# ELECTRICAL AND OPTICAL PROPERTIES OF SEMICONDUCTORS

# ELEKTRICHESKIE I OPTICHESKIE SVOISTVA POLUPROVODNIKOV

# ЭЛЕКТРИЧЕСКИЕ И ОПТИЧЕСКИЕ СВОЙСТВА ПОЛУПРОВОДНИКОВ

# The Lebedev Physics Institute Series

Editor: Academician D. V. Skobel'tsyn

Director, P. N. Lebedev Physics Institute, Academy of Sciences of the USSR

*Proceedings (Trudy) of the P. N. Lebedev Physics Institute*

Volume 37

# ELECTRICAL AND OPTICAL PROPERTIES OF SEMICONDUCTORS

Edited by
## Academician D. V. Skobel'tsyn
*Director, P. N. Lebedev Physics Institute*
*Academy of Sciences of the USSR, Moscow*

Translated from Russian by
Albin Tybulewicz
Editor, "Soviet Physics—Semiconductors"

CONSULTANTS BUREAU
NEW YORK
1968

The original Russian text, published by Nauka Press in Moscow in 1966
for the Academy of Sciences of the USSR as Volume XXXVII of the Proceedings (Trudy) of the P. N. Lebedev Physics Institute, has been cor-
rected by the editor for this edition.

**Электрические и оптические свойства
полупроводников**

*Труды Физического института
им. П. Н. Лебедева*
Том XXXVII

Library of Congress Catalog Card Number 68-18561

ISBN-13: 978-1-4615-8554-1      e-ISBN-13: 978-1-4615-8552-7
DOI:10.1007/978-1-4615-8552-7

# CONTENTS

Electrical and Optical Properties of Electroluminescent Capacitors
Based on ZnS:Cu

# RADIATIVE RECOMBINATION AT DISLOCATIONS
# IN GERMANIUM [†]

## A. A. Gippius

## Introduction

### §1. Potentialities of Radiative Recombination as a Method of Investigation

Interest in radiative recombination in semiconductors has grown considerably in recent years. This is due not only to the practical applications of this phenomenon, such as lasers and semiconductor sources of incoherent radiation [1, 2], but also due to the fact that investigations of recombination radiation yield important information on the properties of semiconducting materials.

A detailed review of the experimental and theoretical papers on radiative recombination in semiconductors would take up too much space and, therefore, we shall consider only certain investigations, using them to illustrate the potentialities of the radiative recombination method.

An analysis of the recombination radiation spectra associated with interband transitions in germanium [3] has confirmed that there are two types of radiative transition: "vertical" transitions for $K = 0$ without phonon participation, and phonon-assisted transitions from the absolute minimum of the conduction band to a maximum of the valence band at $K = 0$.

Since the phonon spectra of germanium and silicon have four branches, the recombination radiation spectrum should contain lines corresponding to transitions accompanied by the emission and absorption of four types of phonon, having wave vectors which correspond to the position of the bottom of the conduction band in the $K$-space. Such lines have been detected in an investigation of the radiative recombination spectra of germanium and silicon at low temperatures [4]. This made it possible to determine the values of the phonon energies, which have been found to agree well with the data on the optical absorption [5] and on the scattering of cold neutrons [6].

In addition to the information on the band structure and on the phonon spectrum, investigations of the recombination radiation spectrum of silicon have yielded information on the mechanism of the interband radiative recombination. The profile of the radiation band has been analyzed for this purpose.

Depending on the recombination mechanism (exciton or electron–hole), the profile of the radiation line is described either by a Boltzmann distribution or by an energy dependence of the expression for the recombina-

---

[†] Thesis for the degree of Candidate of Physico-Mathematical Sciences. Defended at the P. N. Lebedev Physics Institute on March 9, 1964.

tion velocity $np\sigma v$, where n and p are the electron and hole densities, $\sigma$ is the cross section for radiative recombination, and v is the relative velocity of carriers. It has been shown in [7] that all the radiation observed in silicon at $T = 18°K$ is due to the annihilation of excitons. At $T = 83°K$ some of the radiation is due to the recombination of free electrons and holes. Thus, as the temperature is increased, the relative importance of the electron—hole recombination increases. It is found that the binding energy of excitons in silicon is $8 \cdot 10^{-3}$ eV and the radiative lifetime is $6 \cdot 10^{-5}$ sec. It should be mentioned that the experimentally observed line profile is in very good agreement with the profile calculated from the data on the absorption [8] using the principle of detailed balancing [9].

In addition to the radiation due to the recombination of free electrons and holes, silicon emits radiation which is due to the recombination of electrons and holes which are part of complexes consisting of a hole bound to a positive donor ion by a pair of electrons [10]. Such radiation is concentrated in very narrow lines, because recombining electrons and holes are at rest.

Investigations of the impurity recombination radiation spectra make it possible to determine with high accuracy the positions of local levels contributed by impurities to the forbidden band. The ionization energies of boron, gallium, indium, and arsenic in silicon have been obtained in this way in [11]. The values obtained are, in the majority of cases, in good agreement with the values of the ionization energy calculated by other methods.

In addition to the positions of local levels, investigations of the impurity radiative recombination make it possible to obtain other important information on the properties of impurities. Thus, it has been shown in [12] that practically all events involving electron capture by neutral indium atoms in silicon are accompanied by radiation. This, in the opinion of the investigators, is due to the fact that the probability of a many-phonon process in the transfer of about 1 eV of energy is very small, and a phonon cascade process is also not very likely either because the capture takes place at a neutral acceptor. Similar results have been obtained also for other group III impurities in silicon [13].

An investigation of the role of phonons in the process of radiative electron capture by atoms of group III impurities in silicon has shown [13] that as the ionization energy of the impurity increases, so does the fraction of radiation corresponding to transitions without phonon participation (more exactly, without the participation of those phonons which are active in the interband radiative recombination). This is because the description of an electron state at a local level by means of electron wave functions in a band becomes less accurate the greater the distance of this level from the band.

These examples show that investigations of radiative recombination make it possible to obtain valuable information on the band structure, the phonon spectrum, and the properties of impurities. In view of this, it seems very attractive to use recombination radiation to investigate the properties of little-studied systems as deep traps.

In contrast to the shallow levels contributed by elements of groups III and V to the forbidden bands of germanium and silicon, the recombination radiation associated with transitions to deep levels has been studied much less. This may be due to the fact that the probability of radiative recombination is, in this case, relatively small.

In some cases, it is not possible to ascribe definitely the observed radiation to particular impurities [14] or defects [15]. In the case of germanium, attempts to investigate the radiation associated with impurities contributing deep levels have not yet been very successful. Gold, manganese, nickel, iron, and copper have been introduced into germanium [15], but only the radiation associated with copper has been observed, and the level energies obtained have not been found to coincide with any of the known copper levels. The recombination radiation due to the transitions of carriers to energy states associated with dislocations has also been investigated [16, 17]. These investigations will be considered in the next section.

## §2. Energy Structure of Dislocations in Germanium Crystals

There have been quite a few experimental and theoretical investigations of the effect of dislocations on the electrical properties of germanium and of the electron spectrum of dislocations. We shall consider only those investigations which report information on the energy structure of dislocations in germanium.

Measurements have been reported in [18] of the carrier density and mobility in n- and p-type germanium of 15 $\Omega \cdot$ cm resistivity, subjected to plastic bending at 650°C, which has introduced dislocations at a density of ~$10^6$ cm$^{-2}$. It has been found that the introduction of dislocations into p-type samples has practically no effect on the density and mobility of the carriers, whereas, in n-type samples, the density and mobility decrease. It follows from the results of this investigation that dislocations contribute energy states of the acceptor type to the upper half of the forbidden band. The acceptor action of dislocations is associated with the presence of free "dangling" bonds in the case of edge dislocations. In the opinion of the investigators, a local dislocation level is separated by 0.2 eV from the bottom of the conduction band. The determination of the position of this level is difficult because of the absence of a definitely linear part in the curves representing the temperature dependence of the carrier density.

Read [19-21] has analyzed the effect of dislocations on the electrical properties of semiconductors with diamond-type lattices. He has found that dislocations give rise to a series of closely spaced acceptor centers, which have the same level, corresponding to the energy of the capture of an electron by a dangling bond. In an n-type material, electrons may be captured by these acceptor centers, and this reduces the electrical conductivity. The distances between the nearest dangling bonds may be of the order of several angstroms. Therefore, we cannot neglect the Coulomb interaction between electrons captured by a dislocation. Consequently, the extent of the electron capture by dangling bonds depends not only on the position of the dislocation level and the Fermi level, but also on the Coulomb interaction energy. Thus, the usual Fermi statistics are, in this case, inapplicable. Moreover, if the distance between the electrons captured by a dislocation is less than the average distance between chemical donors or acceptors, the dislocation acts as a negatively charged line. A cylindrical region of positive space charge forms around it.

Read has developed several approximate methods for the statistical calculation of the population of dislocation acceptor centers allowing for the interaction of captured electrons. He has found, in particular, that, in germanium with a concentration of uncompensated donors amounting to $10^{15}$ cm$^{-3}$, only 0.1 of the dislocation levels (lying 0.225 eV below the conduction band) are filled at T = 0°. The distance between the electrons captured by a dislocation is, in this case, 36 Å, the radius of the cylindrical region around a dislocation is 2820 Å, and the average distance between donors is 1000 Å.

Read's ideas have been used to interpret the results of measurements of the density and mobility of carriers in germanium crystals with a high dislocation density [22]. In particular, it has been shown that the experimental temperature dependence of the population of dislocation centers is in good agreement with the theoretical dependence calculated on the assumption that the dislocation level lies 0.18-0.20 eV below the conduction band.

In addition to the Hall effect measurements, the energy structure of dislocations has been studied by measuring the nonequilibrium carrier lifetime.

The temperature dependence of the carrier lifetime in p- and n-type germanium crystals with a dislocation density of ~$10^6$-$10^7$ cm$^{-2}$ has been reported in [23]. The temperature dependence of the radius of electron capture by dislocations has been determined for the p-type material: $\rho \propto T^{-3}$. The activation energy (~0.4 eV) has been found from the slope of the approximately linear part of the $\log \tau = f$ (1/T) curve.

Similar measurements have been reported in [24-26]. The activation energies, determined from the slope of the linear part of the $\tau$(T) curve, are: 0.15-0.20 [24], 0.14 [25], and 0.21-0.27 eV [26].

Thus, investigations of the temperature dependence of the lifetime do not give reliable information on the energy structure of dislocations. The values obtained for the activation energy range from 0.14 to 0.4 eV, and very few of the investigators have analyzed the causes of such a large scatter in the values.

It is mentioned in [20, 27] that the assumption of a single acceptor level due to dislocations is not always justified. In particular, Read [20] has used the "harmonic oscillator approximation" to consider the possibility of the motion of an electron in a band formed as a result of the overlap of the wave functions of neighboring dangling bonds.

The idea of dislocation bands has also been used in [28] to interpret the temperature dependence of the carrier density and mobility reported in [22]. It has been assumed that the dislocation band lies 0.06 eV above the valence band [this value has been obtained earlier in measurements of the capacitance and the conductivity of germanium bicrystals, which have walls with closely spaced (~60 Å) dislocations]. It should be mentioned that such a small separation between the dislocation band and the valence band does not agree with the results reported in [18], where it has been shown that dislocations have practically no influence on the carrier density in p-type germanium.

The quantum-mechanical problem of the influence of edge dislocations on the energy spectrum of the electron system in a semiconductor has been considered in [29]. It has been shown that a quasi-continuous spectrum is associated with the motion of electrons along a dislocation. The states of electrons at a dislocation are described by an energy band or several such bands. The widths of dislocation bands may be comparable with the width of the conduction band, and the dislocation bands may overlap the "intrinsic" bands of a crystal. The density of levels in the forbidden band is usually very low, so that we can speak of a continuous spectrum in the form of a thin "veil" stretched across the forbidden band.

Since the equation which is used to determine the energy eigenvalues contains the screened potential of a dislocation and this potential includes the charge per unit length of the dislocation and the screening radius, both of which depend on temperature, we find that the edges of the dislocation bands also depend on temperature.

Electron and hole statistics for semiconductors with dislocations have been considered in [30] on the assumption that the energy spectrum of dislocations represents an array of one-dimensional bands. In sufficiently heavily doped samples there is only one dislocation band at not too low temperatures. A formula has been deduced for the calculation of the electron population of a dislocation, which depends on temperature and on the Fermi level position. The parameters which occur in this formula, in particular, the position of the bottom of the dislocation band $E_d$, can be calculated from the experimental data. Using the results reported in [22], the temperature dependence of the bottom of the dislocation band has been found to obey $E_d = E_c - (8 \cdot 10^{-4})T$, in accordance with the following quantitative considerations: when the negative charge of a dislocation is reduced, the hole levels in the field of the dislocation should become shallower. The same investigator [30] has also considered statistics of the recombination of carriers at dislocations and has shown that, in the case of low injection levels, the lifetime is given by an expression similar to the well-known Shockley—Read expression. In particular, it is found that, in a sufficiently heavily doped n-type material, the temperature dependence of the lifetime is governed by the temperature dependence of the capture radius.

The results reported in [30] have been used to interpret the temperature dependence of the lifetime in plastically deformed germanium crystals [31]. It has been shown that, in the temperature range 180-300°K, we may assume that $\tau = (1/C_p)$, where $C_p$ is the hole capture coefficient. In the 110-200°K range, the temperature dependence of $C_p$ is a power law: $C_p \propto T^{-4}$. The change in the nature of the dependence $\tau(T)$ at $T > 200°K$ is, in the opinion of this investigator [31], associated with the fact that at $T = 200°K$ the quasi-Fermi level intersects the bottom of the dislocation band and, at higher temperatures, lies within this band. The data on the position of the bottom of the dislocation band, reported in [30], have been used in this treatment.

Investigations of the radiative recombination at dislocations yield important information on the energy structure of dislocations. It has been reported in [16] that the plastic deformation of germanium crystals results in the appearance of a radiation band with a maximum near 0.5 eV. The intensity of this band increases as the degree of deformation is increased, but it is little affected by annealing, which destroys acceptor levels of 0.1 eV energy, as observed earlier in [32]. On this basis, it has been concluded that the recombination radiation results from the capture of carriers by centers associated with dislocations. The corresponding level should, in the opinion of the investigator, lie 0.2 eV from one of the bands. The dislocation radiation band has been found to be fairly broad; in some cases, it has obviously consisted of several components.

An attempt to analyze the broad dislocation radiation band into these individual components has been reported in [17]. At T = 63°K, four components of energies 0.5, 0.545, 0.61, and 0.68 eV have been identified. The following may be said about this analysis and its interpretation:

1. The investigator has not made any assumptions about the profiles of the individual radiation components.

2. In the process of the analysis, the investigator has used the experimental method based on a comparison of the profiles of the spectra obtained at various values of the injection current. Thus, this method is basically incapable of separating the components of the radiation which depend in the same way on the injection current. Therefore, the analysis of the spectrum reported in [17] is fairly arbitrary.

3. In the opinion of the investigator, the individual components of the radiation are related to electron transitions between states associated with local narrowing or broadening of the forbidden band near dislocations. The magnitudes of the narrowing and broadening have been calculated in [33]. However, one should treat the results reported in [33] with great caution, since the calculation of changes in the forbidden band width reported in that paper have been carried out by the deformation potential method in the range where this method is essentially inapplicable. In particular, according to [33], a change in the forbidden band width at a distance of one atomic radius from a dislocation amounts to 0.3 eV, and at a distance of three atomic radii it amounts to 0.1 eV. Thus, the application of the deformation potential method to the regions of a crystal in the direct vicinity of a dislocation gives results which contradict the main approximation of the method, i.e., the smooth variation of the band curvature.

This brief analysis of the published literature shows that, in spite of a large number of papers dealing with the energy structure of dislocations, and in spite of the use of different investigation methods, the problem of the spectrum of electron states contributed by dislocations to the forbidden band of germanium has not yet been fully solved. Therefore, the present paper describes an attempt to obtain more accurate information on the energy structure of dislocations and, in particular, to determine the mechanism of radiative recombination at dislocations.

CHAPTER I

# Experimental Method

The main part of this investigation has been concerned with the recombination radiation spectra of germanium crystals having a dislocation density of $10^3$-$10^6$ cm$^{-2}$. To carry out this investigation, it was necessary to prepare crystals with known dislocation densities and to assemble a sufficiently sensitive apparatus for the recording of the radiation spectra. Crystals with dislocation densities of $10^3$-$10^4$ cm$^{-2}$ were prepared by the standard Czochralski method. To prepare crystals with dislocation densities of $10^5$-$10^6$ cm$^{-2}$, we used the special method described below.

## §1. Preparation of Crystals with High Dislocation Densities

To prepare germanium crystals with high dislocation densities, we used the method of growing crystals from plastically deformed seeds [34]. This method was selected because the alternative method of high-temperature plastic deformation produced, in addition to dislocations, new recombination centers (probably impurities which had diffused into a crystal during deformation) which greatly reduced the carrier lifetime.

The pulling of a crystal, using a plastically deformed seed, makes it possible to prepare crystals of a given resistivity with a high density of dislocations. In crystals prepared in this way, dislocations play the dominant role in the recombination of nonequilibrium carriers.

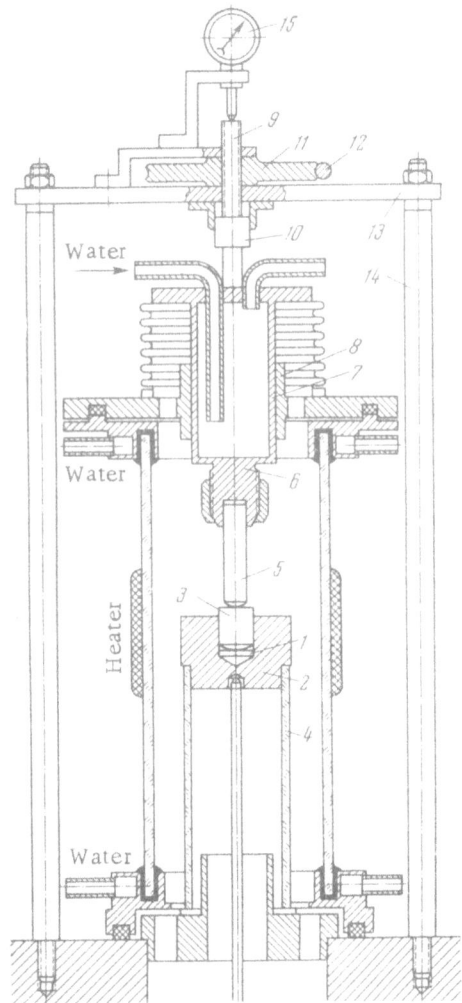

Fig. 1. Cross section of the apparatus used for the plastic deformation of germanium crystals.

To prepare the seeds, we cut germanium slabs of 30 × 20 × 4 mm dimensions. The crystals from which these slabs were cut were oriented by the method of reflectograms, i.e., light figures, which were obtained when light was reflected from a specially etched surface of a crystal [35]. Using this method, it was possible to orient crystals with an accuracy of 30'. The oriented germanium slabs were subjected to bending along an axis coinciding with the longest side.

Plastic deformation was carried out in the apparatus shown in Fig. 1. A germanium slab 1 was placed in a graphite cassette 2, which was heated by means of a resistance heater to the required temperature. The magnitude of the deformation was governed by the curvature of a graphite former 3. A quartz tube 4 was used to support the graphite cassette. The pressure was transmitted to the former 3 by an optical quartz rod 5, clamped (6) to a piston 7. The piston 7 could move vertically in a water-cooled cylinder 8 under the action of a screw 9 to which the cylinder was joined by a sleeve 10. The screw 9 was moved vertically by the rotation of a wheel 11 driven by a worm gear 12. The top plate 13 was supported by strong legs 14. In assembling this apparatus, measures were taken to prevent distortion of the various parts.

The temperature of the graphite cassette was measured by means of a platinum—platinorhodium thermocouple. The deformation was carried out at 750°C. The quartz rod 5 was applied to the former 3 only when the required temperature was reached. The moment of contact between the rod and the former was established by the closing of the electric circuit formed by the thermocouple, the graphite cassette, and a layer of silver deposited on the rod 5 in the form of a strip approximately 2 mm wide.

The deformation took 2 min and was stopped when the necessary degree of bending, measured with a gauge 15, had been reached.

To reduce the contamination of the samples during the high-temperature deformation, the parts of the apparatus which were in the hot zone were carefully cleaned before the measurements and the germanium crystals were etched and coated with a film of gold to facilitate the extraction of the rapidly diffusing impurities.

The density of the dislocations introduced by plastic bending is

$$N_d = \frac{1}{r\,|b|\cos\theta}\,,$$

where r is the radius of curvature, b is the Burgers vector ($\sim4 \cdot 10^{-8}$ cm), and $\theta$ is the angle between the slip plane and the neutral plane of the slab. In this investigation, we used the value r = 10 cm, which corresponds to a dislocation density $N_d \approx (2-3) \cdot 10^6$ cm$^{-2}$.

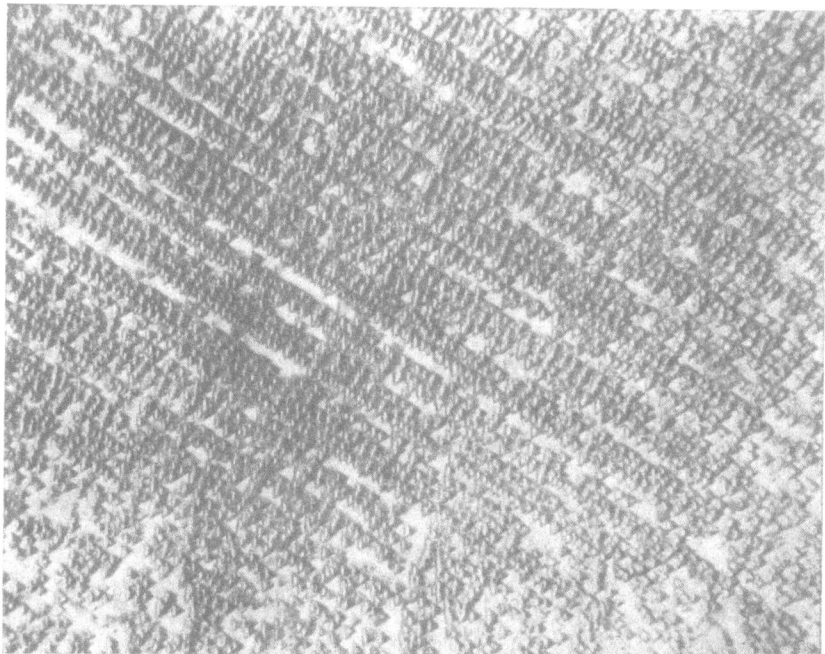

Fig. 2. Distribution of dislocations on the end of a plastically deformed seed.

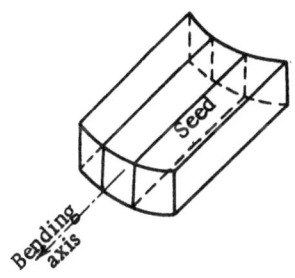

Fig. 3. Method of cutting a seed from a deformed germanium slab.

The dislocation density, deduced from the number of etch pits on the (111) plane of the deformed samples, was $2 \cdot 10^6$ cm$^{-2}$, in agreement with the calculated value. Examination of a cross section of a crystal bent along the [111] axis, shown in Fig. 2, indicated that the dislocations emerged in parallel rows along traces of active slip planes.

The seed crystals were cut from deformed slabs in the way shown in Fig. 3.

We used seeds whose axes coincided either with the [111] or the [211] direction.

The dislocation densities in germanium crystals prepared by growing from such seeds were $10^5$-$10^6$ cm$^{-2}$; the distribution of these dislocations across an ingot section was inhomogeneous. Regions with a high density of uniformly distributed dislocations were interspersed with regions in which dislocation walls predominated (Fig. 4). A metallographic examination of the ingots showed that a high density of etch pits, associated with dislocations, was observed not only on the (111) plane, perpendicular to the growth axis (for crystals grown along the [111] direction), but also on other planes belonging to the {111} family. Thus, the distribution of the dislocations in a crystal did not have a definite anisotropy, and a study of the distribution of etch pits on one of the {111} planes did not give full information on the nature of the dislocation structure in the interior. An investigation of the distribution of dislocations on two planes of the {111} family, near the line of intersection of two such planes, showed that, in the majority of cases, the nature of the distribution of the dislocations in neighboring regions of two {111} planes was the same. Therefore, an investigation of the distribution of points of emergence of dislocations on one of the planes of the {111} family made it possible to draw some conclusions about the distribution of the dislocations in the interior. In particular, if one of the planes included a dislocation wall, a similar wall was frequently observed on the other plane, and the traces of dislocation walls on two {111} planes intersected at a point lying on the line of intersection of these planes. Thus, in many cases, dislocation walls were networks of two intersecting arrays of dislocations.

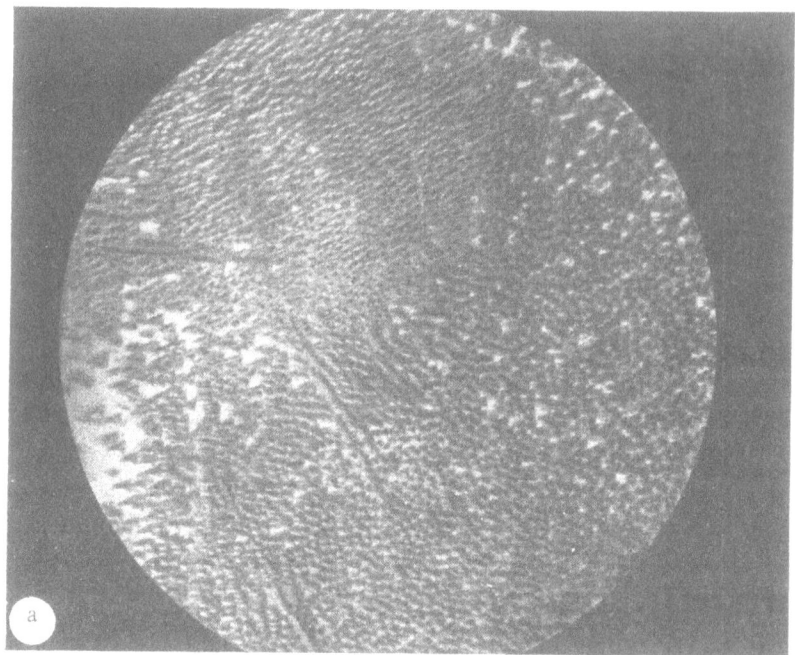

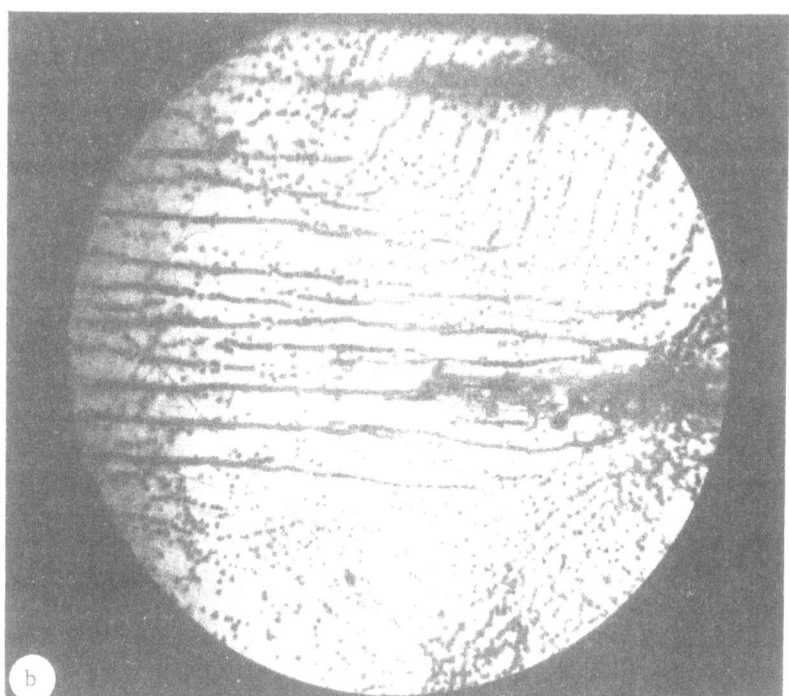

Fig. 4. Distribution of the dislocations in crystals grown using plastically deformed seeds. a) Uniform distribution of dislocations; b) dislocation walls.

## §2. Preparation of Samples

We used germanium samples in the form of Weierstrass spheres, 8 mm in diameter, in order to reduce the losses due to the total internal reflection. In the majority of cases, we knew the orientation of the optical axis of a sample with respect to its crystallographic axes. In those cases when the flat part of the sample coincided with the (111) plane, the dislocation density was determined directly in the sample under investigation and the central part of the flat surface, where carriers were injected, was photographed. Thus, we could compare the observed recombination radiation spectrum with the dislocation structure.

The recombination radiation was excited by electrical injection from a metal pressure contact placed in the center of the flat surface, along whose circumference an ohmic contact was deposited. It was found that the form of the recombination radiation spectrum was independent of the contact material. In most cases, the injection took place from an aluminum contact. A soft aluminum wire, 1 mm thick, was sharpened and pressed against the surface of a sample by a phosphor bronze spring. The point of the wire became flattened and provided a contact of ~0.02 mm² area.

We used a metal pressure contact and not a p−n junction because dislocations form near a p−n junction prepared by the fusion method (at 500-600°C), as reported by Newman [16]. Since, in some cases, we were interested in the recombination radiation spectrum associated with a definite dislocation structure, it was undesirable to use p−n junctions because the formation of such junctions altered the dislocation structure of the crystal.

We did not apply the "forming" treatment to the metal contact since it was found that the forming produced radiation in the impurity part of the spectrum, evidently associated with dislocations, which appeared near a contact under the action of local heating and pressure, applied through the contact. (These control experiments were carried out on dislocation-free samples which did not emit impurity radiation before forming.)

The ohmic contact, in the shape of a ring, was provided either by the electrolytic deposition of copper or by the fusing-in of tin at 250°C.

The surface through which carriers were injected was treated in various ways. In those cases when the surface was etched, we used a special Teflon holder to protect the polished spherical surface.

## §3. Apparatus for Measuring Recombination Radiation Spectra

The recombination radiation spectra were measured using the apparatus a block diagram of which is given in Fig. 5. The apparatus consisted of a circuit used to excite recombination radiation, a cryostat, a monochromator, and a recording circuit.

a) Excitation Circuit. Recombination radiation was excited electrically. Since a lead sulfide photoresistor was used as the detector, it was necessary to modulate the recombination radiation flux. This was done by modulating the current through the injecting contact.

In the majority of cases, an alternating voltage of 60 cps was applied to the contact. A ZG-10 oscillator was used as the source of this voltage and a power amplifier based on 6P3S tubes was connected to the oscillator output. The constant component of the current was measured with an M-104 instrument. We shall show later that there were grounds for assuming that the injection coefficient was very little affected by changes in the forward current. Therefore, we called this current the "injection current," assuming that the injection coefficient was close to unity. The working range of the injection currents had a lower limit set by the sensitivity of the apparatus and an upper limit set by the heating of the contact region in the sample. Using sinusoidal modulation it was possible to work in the range of injection currents from 100 $\mu$A to 150 mA. In some measurements, it was necessary to increase the injection current. In order to prevent the heating of the sample, we injected strong currents in the form of 100-$\mu$sec pulses of 60 cps repetition frequency. Voltage pulses from a GIS-2 generator, synchronized by the sinusoidal voltage from the ZD-10 oscillator, were applied to the input of a power amplifier based on P-210 transistors. Thus, current pulses of up to several amperes could be obtained. To determine the value of the current in these cases, we measured (using a pulse voltmeter or an oscillograph) the voltage drop across a calibrated resistor.

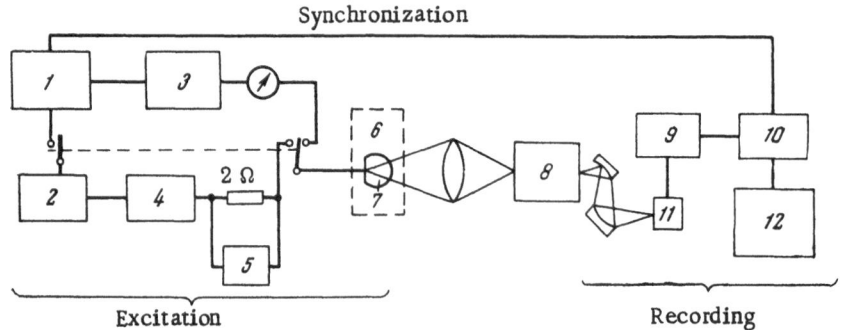

Fig. 5. Block diagram of the apparatus used to determine the recombination radiation spectra. 1) ZG-10; 2) GIS-2; 3,4) power amplifiers; 5) MVI-1; 6) cryostat; 7) sample; 8) monochromator; 9) selective amplifier; 10) synchronous detector; 11) PbS; 12) ÉPP-09.

TABLE 1

| Monochrom. type | $n$, mm | $d/f$ | $\beta$ | $D/f$ | $f$, mm | $D$, mm | Detector size, mm$^2$ | $\beta_1$ | $D_1/f_1$ | $f_1$, mm | $D_1$, mm |
|---|---|---|---|---|---|---|---|---|---|---|---|
| Prism . . . | 18 | 1/6 | 3 | 1/1,5 | 150 | 100 | 6·1 | 3 | 1/1,5 | 150 | 100 |
| Grating . . | 6 | 1/2 | 1 | 1/1 | 120 | 120 | 6·1 | 1 | 1/1 | 65 | 65 |

b) Cryostat. The investigation of the recombination radiation was carried out in the temperature range 78-300°K. At low temperatures, the sample was placed in a cryostat, two variants of which were used. In one, the sample, which had been soldered (with tin in vacuum) to a nickel mandrel 0.2 mm thick, was placed in direct contact with liquid nitrogen. In most cases, the joint was vacuum-tight. The nickel mandrel holding the sample was soldered with Wood's alloy to a flange which was pressed against another flange through an annealed copper ring of rhombic cross section. Both flanges and the tightening bolts were made of phosphor bronze. This form of seal, originally described by Strelkov [36], was found to work well over a wide range of temperatures. The space between the flanges was connected to an enclosure which contained 0.5 liters of liquid nitrogen. Such a construction of the cryostat ensured very good heat transfer and made it possible to work at high injection currents (up to 500 mA) without significant overheating of a sample.

On the other hand, this construction had a number of disadvantages: the placing of a sample in the cryostat took a fairly long time; the joint between the sample and the mandrel was sometimes not vacuum-tight; the conditions on the surface of the sample were unstable when the measurements were repeated. Therefore, the majority of the measurements were carried out in a cryostat of different construction. In this case, the sample was held in a massive copper holder. The minimum temperature of the sample was 85°K. The temperature was measured with a copper—constantan thermocouple. To obtain intermediate temperatures (85-300°K), nitrogen vapor from a Dewar flask was blown through the inner enclosure of the cryostat. The flow rate of the nitrogen vapor was controlled with a system developed by V. D. Kopanev at the Physics Institute, Academy of Sciences of the USSR.

c) Monochromator. The recombination radiation which was generated in a sample was collected in the plane of the entry slit of a monochromator by a NaCl lens. A lens was more convenient than a mirror for the particular monochromator employed. The lens material was selected to reduce the chromatic aberration. The dispersion of NaCl in the investigated spectral range (1-3 μ) was small, and the chromatic aberration was therefore slight.

We used two types of monochromator. Initially, we employed a mirror monochromator with a quartz prism; later, we used a monochromator with a diffraction grating, which had a much higher illumination efficiency.

The gain in the efficiency resulting from the replacement of a prism monochromator with a grating instrument could be calculated from the following considerations. Take a monochromator whose entry and exit collimator focal lengths are $f_1 = f_2 = f$, and whose angular magnification is equal to unity, i.e., $S_1 = S_2 = S$. Here, $S_1$ and $S_2$ are the cross-sectional areas of a beam falling on and emerging from a prism or a grating. These conditions were satisfied by both monochromators used in the present investigation.

In the case of a continuous spectrum, the light flux per wavelength interval $\Delta\lambda$ accepted by a monochromator is proportional to the source brightness $B_\lambda$, the spectral interval $\Delta\lambda$, the entry slit area $a_1 h_1$, and the solid angle supported by the effective stop at the center of the entry slit:

$$\Phi = B_\lambda \Delta\lambda a_1 h_1 \frac{S}{f_1^2} \,, \tag{1}$$

where $a_1$ is the width and $h_1$ the height of the entry slit.

The light flux through the exit slit, whose spectral width is $a_\lambda = \Delta\lambda$, is (for a sufficiently narrow entry slit)

$$\Phi = \tau_\lambda B_\lambda \Delta\lambda a_1 h_1 \frac{S}{f^2} \,, \tag{2}$$

where $\tau_\lambda$ is the transmission of the monochromator.

Usually, the width of the entry slit of a monochromator is made equal to the width of the exit slit: $a_2 = a_1$. Taking into account the relationship between the geometrical and spectral slit widths

$$a_1 = \frac{d\varphi}{d\lambda} \Delta\lambda f, \tag{3}$$

where $d\varphi/d\lambda$ is the angular dispersion, we can write the expression for the light flux through the entry slit in the form

$$\Phi = \tau_\lambda B_\lambda (\Delta\lambda)^2 \frac{h}{f} S \frac{d\varphi}{d\lambda} \,. \tag{4}$$

The efficiency of a monochromator is defined as the light flux at the exit for $B_\lambda = 1$ and $\Delta\lambda = 1$:

$$L = \tau_\lambda \frac{h}{f} S \frac{d\varphi}{d\lambda} \,. \tag{5}$$

Thus, the efficiency of a monochromator is proportional to the transmission of the instrument $\tau$, the angular height of the slit $\beta^* = (h/f)$, the dimensions of the dispersing element S, and the angular dispersion $d\varphi/d\lambda$. (Coefficients of the order of unity, which relate the dimensions of the dispersing element to the size of the effective stop, are ignored.)

Comparison of the efficiency of the prism and grating instruments has been carried out in [37]. The value of $\tau_\lambda$ for prism monochromators is about 50%. The transmission of grating monochromators is also usually not less than 50% for gratings with grooves inclined at the blaze angle.

The quantity $\beta^* = (h/f)$ is governed mainly by the aberrations of the focusing system of the monochromator and is practically the same for prism and grating monochromators. (In any case, the maximum value of $\beta^*$ of grating monochromators is not less than that of prism instruments.)

Thus, comparing the two types of monochromator, we may assume that

$$(\tau\beta^*)_{\text{prism}} = (\tau\beta^*)_{\text{grating}}. \tag{6}$$

The efficiencies of grating and prism monochromators differ mainly because of the difference in the term $S(d\varphi/d\lambda)$. In a rough estimate, we may assume that $S_{grating} = S_{prism}$. Then, the ratio of the efficiencies of prism and grating monochromators is given by the expression

$$\gamma = \frac{\left(\dfrac{d\varphi}{d\lambda}\right)_{prism}}{\left(\dfrac{d\varphi}{d\lambda}\right)_{grating}}. \tag{7}$$

As shown in [37], the value of this ratio does not exceed 0.15 in the wavelength range 1.0-15 $\mu$. Thus, the efficiency of the grating instrument is more than 6 times greater than the efficiency of a prism instrument.

The value obtained for the gain in the efficiency should be regarded as an average value, and it is necessary to use Eq. (5) in the comparison of actual instruments.

In the present investigation, we used monochromators with the parameters listed below:

A prism monochromator, Wadsworth mounting: $h = 18$ mm, $f = 310$ mm, $\beta* = (h/f) = 0.058$, $(d\varphi/d\lambda) = 0.0258\ \mu^{-1}$, $S = 19.6$ cm$^2$. (The effective stop area S of this monochromator was governed by the dimensions of the collimator mirror.)

A grating monochromator of the autocollimator type: $h = 6$ mm, $f = 270$ mm, $\beta* = (h/f) = 0.022$, $(d\varphi/d\lambda) = 0.218\ \mu^{-1}$, $S = 132$ cm$^2$. (In this monochromator, the value of S was governed by the dimensions of the grating.)

Substituting these values into Eq. (5), we find that $L_{grating}/L_{prism} = 22$. The value of the ratio $L_{grating}/L_{prism}$, found experimentally, was close to the calculated value.

Thus, the replacement of a prism monochromator with a grating instrument increased the light flux by a factor of more than 20, which corresponded approximately to a fivefold increase in the resolving power (which was limited by the sensitivity of the radiation detector).

As already mentioned, the entry slit was illuminated through a lens made of NaCl. The radiation emerging from the monochromator was focused on the surface of the detector by a spherical mirror. The parameters of the lens were selected so that the image of the source on the entry slit completely covered the slit and the angular aperture of the monochromator was completely filled.

The magnification of an optical system depends on the ratio of the dimensions of the source and the slit: $\beta = (h/l)$, where $l$ is the size of the light source. The relative aperture of the lens $D/f$ is related to the relative aperture of the collimator $d/f$ by the expression [38]

$$\frac{D}{f} = (\beta + 1)\frac{d}{f}. \tag{8}$$

Thus, the dimensions of the source of light and the parameters of the spectroscopic instrument govern the magnification of the system $\beta$ and the relative aperture of the lens $D/f$. The focal length is selected in accordance with the construction of the instrument. The parameters of the optical system which focuses the image of the exit slit onto the surface of the radiation detector are selected on the basis of similar considerations.

In our case, the source of light was a direct virtual magnified image of the luminous region produced by the Weierstrass sphere. Its size was 4-6 mm. Table 1 lists the parameters of the monochromators used in the present investigation and the parameters of the lenses and mirrors employed to illuminate the entry slit and the radiation detector.

Subscripts are used to denote the quantities which refer to the optical system employed to focus the radiation onto the detector.

The prism monochromator was calibrated using the emission lines of mercury vapor and the absorption bands of atmospheric $H_2O$ vapor. To calibrate the grating monochromator, we used sodium and mercury lamps. In this case, we observed six orders (from 4 to 9) of the Na line at $\lambda = 5896$ Å and seven orders (from 4 to 10) of the Hg line at $\lambda = 5461$ Å.

d) Recording of Radiation. The radiation detector was a lead sulfide photoresistor cooled to $-80°C$. The photoresistor was made in the form of a miniature Dewar flask with a photosensitive layer deposited on its inner wall.

The photoresistor was connected in series with a load resistor and supplied with a voltage of 20 V from a battery shunted by a capacitor in order to reduce voltage fluctuations. The whole circuit, together with the photoresistor, was screened in order to reduce induced strays. The electrical signal from the photoresistor, due to illumination with a modulated radiation flux, was applied to the input of a selective narrow-band amplifier ($f_0 = 60$ cps, $K_0 = 10^6$, $\Delta f = 1$ cps) which amplified the first harmonic of the signal (both for sinusoidal modulation and for pulse modulation through a rectifying contact).

The signal from the amplifier output was applied to a synchronous detector. The time constant of the synchronous detector was variable within the limits 1-100 sec, which represented a transmission band $\frac{1}{3}$-$\frac{1}{300}$ cps. A reference voltage was provided by the anode of the second 6Zh4 tube of the ZG-10 oscillator (this ensured that the value of the reference voltage was independent of the signal taken from the oscillator). The ZG-10 oscillator was used as a synchronizing element in the sinusoidal and pulse modulation (in the latter case, the GIS-2 pulse generator was synchronized by a signal from the ZG-10 oscillator).

A constant voltage from the synchronous detector was applied to an ÉPP-09 automatic recorder. Synchronous motors were used to drive the chart paper of the recorder and to rotate the prism or grating. Thus, the recombination radiation spectra were recorded automatically.

Special attention was paid to the stability of the recording system; in particular, all the electronic units were supplied with a stabilized voltage. A battery was used to provide the heating current for the electron tubes of the narrow-band amplifier. The radiation detector was fitted with a device which kept solid carbon dioxide pressed against the bottom of the enclosure during the evaporation of the carbon dioxide. These measures ensured that the instability of the system did not exceed 1%.

It should be mentioned that the form of the radiation band recorded on the chart differed from the true spectral distribution because of the distortions introduced by the optical and detector—recorder systems. These distortions were due to the selectivity of the radiation detector, the dependence of the transmission of the monochromator on the wavelength, the presence of atmospheric absorption bands, and the chromatic aberration of the lenses used. Consequently, the system as a whole had a certain spectral sensitivity which was more or less selecttive. This sensitivity curve was determined using a "black body" whose spectrum was described by Planck's formula; the curve was used later to correct the spectra recorded on the chart.

The recording system, characterized by a time constant $\tau$, reduced the amplitude, and broadened and shifted the radiation band maximum; these distortions were amplified by an increase in $\tau$ and in the rate of recording V. The accuracy of the reproduction of the radiation band profile by the monochromator was governed by the value of the spectral slit width or the resolution $\Delta \lambda$. By analogy, the accuracy of the reproduction of the band by the detector—recorder system was characterized by a "time resolution" $V\tau$, i.e., by the value of the spectral interval

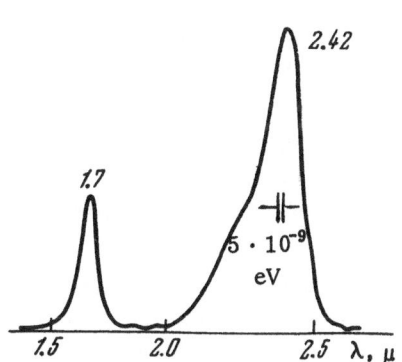

Fig. 6. Recombination radiation spectrum, recorded on a chart, of an n-type germanium sample having an electron density of $5 \cdot 10^{13}$ cm$^{-3}$ and a dislocation density of $5 \cdot 10^3$ cm$^{-2}$. The injection current was 50 mA.

scanned in a time $\tau$. The spectra were usually recorded using a slit width of 0.017 $\mu$. The time resolution was ~0.02 $\mu$. Thus, the accuracy of the recording of the spectra was limited by the monochromator resolution.

CHAPTER II

# Radiative Recombination in Germanium Crystals with Dislocation Densities of $10^3$-$10^4$ cm$^{-2}$

In this chapter, we shall report the results of measurements of the recombination radiation spectra of n-type germanium crystals with dislocation densities of $10^3$-$10^4$ cm$^{-2}$ and carrier densities of $5 \cdot 10^{13}$-$9 \cdot 10^{15}$ cm$^{-3}$. A study of the recombination radiation of these crystals made it possible to determine some important characteristics of the process of radiative recombination at dislocations in germanium.

At room temperature, the recombination radiation spectrum of crystals containing dislocations consisted of a band with a maximum at 1.85 $\mu$ (0.672 eV), corresponding to indirect band−band transitions (the intrinsic radiation band). Cooling to liquid nitrogen temperature narrowed this band and shifted it toward higher energies (1.71 $\mu$, i.e., 0.73 eV); moreover, a new band appeared at 2-2.5 $\mu$ (Fig. 6). This band was found in crystals with dislocations but was absent from the dislocation-free material; it was therefore called the dislocation band. More detailed data on the dependence of the intensity of this band on the dislocation density will be given in Chapter III.

The position of the intrinsic band was governed by the forbidden band width of the semiconductor and, therefore, it could be used as an additional check of the temperature of the contact region in the sample. The accuracy of the determination of the temperature from the position of the intrinsic band was about 3°. The temperature remained constant within this accuracy in the range of injection currents employed in our investigation.

The position and profile of the intrinsic band were the same for all the injection currents and all samples. The position and profile of the dislocation band varied with the degree of doping, the injection current, and the treatment of the surface through which carriers were injected.

## §1.  Dependence of the Dislocation Band Profile on the Degree of Doping

As mentioned in the Introduction, investigations of the radiative recombination at dislocations in germanium have established that the dislocation band has several components. It is not clear which electron transitions correspond to the individual components. We attempted to answer this question by measuring the recombination radiation spectra for several series of germanium samples having different positions of the Fermi level.

The dependence of the dislocation band profile on the degree of doping is shown in Fig. 7. In a high-resistivity n-type material, the radiation was concentrated within a fairly narrow band which had a maximum at $\lambda$ = 2.42 $\mu$. When the electron density was increased, the spectrum acquired components which corresponded to electron transitions of higher energy.

The presence of several components indicated that the radiative recombination took place simultaneously at several levels. Assuming that the observed radiation was due to carrier transitions from an energy band to local levels, we could determine, in principle, the positions of these local levels from the positions of the maxima in the spectrum.

However, we shall show in Chapter III that there are grounds for assuming that, in the case of recombination radiation associated with dislocations, we are, in fact, dealing with transitions between the energy states of a local center, so that the relationship between the radiation band profile and the positions of the levels is not as simple as in the case of transitions of carriers from an energy band.

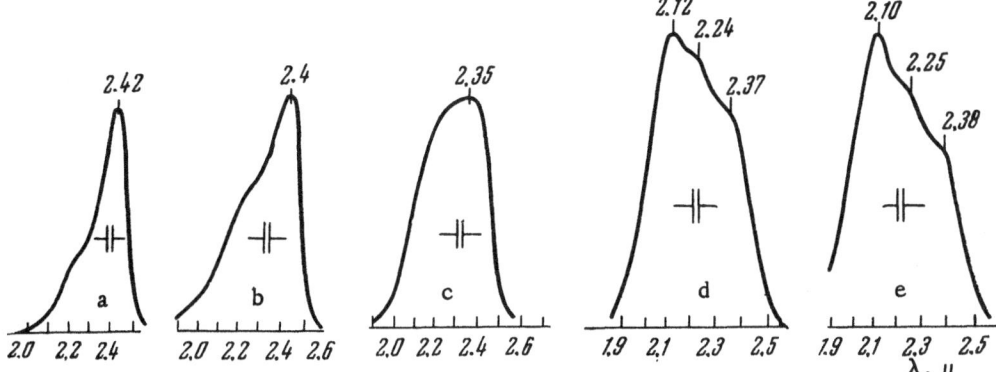

Fig. 7. Dependence of the dislocation band profile on the degree of doping. Injection current in all cases, 100 mA. Carrier densities n = 5 · 10$^{13}$ (a), 2 · 10$^{14}$ (b), 5 · 10$^{14}$ (c), 3.4 · 10$^{15}$ (d), 9 · 10$^{15}$ cm$^{-3}$ (e).

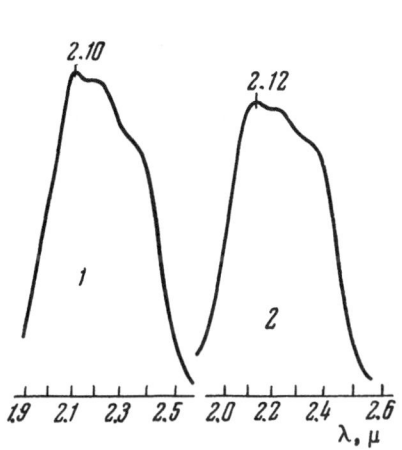

Fig. 8. Dependence of the dislocation band profile on the injection current. The injection current for curve 1 was 10 times that for curve 2.

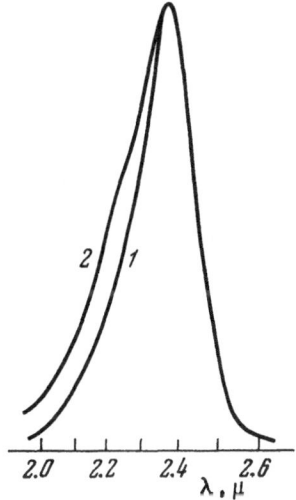

Fig. 9. Influence of the surface treatment on the dislocation band profile. 1) Grinding; 2) etching.

It should be mentioned that when the electron density was increased from 5 · 10$^{13}$ to 9 · 10$^{15}$ cm$^{-3}$, the gap between the Fermi level and the conduction band (at liquid nitrogen temperature) changed from 0.08 to 0.04 eV. Since this increase in carrier density markedly altered the form of the spectrum, we concluded that the levels lay in the upper half of the forbidden band.

The observed radiation was due to transitions of holes to a system of levels associated with dislocations, and, as the electron density was increased, i.e., as the Fermi level rose, the population of the levels lying close to the conduction band increased and these levels began to play a more important role in the recombination, which resulted in the enhancement of the short-wavelength component of the radiation spectrum. The problem of the exact positions of these levels will be considered in Chapter III.

## §2.   Dependence of the Dislocation Band Profile on the Nonequilibrium Carrier Density

The steady-state density of nonequilibrium carriers is given by the formula $\Delta n = g\tau$, where g is the generation rate and $\tau$ is the lifetime of carriers. Thus, $\Delta n$ can be varied by varying either g (i.e., the injection current) or $\tau$ (i.e., for example, by treatment of the surface through which carriers are injected).

The dependence of the dislocation band profile on the injection current is shown in Fig. 8. As the current was increased, the short-wavelength component of this spectrum became stronger. This was interpreted as an increase in the population of the levels lying close to the conduction band when the density of nonequilibrium carriers was increased.

The treatment of the surface through which carriers were injected also affected the dislocation band profile. When etching replaced grinding, the relative intensity of the short-wavelength component increased in a manner similar to that observed when the injection current was raised. Etching reduced the surface recombination velocity, and this increased the effective lifetime in the injection region and, consequently, increased $\Delta n$ at a fixed injection current. The corresponding increase in the population of the levels lying close to the conduction band enhanced the short-wavelength component, as shown in Fig. 9.

The dependence of the dislocation band profile on the surface treatment could not be due to radiative recombination at the surface because the changes observed in the spectra were an order of magnitude greater than the intensity of the impurity radiation observed for dislocation-free crystals subjected to the same surface treatments.

When the optical system was focused on a region far from the contact, the short-wavelength component of the dislocation band became weaker because of a decrease in $\Delta n$. All the changes in the spectra which were observed when the injection current or surface treatment was varied could be explained by an increase or decrease in $\Delta n$ and the associated change in the level population. There was a definite correlation between the relative intensity of the short-wavelength component of the dislocation band and the intensity of the intrinsic band, which could be used as a measure of the value of $\Delta n$.

## §3.   Dependence of the Dislocation Band Intensity on the Injection Current

The dependence of the intensities of the intrinsic and dislocation bands on the injection current, plotted on a double logarithmic scale, is shown in Fig. 10. In the intrinsic band case, this dependence had a form of $I \propto i^m$, where m is close to 2; hence, we can conclude that the injection level was sufficiently high and that the inequality $\Delta n \geq n_0$ was satisfied. The intensity of the dislocation band was a linear function of the injection current for low values of this current, but at high currents this intensity was proportional to $i^\alpha$, where $\alpha$ could be considerably less than unity (in the case presented in Fig. 2, $\alpha = 0.2$ for large currents).

This departure from linearity was not associated with a departure from the proportionality between $\Delta n$ and i because of a change in the carrier lifetime and the efficiency of the injection contact. Since we used high injection levels, the lifetime was constant at those values of the current at which departure from linearity was observed. The assumption of the constancy of the carrier lifetime or, at least, the constancy of the product $\tau\gamma$ ($\gamma$ is the efficiency of the injecting contact) was confirmed by the fact that the intrinsic band was independent of the injection current in this range of currents.

The change in the number of effective recombination centers, due to an increase in the population of the levels when the injection current was increased, would only strengthen the dependence of the intensity on the current (our experiments satisfied the condition $N_t \ll \Delta n$, so that a change in the population of the recombination centers could affect the carrier density in the bands only through a change in $\tau$).

A local increase in the temperature near the contact also could not explain the observed effect, because the temperature dependence of the intensity of the dislocation band indicated that, in this case, the temperature in the contact region would have to rise by about 150°, but we have already established that the increase

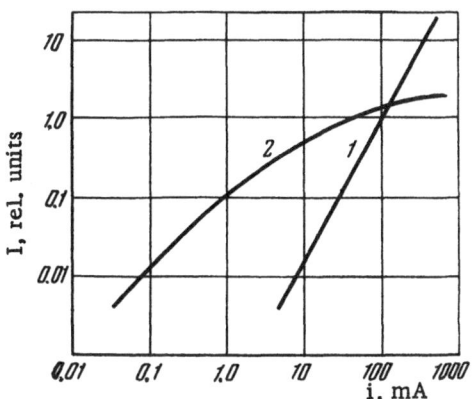

Fig. 10. Dependence of the recombination radiation band intensity on the injection current; n-type sample, $n_0 = 5 \cdot 10^{13}$ cm$^{-3}$. 1) Intrinsic band; 2) dislocation band.

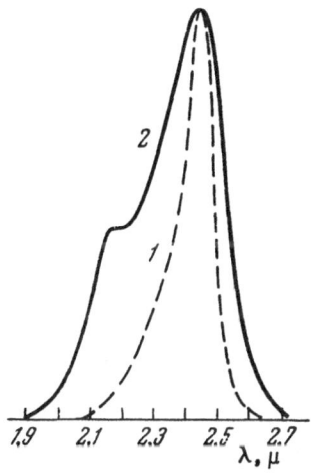

Fig. 11. Enhancement of the intensity of the long-wavelength components due to an increase in the electron density; n-type samples, $N_d \approx 10^5$ cm$^{-2}$. 1) $n_0 = 5 \cdot 10^{13}$ cm$^{-3}$; 2) $n_0 = 2 \cdot 10^{15}$ cm$^{-3}$.

in the temperature was not more than 3°. This forced us to look for a different explanation of the observed departure from linearity. This departure was probably associated with some processes within the centers which took place during recombination. The effective time for these processes limited the number of recombination events which could take place at a given center per unit time. According to our estimates, the average rate of generation of nonequilibrium carriers in the radiating region reached $10^{22}$ cm$^{-3} \cdot$ sec$^{-1}$ (for an injection current of 100 mA) and, therefore, the "saturation" of recombination centers was probably responsible for the observed nonlinearity. When "saturation" occurred, the proportionality between the number of radiative transitions per unit time and $\Delta n$ was no longer observed, and this caused the departure from linearity.

The clear difference between the dependences of the intensities of the dislocation and intrinsic bands on $\Delta n$ meant that a change in the effective lifetime affected in different ways the intensities of these two bands. When the surface was etched instead of ground, the intensity of the intrinsic band increased by a factor of 10-100, while the intensity of the dislocation band (for the same injection current) changed by much less (by a factor of 1.5-2).

Similarly, irradiation with fast electrons greatly altered the intensity of the intrinsic band but had relatively little effect on the intensity of the dislocation band.

Thus, an investigation of the recombination radiation of germanium crystals with dislocation densities of $10^3$-$10^4$ cm$^{-2}$ has established the following important characteristics of the process of radiative recombination at dislocations:

1. The presence of several components in the spectrum indicates that the radiative recombination takes place simultaneously at several levels.

2. The dependence of the radiation band profile on the position of the equilibrium or steady-state Fermi level shows that the radiation is associated with transitions of holes to states lying in the upper half of the forbidden band.

3. The sublinear dependence of the intensity of the dislocation band on the injection current indicates the existence of some intra-center processes responsible for the "saturation" effect.

The investigation has failed to determine the transitions to which the various components correspond, the profiles of these components, and radiative recombination mechanisms. These problems will be considered in Chapter III.

Fig. 12. Weakening of the intermediate-wavelength components due to a reduction of the injection current. The injection current for curve 1 was 10 times that for curve 2.

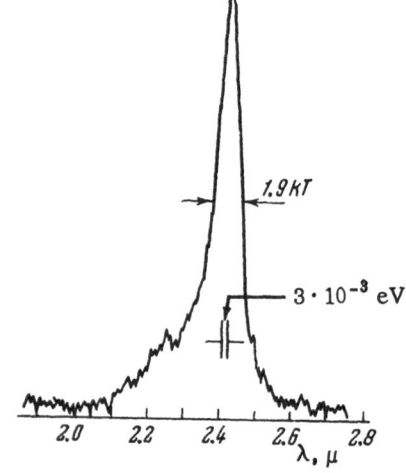

Fig. 13. Automatically recorded recombination radiation spectrum of an n-type germanium sample with an electron density of $1.3 \cdot 10^{13}$ cm$^{-3}$ and a dislocation density of $10^5$ cm$^{-2}$. Injection current was 100 $\mu$A.

CHAPTER III

# Radiative Recombination in Germanium Crystals with High ($10^5$-$10^6$ cm$^{-2}$) Dislocation Densities. Radiative Recombination Mechanism

## §1.   Dependence of the Form of the Spectrum on the Position of the Fermi Level and on the Injection Current

The dependence of the dislocation band profile on the density of equilibrium carriers and on the injection current in crystals with high dislocation densities was, in general, similar to that observed for crystals with $N_d \approx 10^3$-$10^4$ cm$^{-2}$. An increase in the equilibrium electron density or an increase in the injection current enhanced selectively the short-wavelength components of the spectrum. In some cases, it was found that an increase in $n_0$ caused the appearance of not only the short-wavelength, but also the long-wavelength components, as shown in Fig. 11. Moreover, it was found that, sometimes, when the injection current was reduced, the usual gradual decrease in the short-wavelength component intensity was replaced by a fairly strong reduction in the intensity of the intermediate-wavelength components (Fig. 12).

As shown in Chapter II, the changes in the spectra which usually accompanied the changes in the position of the equilibrium or steady-state Fermi level could be explained on the assumption that the dislocation band was due to transitions of holes to levels lying in the upper half of the forbidden band. If we assumed that the radiation was due to such transitions to these levels from the valence band, it was difficult to explain the observed anomalies in the influence of the degree of doping and of the value of the injection current on the dislocation band profile (Figs. 11 and 12). Although the reduction in the intensity of the intermediate-wavelength components (Fig. 12) could be explained by the fact that, in principle, the steady-state population of local levels was not always described by a common quasi-Fermi level, the appearance of the long-wavelength components in the spectrum when the equilibrium electron density was increased was inexplicable.

Thus, an investigation of the recombination radiation spectra of crystals with high dislocation densities established new experimental facts, which indicated that the usually accepted representation of the radiative recombination mechanism would have to be refined.

## § 2. Analysis of the Radiation Band Profile

To obtain information on the radiative recombination mechanism it would be desirable to separate one radiation component and analyze its profile. When the equilibrium carrier density or the injection current was increased, the relative intensity of the short-wavelength components increased and this broadened the radiation band. The short-wavelength components were weakest for a high-resistivity material at low injection currents. Under these conditions, we observed a fairly narrow radiation band whose width was 0.014 eV. Figure 13 shows the spectrum of a germanium sample with an equilibrium electron density $n_0 = 1.3 \cdot 10^{13}$ cm$^{-3}$ for an injection current of 100 $\mu$A. The spectrum was recorded with a sufficiently high resolution so that, in this case, we could draw some reliable conclusions not only about the width but also about the profile of the radiation band.

Assuming that, under these conditions, the spectrum had only one component, we compared the parameters of the experimental curve with the parameters of the theoretical curve, making certain assumptions about the radiative recombination mechanism. The parameters were the width of the curve at half the amplitude $\Delta E$ and the position of the maximum $E_m$ relative to a point obtained by extrapolating the long-wavelength edge to zero.

If the observed radiation is associated with carrier transitions from a band to a local level, with or without the participation of one type of phonon, the line profile should be governed by the energy dependence of the expression for the capture rate $pv\rho_{rad}N_d$, where $p = p(E)$ is the density of carriers of energy E in the band, v is the velocity of the carriers, $\rho_{rad}$ is the radiative capture radius, and $N_d$ is the dislocation density. The parameters of the theoretical curve, calculated on the assumption that $\rho_{rad}$ = const, were as follows: $\Delta E = 2.45$ kT, $E_m$ = kT. When we introduced the dependence of the capture radius on the velocity in the form $\rho_{rad} \propto v^{-\alpha}$, then $\Delta E$ and $E_m$ were found to be smaller. In particular, if $\rho_{rad} \propto v^{-1}$, then $\Delta E = 1.8$ kT and $E_m = kT/2$.

The parameters of the experimental curve, shown in Fig. 13, were as follows: $\Delta E = (1.9 \pm 0.05)$ kT and $E_m = (1.7 \pm 0.15)$ kT.

Thus, the experimentally observed radiation spectrum differed from the theoretical profile calculated on the assumption that the radiation was associated with carrier transitions from a band to a local level without the participation of phonons, or with the participation of phonons of one type only. It should be mentioned that this assumption was not the only possible one. We shall now consider the role of phonons in radiative recombination. The disagreement between the experimental and theoretical curves can be explained, in principle, by assuming that several types of phonon take part in the recombination. The spectrum observed is then the result of the superposition of several radiation lines, each of which corresponds to a transition of a carrier from a band to a local center in which a particular phonon participates. In order to explain the small value of $\Delta E$ for the experimental curve, it is necessary to make an important assumption that these lines are narrow, which means that the dependence of $\rho_{rad}$ on v is strong.

In the interpretation of our results, we have so far assumed that the radiation is due to transitions to local levels introduced by dislocations into the forbidden band. However, there is an alternative explanation based on the concept of dislocation bands, developed in [29, 30]. The interpretation of the radiative recombination spectra on the basis of this concept is difficult because we do not know the width of the dislocation band (or bands — if there are several) or the density-of-states function for these bands. Nevertheless, the radiation line associated with a radiative transition of a carrier from the valence band to the dislocation band can be quite narrow because, firstly, the law of conservation of the components of crystal momentum of a carrier along the dislocation axis should be conserved in the capture event and, secondly, there may be a dependence of $\rho_{rad}$ on v. Therefore, the experimentally observed radiation band (cf. Fig. 13) can obviously be represented

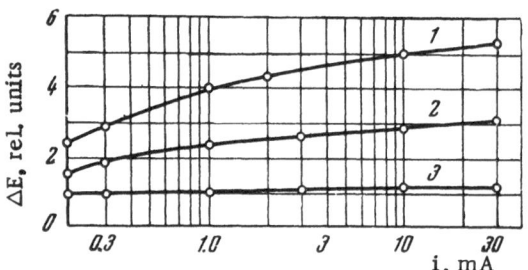

Fig. 14. Dependence on the injection current of the radiation band width at 0.3 (1), 0.5 (2), 0.8 (3) of the maximum amplitude.

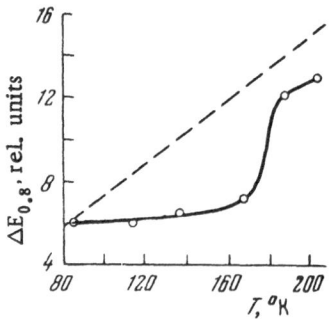

Fig. 15. Temperature dependence of the radiation band width.

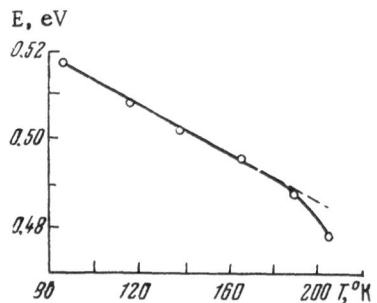

Fig. 16. Temperature dependence of the energy of a transition corresponding to the maximum of the radiation band.

by a superposition of lines corresponding to transitions of carriers from the valence band to the dislocation band in which various phonons participate.

Thus, although the experimental curve does not agree with the theoretical band profile based on the assumption of radiative transitions from the valence band to local levels without phonon participation or with the participation of phonons of one type only, it can probably be represented by the superposition of radiation lines corresponding to transitions in which phonons of various types participate. Individual lines are then associated with the capture of a carrier from the band by a local center in the presence of a strong dependence of $\rho_{rad}$ on v, or are due to transitions from the valence band to the dislocation band.

The results of an investigation of the temperature dependence of the width of the observed radiation band forced us to reject both these explanations.

§3. Temperature Dependence of the Radiation Band Width. Mechanism of Radiative Recombination at Dislocations

Before we consider the temperature dependence of the band width, we must mention that the intensity of the dislocation radiation band decreases when the temperature is increased. The spectrum shown in Fig. 13 was recorded for a current of 100 $\mu$A. An investigation of the temperature dependence of the band width over a wide range of temperatures using this value of the current was impossible because, when the temperature was increased, the accuracy of measurement of the spectra decreased considerably because of the reduction in intensity. Therefore, to carry out this investigation, it was necessary to increase the injection current. This broadened the band, but it was found that the broadening was different at different heights of the profile.

Figure 14 shows the dependence on the injection current of the band width at levels representing 0.3, 0.5, and 0.8 of the maximum amplitude. We can see that the band width at the 0.8 level is practically independent of the injection current.

Thus, it is natural to assume that the spectrum observed for strong injection currents consists of several components whose profiles are independent of the current, and that the energy of the component with the longest wavelength differs from the energy of the other components so much that, when the current is increased, the weak components overlap the strong component relatively little and, therefore, the width of the top part of the composite curve, governed by the width of the strong component, remains unchanged. The temperature dependence reported here refers to the width of this component (denoted by $\Delta E_{0.8}$) at 0.8 of the maximum amplitude.

Figure 15 shows the temperature dependence of $\Delta E_{0.8}$ in the range 80-200°K. It is worth noting that the value of $\Delta E_{0.8}$ is constant in the range 80-170°K. This constancy of $\Delta E_{0.8}$ cannot be explained by assuming that the radiation is due to transitions of holes from the valence band to energy states in the forbidden band, irrespective of the nature of these states. For carrier transitions from an energy band, the width of the radiation band would be proportional to T unless we make the very unlikely assumption of a strong temperature dependence of the function $\rho(v)$. A line representing a linear dependence on T is shown dashed in Fig. 15.

This lack of dependence of the radiation band width on temperature can be easily explained on the basis of the following mechanism: initially, a hole is captured by an excited state of a center, and then a radiative transition takes place within the center. In this case, the shape of the radiation band should be independent of the carrier energy distribution in the energy band, as observed experimentally.

If we make this assumption about the radiative recombination mechanism, the width of the band should be governed by the internal properties of centers and by the energy of phonons which may take part in the radiative transitions. The role of phonons in the radiative recombination at dislocations is not quite clear, but some conclusions about their energy (on the assumption that the line width is governed by the carrier interaction with phonons) can still be made. Since the experiments were carried out at not too low temperatures ($T \approx 100°K$) we could have, in principle, transitions accompanied by the emission or absorption of phonons [4]. Therefore, knowing the line width, we could estimate the energy of phonons participating in the radiative recombination. As already mentioned, the radiative recombination band was found to be fairly narrow (E = 0.014 eV), which indicated that the phonon energy could be only several thousandths of an electron-volt.

Let us now consider Fig. 15. The value of $\Delta E_{0.8}$ remains constant up to about 170°K. When the temperature is increased beyond this point, the radiation band rapidly broadens. As already mentioned, an increase of temperature is accompanied by a reduction in the radiation intensity. At $T \approx 170°K$, the intensity of the investigated band becomes comparable with the intensity of another band which is not noticeable at lower temperatures. Since we know nothing about the latter band, we can only assume that it has a temperature dependence different from the temperature dependence of the band observed at T < 170°K. Because of this difference between the temperature dependences, the intensities of these two bands become comparable at $T \approx 170°K$ and the width of the combined bands is no longer governed by the width of the component observed at low temperatures. This explanation is confirmed by the temperature dependence of the photon energy, corresponding to the maximum of the radiation band (Fig. 16). Up to $T \approx 170°K$, the transition energy depends linearly on temperature and the coefficient of proportionality is $-2.6 \cdot 10^{-4}$ eV/°K, which is close to the temperature coefficient of the forbidden band width ($-3.3 \cdot 10^{-4}$ eV/°K). At $T \approx 170°K$, this dependence is no longer obeyed. The simultaneous change in the nature of the temperature dependences of the band width and of the transition energy indicates that, at $T \approx 170°K$, a new radiation band appears, representing electron transitions which do not play a significant role at low temperatures.

Thus, an investigation of the temperature dependence of the radiation band width shows that the radiation corresponding to one of the components of the spectrum is due to electron transitions within recombination centers and not due to transitions of carriers from a band to such centers. If we extend this conclusion to all components of the spectrum, we find that we are able to account for many experimental observations which are otherwise inexplicable.

## §4. Interpretation of Experimental Results by Means of Intra-Center Transitions

a) Dependence of the Radiation Band Profile on the Fermi Level Position. We shall consider first the dependence of the radiation band profile on the density of equilibrium electrons and on the injection current. If we start with the hypothesis of intra-center radiative transitions, then different components of the spectrum should be ascribed to different local centers. (The nature of these centers

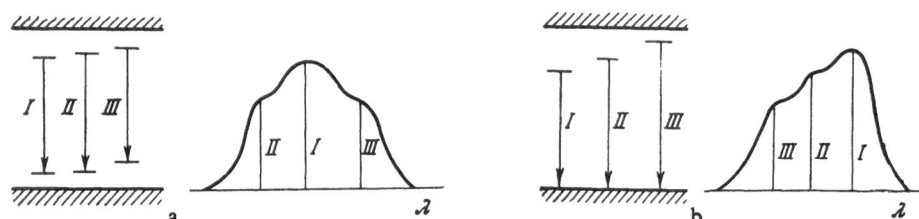

Fig. 17. Radiative transition scheme. a) In the presence of excited states; b) when carriers are captured from a band.

will be discussed in subsections "b" and "c" of the present section). It is natural to assume that different local centers may have different excited hole states. A radiative transition scheme for this case is presented in Fig. 17a. It is evident from this figure that a higher level may correspond to a transition energy III which is less than that for a lower level (transition II). When the Fermi level rises, long-wavelength components appear in the spectrum, as observed experimentally in some cases (Fig. 11). The same explanation can be used to account for the weakening of the intermediate-wavelength components when the injection current is reduced (Fig. 12).

In the case of radiative transitions of carriers from a band to local centers (Fig. 17b), the position of the local level can be found from the radiative recombination data provided we know the energy of the emitted quantum and the energy of the associated phonon, if phonons participate in the radiative recombination. In the case of intra-center transitions, we have to know, moreover, the energy of the excited state. This results in an additional complication in the determination of the level positions from the radiative recombination spectra, because the excited-state energy must be determined independently and there are as yet no reliable data on this energy.

b) Dependence of the Intensity of the Intrinsic and Dislocation Radiation Bands on the Injection Current. If the observed radiation is due to intra-center transitions, then the hole lifetime in the excited state will limit the number of radiative transitions which may take place at a given center per unit time. Thus, we find a natural explanation for the phenomenon of the "saturation" of the radiation intensity when the injection current is increased (Chapter II). In crystals with higher dislocation densities, the "saturation" is weaker. As reported in Chapter II, the dependence of the dislocation radiation band on the injection current (at high currents) has the form $I \propto i^{0.2}$ for crystals with $N_d \approx 10^3$-$10^4$ cm$^{-2}$. When the dislocation density is increased up to $10^5$-$10^6$ cm$^{-2}$, this dependence (at the same current) becomes $I \propto i^{0.6-0.8}$. This change in the nature of the dependence is evidently associated with the fact that, when the concentration of centers is increased, fewer events take place at each of these centers (at a fixed value of the generation rate) and this weakens the "saturation."

The phenomenon of the "saturation" of the intensity is observed, in particular, in crystals with dislocation densities of $10^6$ cm$^{-2}$. In such crystals, the dislocations play the dominant role in the nonequilibrium carrier recombination, i.e., the dislocation recombination channel is practically the only radiative channel (the fraction of band—band radiative transitions is then very small). If the "saturation" would have been applied to the whole dislocation recombination channel, this would have affected the dependence of the intensity of the intrinsic radiation band on the injection current. In fact, the quadratic dependence of the intensity of the intrinsic radiation band on the injection current, observed in the majority of cases at high currents, means that, firstly, $\Delta n > n_0$ and, secondly, $\tau$ = const. The value of $\tau$ of such crystals is governed by the recombination efficiency of dislocations. The saturation of the dislocation recombination channel would have increased $\tau$, i.e., would have altered the dependence of the intensity of the intrinsic radiation band on the injection current (in particular, this dependence should become stronger if the fraction of radiative band—band transitions is small).

Since, in the range of currents at which there is a transition from a linear to a sublinear dependence of the dislocation radiation band intensity on the current, the dependence of the intensity of the intrinsic radiation band remains unaffected, we may conclude that when the radiative component of the dislocation recombination channel becomes saturated, the dislocation channel as a whole (radiative and nonradiative) is still far from saturation. Hence, we may conclude that the "radiators" associated with dislocations have an electron structure different from the electron structure of a dislocation as a whole. If we bear in mind that (Chapter IV) only an $\sim 10^{-4}$ fraction of recombination events at dislocations produces radiation at T = 80°K, it is natural to assume that radiative recombination events may take place not over the whole length of a dislocation but at some parts of it. Such parts may be various irregularities and defects in the dislocation structure. Although this conclusion cannot be justified rigorously, it is supported indirectly by several experimental observations.

c) Dependence of the Intensity and Profile of the Dislocation Radiation Band on the Dislocation Structure of a Crystal. Investigation of the dependence of the intensity of the dislocation radiation band on the density of randomly distributed dislocations showed that when $N_d \approx 10^5$-$10^6$ cm$^{-2}$, the intensity depended very weakly on $N_d$ (at a fixed injection current). The causes of this lack of dependence will be considered in subsection "d" of the present section. It was found that the intensity of the radiation of crystals with dislocation walls was considerably stronger (by a factor of 3-4) than the intensity of radiation of crystals with a high density of randomly distributed dislocations. Bearing in mind the nature of the dependence of the intensity on $N_d$, it was difficult to account for this strong difference simply by an increase in the average dislocation density. As mentioned in Chapter I, the dislocation walls observed in the investigated crystals were frequently in the form of networks of two intersecting dislocation arrays. In this case, the dislocations obviously have a large number of jogs, formed by the intersection of moving dislocations, or nodes (the latter are formed when the planes in which the dislocation arrays are located coincide exactly; such coincidence of the planes could not be confirmed by metallographic investigations because of their insufficient accuracy).

In view of this, the higher radiation intensity observed in crystals with dislocation walls could be explained by assuming that the jogs or nodes on dislocations were the "radiators."

Since the radiation spectrum associated with dislocations consisted of many components, it was natural to attempt to find what these individual components represent. Initially, an attempt was made to ascribe individual components to different types of dislocations. In this connection, it should be mentioned that a reliable analysis of the structure of germanium, particularly at high dislocation densities, is very difficult. This is because we cannot use the method of dislocation decoration [40]. Since we were unable to analyze in detail the dislocation structure, we carried out measurements on two series of samples with very different dislocation structures: in the samples of one series, the dislocations were concentrated in walls corresponding to the {110} planes, while in samples of the other series, the dislocations were distributed at random. It was found that the radiation spectra of samples of these two series differed considerably (Fig. 18). The spectra of the samples with dislocation walls had strong components of wavelength 2.4 μ. Special control experiments carried out showed that the difference between the spectra was not due to different electron densities in the regions with a high concentration of dislocations.

It is known [41] that dislocations forming low-angle boundaries in the {110} planes lie along the [211] direction. Moreover, it is known [41, 42] that crystals pulled along the [111] axis contain randomly distributed dislocations along the directions [111], [110], [211]. From these data, it was natural to conclude that the difference between the spectra of the samples of the two series was due to the fact that the samples with dislocation walls contained mainly dislocations of the [211] type, which gave rise to the radiation of wavelength 2.4 μ. However, this conclusion was contradicted by the results obtained for crystals pulled along the [211] axis, in which, in spite of the predominance of dislocations of the [211] type, the component of the radiation of wavelength 2.4 μ had a lower relative intensity. Moreover, the assumption that the [211] dislocations in the walls had an electron structure different from the structure of randomly distributed [211] dislocations had to be rejected because it was found that the spectrum of crystals containing dislocation walls was independent of the distance between dislocations in the walls. (This distance was of the order of several microns, and, therefore, we could hardly expect that the interaction between the dislocations could have affected their electron structure.)

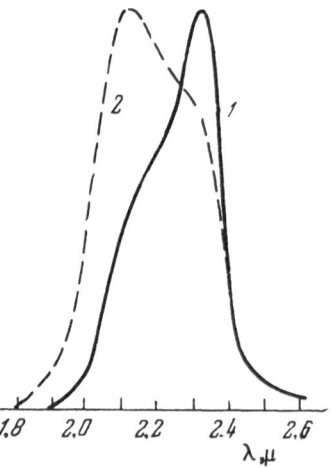

Fig. 18. Dependence of the radiation band profile on the dislocation structure. 1) Dislocation walls; 2) random distribution of dislocations.

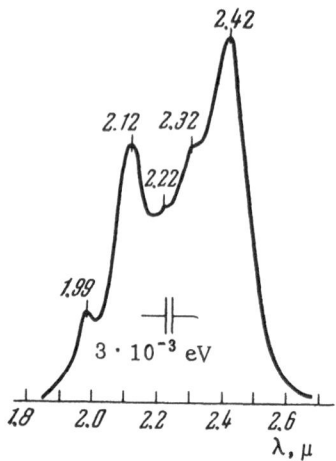

Fig. 19. Radiative recombination spectrum of an n-type germanium sample with $n_0 = 10^{15}$ cm$^{-3}$, $N_d \approx 10^6$ cm$^{-2}$.

Thus, an attempt to assign different components of the radiation to dislocations of different types met with considerable difficulties. However, this difference between the spectra could be easily explained by assuming that individual components of the radiation corresponded not to different dislocations but to different types of imperfection in the dislocation structure. We should bear in mind that the type of dislocation also plays a role, since, for example, the microstructure of a jog depends on the nature of the intersecting dislocations. From this point of view, the observed difference between the spectra of samples having different dislocation structures is due to the fact that the component of wavelength 2.4 $\mu$ corresponds to local levels associated with jogs or nodes, which are very likely to be present in the [211] dislocations.

The reported experimental data support the hypothesis of the mechanism of radiative recombination at dislocations in which radiative recombination events take place not over the whole dislocation but at certain points on it. Such points are various imperfections of the dislocation structure, such as jogs or nodes. Such jogs or nodes, like noninteracting isolated objects, have local levels, while dislocations as a whole have a band spectrum of energy states due to the periodicity of the dislocation structure [29]. The radiation appears due to intra-center transitions after the capture of a hole into an excited state.

It has already been mentioned that intra-center transitions prevent us from determining the positions of local levels from the radiative recombination spectra because it is necessary to know the energy of the excited states (and, in general, the energy of a phonon which may take part in a radiative recombination event). Nevertheless, some information on the positions of local levels associated with "radiators" on dislocations can still be obtained from the dependence of the radiation band profile on the density of equilibrium electrons. As mentioned in Chapter II, this dependence is due to a change in the degree of population of the levels. It should be stated here that the ratio of the probabilities of population of two levels is independent of the Fermi level position if this level is sufficiently far from the two levels in question (in practice, the Fermi level must be separated by a gap greater than 3 kT). The fact that when the Fermi level varies within an energy range of 0.04 eV (when the carrier density is increased from $5 \cdot 10^{13}$ to $9 \cdot 10^{15}$ cm$^{-3}$) the ratio of the degree of population of various levels changes very considerably (Fig. 7), shows that the local levels associated with "radiators" are concentrated in a fairly narrow (not more than 0.1 eV wide) range of energies. It is fairly difficult to say anything definite about the separation between these levels and the conduction band, in spite of the fact that we know the position of the Fermi level (0.08-0.04 eV from the bottom of the conduction band) because, in the calculation of the population of the "radiator" levels, we cannot use the position of the Fermi level in the bulk of a sample in view of the presence, around a dislocation, of a cylindrical region of positive space charge. The number of the

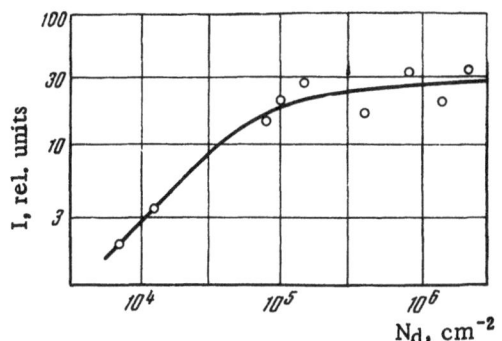

Fig. 20. Dependence of the intensity of the dislocation radiation band on the density of dislocations. Injection current was 100 mA.

"radiator" levels in the forbidden band is fairly high (Fig. 19). However, it is not always possible to associate individual components of the spectrum with particular types of "radiator" because the methods for the detailed analysis of the dislocation structure of germanium are not yet fully developed.

d) Dependence of the Intensity of the Dislocation Radiation Band on the Injection Current and Dislocation Density. If the radiation is due to intra-center transitions, its intensity can be written in the form

$$I = \frac{p_d}{\tau_r} , \qquad (9)$$

where $p_d$ is the density of holes in the excited state and $\tau_r$ is the lifetime of hole in the excited state in the case of radiative transitions.

To calculate $p_d$, it is necessary to know the concentration of "radiators" and their properties, in particular, the excited state energy. Since these quantities are not known, the calculations given below can be considered simply as an illustration of the way in which Eq. (9) can be used to explain the observed dependence of the radiation intensity on the injection current and dislocation density.

The change in the hole density in an excited state per unit time can be written thus:

$$\frac{dp_d}{dt} = - \frac{p_d}{\tau_r} - \gamma_p p_d P_{vM} + \gamma_p p (M - p_d), \qquad (10)$$

where M is the concentration of "radiators"; $\gamma_p$ is a coefficient representing the capture of holes into the excited state; $P_{vM} = P_v^{-\frac{E_M}{kT}}$ is the reduced density of states; $P_v$ is the density of states in the valence band; and $E_M$ is the energy of the excited state, measured from the top of the valence band.

The first term on the right-hand side of Eq. (10) represents radiative transitions, the second term represents the thermal emission of holes from the excited state, and the third term represents the capture of holes into the excited state.

We shall assume that the first term is much smaller than the other two terms. This assumption is valid if

$$\tau_r \gg \frac{1}{\gamma_p P_{vM}} ,$$

i.e., if the radiative lifetime of a hole is much longer than the lifetime for thermal transitions. Then, under steady-state conditions, we have

$$\gamma_p p_d P_{vM} = \gamma_p p (M - p_d),$$

whence

$$p_d = \frac{pM}{p + P_{vM}} . \qquad (11)$$

Substituting the expression for $p_d$ into Eq. (9), and noting that $M = kN_d$, where k is the number of "radiators" per unit length of a dislocation, we obtain an expression for the radiation intensity in the form

$$I = \frac{p}{p + P_{vM}} \frac{kN_d}{\tau_r}. \qquad (12)$$

Using this expression, we shall consider first the dependence of the intensity I on the injection current, assuming that p is proportional to this current. At low injection currents, $p \ll P_{vM}$, and Eq. (12) becomes

$$I = \frac{p}{P_{vM}} \frac{kN_d}{\tau_r}, \qquad (13)$$

in agreement with the experimentally observed linear law. At high currents, $p \propto P_{vM}$ and a deviation from the linear law is observed. If $p \gg P_{vM}$, then $p_d = M$ and $I = kN_d/\tau_r = const$.

We shall now consider the dependence of I on the dislocation density at a fixed value of the injection current, and we shall assume that the injection current is sufficiently high, i.e., we shall assume those conditions under which the experimental dependence $I(N_d)$, shown in Fig. 20, was investigated.

At low ($\sim 10^3$ cm$^{-2}$) dislocation densities, dislocations do not dominate the recombination and, therefore, p is practically independent of $N_d$. In this case,

$$I = \frac{p}{p + P_{vM}} \frac{kN_d}{\tau_r} = const\, N_d,$$

i.e., we obtain a linear dependence in agreement with the experimental results. At high values of $N_d$ ($\sim 10^5$ cm$^{-2}$), dislocations represent the main recombination channel and, in this case, $\tau = 1/\sigma_R N_d$ ($\sigma_R$ is the recombination efficiency of dislocations) and $p = g\tau = g/\sigma_R N_d$. Moreover, since the lifetime is short, we have $p \ll P_{vM}$ and

$$I = \frac{p}{P_{vM}} \frac{kN_d}{\tau_r} = \frac{g}{P_{vM} \sigma_R N_d} \frac{kN_d}{\tau_r} = \frac{gk}{P_{vM} \sigma_R \tau_r} = const.$$

At high dislocation densities, the radiation intensity is independent of $N_d$ (at a fixed injection current), again in agreement with the experimental data.

Moreover, the formulas obtained predict the experimentally observed dependence of the degree of "saturation" on the dislocation density at a fixed current (cf. subsection "b"). In fact, at low values of $N_d$, dislocations do not play an important role in recombination, $\tau$ is large, and, consequently, at high currents, $p \approx P_{vM}$, and the dependence of I on p is nonlinear, i.e., saturation is observed. At high values of $N_d$ we have, as already mentioned, $p \ll P_{vM}$, and, consequently, $I \propto p$, i.e., there is no "saturation." In fact, we observed experimentally only a weakening of the "saturation" for crystals with higher dislocation densities.

We shall now try to obtain some quantitative estimates. As already stated, the strong "saturation" of the dependence of the radiation intensity on the injection current is observed for crystals with $N_d \approx 10^4$ cm$^{-2}$. For such crystals, $\tau \approx 10^{-6}$ sec, T = 80°K, and the generation rate (for a current of 100 mA) is estimated to be $\sim 10^{22}$ cm$^{-3} \cdot$ sec$^{-1}$. Hence, we find that $p \approx 10^{16}$ cm$^{-3}$. Since strong "saturation" is observed in this case, it follows that $p \approx P_{vM}$, i.e., $P_{vM}$ is also $\sim 10^{16}$ cm$^{-3}$, and hence $E_M = 0.03$-$0.04$ eV.

From the condition $\tau_r \gg 1/\gamma_p P_{vM}$, assuming that $P_{vM} \approx 10^{16}$ cm$^{-3}$ and $\gamma_p \approx 10^{-8}$ (i.e., $\sigma_p \approx 10^{-15}$ cm$^2$, which is quite reasonable for excited states), we obtain $\tau_r \gg 10^{-8}$ sec.

Thus, the dependence of the radiation intensity on the injection current and dislocation density obtained in our treatment is in qualitative agreement with the experimental data if reasonable values are used for the parameters.

e) Role of Impurity Atmospheres. In discussing the nature of "radiators" associated with dislocations, the question arises whether these "radiators" are perhaps impurity atoms at dislocations. Dislocations in a crystal are surrounded by an uncontrolled impurity atmosphere. In principle, the dislocation radiation may be associated not with a dislocation itself, but with atoms in an impurity atmosphere around it. However, some data suggest that this hypothesis should be rejected. In particular, if a current of about 2 A is passed for a short time (of the order of one second) through the contact, the resultant strong local heating and the pressure applied by the contact produce, near the contact, dislocations which are free of impurity atmospheres. After such "forming," the intensity of the intrinsic radiation band decreases and the intensity of the dislocation band increases (these experiments were carried out on crystals with an initial dislocation density of $10^3$-$10^4$ cm$^{-2}$). This result confirms that atoms in impurity atmospheres do not make any significant contribution to the radiative recombination at dislocations. However, it should be mentioned that, because of the complexity of the problem of the role of impurity atmospheres, further careful control experiments are needed to obtain the final answer to this question.

Thus, an investigation of the dislocation band profile of samples with low equilibrium electron densities, of the temperature dependence of the band width, and of the dependence of the band intensity on the injection current, allows us to draw certain conclusions about the mechanism of radiative recombination at dislocations:

1. Radiative recombination events do not occur over the whole length of a dislocation but only at certain parts. These parts are the various imperfections of the dislocation structure — for example, jogs or nodes formed at intersections of dislocations. These imperfections are noninteracting isolated objects which have local levels, while the dislocation as a whole is described by a band of energy states associated with the periodicity of its structure.

2. The observed radiation results from intra-center transitions after the capture of a hole into an excited state.

3. From the dependence of the radiation band profile on the Fermi level position, we may conclude that the local levels of "radiators" are concentrated in a fairly narrow (not more than 0.1 eV) range of energies. To determine exactly the level positions, it is necessary to know the energy of the corresponding excited states.

4. The width of the radiation band of a high-resistivity material indicates that only phonons of low energies (of the order of several thousandths of an electron-volt) can take part in radiative recombination events.

CHAPTER IV

# Determination of the Quantum Yield of the Recombination Radiation Associated with Dislocations

The quantum yield of photoluminescence ($\gamma_q$) is defined as the ratio of the number of emitted quanta to the number of absorbed quanta under steady-state conditions. In the case of recombination radiation, irrespective of its method of excitation, the quantum yield is understood to mean the ratio of the emitted quanta to the number of generated electron—hole pairs. To ensure that the value of $\gamma_q$, defined in this way, represents the properties of a given type of recombination center and not the property of a material, it is necessary to measure this quantum yield under such conditions that the recombination at a given type of center is the dominant process. Thus, we shall regard the quantum yield of the recombination radiation to be the ratio of the number of quanta of recombination radiation to the total number of recombination events at a given type of recombination center.

The knowledge of the value of $\gamma_q$ is very important for the understanding of the mechanism of radiative recombination at dislocations. To determine this quantity, it is necessary to measure independently the number of quanta I emitted per unit time per unit volume and the value of the volume generation rate g under such conditions that dislocations dominate the recombination of nonequilibrium carriers. In this case, $\gamma_q = I/g$.

The investigations of the recombination radiation spectra described in Chapters II and III were carried out using electrical injection of carriers from a metal pressure contact into a sample in the form of a

Weierstrass sphere. It is fairly difficult to relate the observed radiation intensity to the quantity I. Moreover, the value of the generation rate can be estimated only very approximately. Therefore, quantitative measurements are best carried out on plane-parallel samples using uniform volume excitation.

To achieve a sufficiently powerful volume excitation, we used bombardment with a beam of fast electrons. A lead sulfide photoresistor was used as the radiation detector. In addition to the values of I and g, we also determined the density of nonequilibrium carriers $\Delta p$ from the conductivity signal due to the bombardment with a beam of fast electrons (we shall call this the $\beta$-conductivity). Knowing the value of $\Delta p$, we used the formal expression for the intensity $I = \Delta p \rho_{rad} v N_d$ to calculate the value of the "radiative capture radius" $\rho_{rad}$ and compare it with the published values of the radius for the capture by dislocations ($\rho$), in the nonradiative transition case. The value of $\rho_{rad}$ obtained in this way cannot be given an explicit meaning (see the beginning of the present section) but, nevertheless, the ratio $\rho_{rad}/\rho$ gives the quantity $\gamma_q$, which is of interest to us.

An electrostatic Van de Graaff generator was used as the source of fast electrons. The electron beam was modulated by applying voltage pulses from a multivibrator to the control grid of the electron gun of the generator. To synchronize the multivibrator, we used pulses from a photomultiplier illuminated with modulated light from an incandescent lamp outside the generator. The lamp was modulated by an apertured disk which was rotated by a synchronous motor. The motor was supplied with a sinusoidal voltage from a ZG-10 oscillator, amplified by a special power amplifier. In this way, we were able to achieve stable modulation and smooth control of the frequency.

The electron beam reaching a sample excited recombination radiation, which was recorded with the cooled lead sulfide photoresistor placed, for the sake of protection from x-ray radiation, inside a lead cylinder. The recombination radiation passed through a germanium filter (with an antireflection coating), which was kept at room temperature and which absorbed practically completely the radiation of $\lambda < 1.8\ \mu$, i.e., it absorbed, in particular, the intrinsic recombination radiation of a sample kept at a temperature close to that of liquid nitrogen.

The signal from the photoresistor was passed through a narrow-band amplifier ($f_0 = 630$ cps, $\Delta f = 8$ cps, $K \approx 10^6$) and measured with a cathode voltmeter. The $\beta$-conductivity signal was fed to a 28IM amplifier and measured with a cathode voltmeter. Both amplifiers responded only to the first harmonic of the signal.

The magnification of the optical system, which focused the recombination radiation onto the photoresistor surface, was selected in such a way that the image of the sample covered the working area of the photoresistor. Then, the photoresistor signal was proportional to the illumination.

The photoresistor was calibrated using a "black body" and filters which transmitted only that part of the spectrum which contained the dislocation band of the recombination radiation (antireflection-coated germanium and Pyrex were used for such filters).

From the value of the signal at the calibrated photoresistor we calculated the number of quanta I of light generated to a unit volume of a sample. The value of I was given by the formula

$$I = \frac{wY\beta^2}{d\delta T}\ ,\qquad\qquad (14)$$

where w is the illumination of the photoresistor when the signal−noise ratio is 1; Y is the signal−noise ratio; $\beta$ is the magnification of the optical system; d is the thickness of the sample; T is the transmission of the germanium filter; and $\delta$ is the fraction of quanta which emerge from a sample within the limits of a solid angle determined by the aperture of the lens. The value of I was calculated allowing for the multiple reflection ot the radiation from the surfaces of a sample.

The rate of volume generation g by the incident fast electrons was calculated from the formula

$$g = \frac{Ej}{\varepsilon d}\ ,\qquad\qquad (15)$$

TABLE 2

| Sample No. | $n_0$, cm$^{-3}$ | $N_d$, cm$^{-2}$ | $j$, $\mu A/cm^2$ | $g$, cm$^{-3} \cdot$ sec$^{-1}$ | $I$, cm$^{-3} \cdot$ sec$^{-1}$ | $\gamma_q$ |
|---|---|---|---|---|---|---|
| 4 | $6.5 \cdot 10^{14}$ | $5 \cdot 10^4$ | 0.5 | $1.7 \cdot 10^{19}$ | $6.2 \cdot 10^{15}$ | $3.7 \cdot 10^{-4}$ |
|   |   |   | 0.75 | $2.5 \cdot 10^{19}$ | $7.7 \cdot 10^{15}$ | $3.1 \cdot 10^{-4}$ |
| 5 | $5.2 \cdot 10^{14}$ | $4 \cdot 10^4$ | 0.5 | $1.7 \cdot 10^{19}$ | $3.8 \cdot 10^{15}$ | $2.2 \cdot 10^{-4}$ |
|   |   |   | 1.0 | $3.4 \cdot 10^{19}$ | $6.4 \cdot 10^{15}$ | $1.9 \cdot 10^{-4}$ |
| 7 | $3.8 \cdot 10^{14}$ | $10^5$ | 1.25 | $4.2 \cdot 10^{19}$ | $8.2 \cdot 10^{15}$ | $1.9 \cdot 10^{-4}$ |
|   |   |   | 2.25 | $7.4 \cdot 10^{19}$ | $1.28 \cdot 10^{16}$ | $1.7 \cdot 10^{-4}$ |

where E is the energy lost by the incident electrons in the thickness of the sample; j is the density of the current of the fast incident electrons; $\varepsilon$ is the average ionization energy; and d is the thickness of the sample.

Analysis of the ionization distribution curves [43] showed that electrons of 1 MeV energy lost about 80% of their energy in a sample 0.5 mm thick. The average ionization energy was assumed to be 3 eV [44].

The density of nonequilibrium carriers in a sample was calculated using a formula applicable to the case when the load resistance was much greater than the resistance of the sample [45]:

$$\Delta p = \frac{V_\beta n_0^2}{iA - V_\beta n_0} \frac{1}{c}, \tag{16}$$

where $V_\beta$ is the $\beta$-conductivity signal; $n_0$ is the equilibrium density of carriers in the sample, found from the Hall measurements; i is the current through the sample; $A = l / Se\mu_n$; $l$ is the length of the sample; S is the cross-sectional area of the sample; $c = 1 + (\mu_p/\mu_n)$; it was assumed that $\Delta n = \Delta p$, since the concentration of recombination centers was low.

In these measurements, we used an n-type germanium sample with a carrier density $n_0 \approx 5 \cdot 10^{14}$ cm$^{-3}$ and a dislocation density of ~$10^5$ cm$^{-2}$ generated by thermal shock during growth.[†] The investigated samples were made in the form of polished plane-parallel plates of 10 × 5 × 0.5 mm dimensions. Table 2 lists the results of measurements for three samples.

The scatter of the values of $\gamma_q$ for the same sample but different currents was about 20%, which was a measure of the accuracy of the determination of this quantity by the method used. The difference between the results for samples Nos. 4 and 5, which was outside the limits of this error, could be due to the characteristic properties of the radiative recombination at dislocations mentioned in Chapter III. Samples Nos. 4 and 5 could have had different dislocation structures and different concentrations of "radiators," which would have accounted for the difference between the values of $\gamma_q$.

In addition to the values of I and g, we measured the nonequilibrium carrier density, which made it possible to determine the value of $\rho_{rad}$ and compare it with the value of $\rho$ known from the recombination experiments. The value of $\rho = 6 \cdot 10^{-6}$ cm at T = 130°K was reported in [31]. In our experiments at the same temperature, we found that $\rho_{rad} = 2 \cdot 10^{-11}$ cm.

Thus, we found that $\rho_{rad}/\rho = 3 \cdot 10^{-6}$. At the same temperature, the value of I/g was found to be $5 \cdot 10^{-6}$. Bearing in mind $\rho_{rad}$ and $\rho$ measured for different crystals, the agreement between the values of $\rho_{rad}/\rho$ and I/g should be regarded as satisfactory.

---

[†] The use of thermal shock, i.e., a sudden change of thermal conditions during crystal growth, made it possible to obtain crystals with a homogeneous distribution of dislocations over a relatively large area. However, it was difficult to reach $N_d > 10^5$ cm$^{-2}$.

The value of $\gamma_q$ was found to decrease when the temperature was increased. Since the recombination radiation spectrum of crystals with $n_0 \approx 5 \cdot 10^{14}$ cm$^{-3}$ consisted of many components (Fig. 7), and we measured the total radiation intensity in our experiments, the nature of the temperature dependence of $\gamma_q$ could not be used to deduce the temperature dependence of the quantum yield of individual components.[†] It was very difficult to use the same method to measure the quantum yield of high-resistivity samples, whose spectra had mainly one component each, because the properties of these samples varied very rapidly when they were subjected to bombardment with electrons of $E \approx 1$ MeV.

Thus, the recombination radiation at dislocations has a very small quantum yield ($10^{-4}$). This very low value of $\gamma_q$ confirms the assumption made in Chapter III on the mechanism of radiative recombination at dislocations and on the nature of "radiators" associated with dislocations. The majority of recombination events at dislocations are nonradiative (more exactly, they take place without the emission of quanta of energy greater than 0.35 eV). The nonradiative recombination channel of dislocations is very efficient ($\rho \approx 6 \cdot 10^{-6}$ cm at $T = 130°$K), which is possibly associated with the energy band spectrum of dislocations [29, 30]. Radiative recombination takes place not over the whole length of a dislocation but at specific parts of it, i.e., at "radiators," which are usually various imperfections of the dislocation structure. Naturally, under such conditions, the quantum yield of recombination radiation would be small.

## Conclusions

The investigations reported here of the recombination radiation of germanium crystals with dislocations have yielded new data on the mechanism of radiative recombination at dislocations and on the energy structure of "radiators" associated with dislocations. The introduction of the concept of intra-center transitions taking place at "radiators" which are various imperfections in the dislocation structure has made it possible to explain various phenomena observed in the investigation of the recombination radiation of crystals with dislocations — in particular, the dependence of the radiation band intensity on the injection current and dislocation density. Analysis of the radiation band profile has yielded some information on the role of phonons in the radiative recombination at dislocations.

Nevertheless, the problem of the mechanism of radiative recombination cannot be regarded as fully solved. Although the concept of intra-center transitions has been found to be very useful, the energy of excited states and the lifetime of a hole in an excited state remain unknown. To determine the values of these parameters, one would need to study the kinetics of recombination radiation. Moreover, the measurements should be carried out at lower temperatures.

In conclusion, the author takes this opportunity to express his deep gratitude to his scientific director, V. S. Vavilov, for his constant attention, help, and support throughout this investigation. The author is also very grateful to B. M. Vul for his interest in this research and for discussing the results obtained; to L. S. Smirnov and A. V. Spitsyn for their help and valuable discussions; and to M. S. Murashov and V. S. Konoplev, with whose help some of the results reported were obtained. The author is also much indebted to L. S. Silonov, B. D. Kopylovskii, and N. V. Zhigalov for the construction, assembly, and adjustment of the radiofrequency apparatus; to V. D. Kopanev, S. I. Vintovkin, and Yu. M. Kolotov for their help in assembling the apparatus; and to laboratory assistant L. N. Egorova for the preparation of the samples used in the measurements.

---

[†] By the quantum yield of an individual component, we naturally understand the value $\gamma_q$ under such conditions that the spectrum consists of only one component.

## Literature Cited

1.  R. N. Hall, G. E. Fenner, J. D. Kingsley, T. J. Soltys, and R. O. Carlson, Phys. Rev. Letters, 9:366 (1962).
2.  C. Hilsum, Brit. Commun. and Electronics, 10:450 (1963).
3.  J. R. Haynes, Phys. Rev., 98:1866 (1955).
4.  J. R. Haynes, M. Lax, and W. F. Flood, J. Phys. Chem. Solids, 8:392 (1959).
5.  G. G. Macfarlane, T. P. McLean, J. E. Quarrington, and V. Roberts, Phys. Rev., 108:1377 (1957).
6.  B. B. Brockhouse, J. Phys. Chem. Solids, 8:400 (1959).
7.  J. R. Haynes, M. Lax, and W. F. Flood, Proc. Internat. Conf. Semiconductor Phys. (Prague, 1960), p. 423.
8.  G. G. Macfarlane, T. P. McLean, J. E. Quarrington, and V. Roberts, Phys. Rev., 111:1245 (1958).
9.  W. van Roosbroeck and W. Shockley, Phys. Rev., 94:1558 (1954).
10. J. R. Haynes, Phys. Rev. Letters, 4:361 (1960).
11. J. R. Haynes and W. Westphal, Phys. Rev., 101:1676 (1956).
12. Ya. E. Pokrovskii and K. I. Svistunova, Fiz. tverd. tela, 3:2820 (1961).
13. Ya. E. Pokrovskii and K. I. Svistunova, Paper presented at Third All-Union Conference on Photoelectric Phenomena in Semiconductors (Kiev, 1963).
14. E. I. Johnson, I. Filinski, and H. Y. Fan, Proc. Internat. Conf. Semiconductor Phys. (Exeter, 1962), p. 375.
15. J. J. Lambe, C. C. Klick, and D. L. Dexter, Phys. Rev., 103:1715 (1956).
16. R. Newman, Phys. Rev., 105:1715 (1957).
17. C. Bénoit à la Guillaume, J. Phys. Chem. Solids, 8:150 (1959).
18. G. L. Pearson, W. T. Read, and F. G. Morin, Phys. Rev., 93:666 (1954).
19. W. T. Read, Phil. Mag., 45:775 (1954).
20. W. T. Read, Phil. Mag., 45:1119 (1954).
21. W. T. Read, Phil. Mag., 46:111 (1955).
22. R. Logan, G. Pearson, and D. Kleinman, J. Appl. Phys., 30:885 (1959).
23. G. K. Wertheim and G. L. Pearson, Phys. Rev., 107:694 (1957).
24. A. Z. Belyaev, V. N. Vasilevskaya, and E. G. Milekhin, Fiz. tverd. tela, 2:227 (1960).
25. M. I. Iglitsyn and L. I. Kolesnik, Fiz. tverd. tela, 2:1542 (1960).
26. I. W. Allen, J. Electron. Control, 1:580 (1955).
27. W. Shockley, Phys. Rev., 91:228 (1953).
28. R. K. Mueller, J. Appl. Phys., 30:2015 (1959).
29. V. L. Bonch-Bruevich and V. B. Glasko, Fiz. tverd. tela, 3:36 (1961).
30. Yu. V. Gulyaev, Fiz. tverd. tela, 3:1094 (1961).
31. L. I. Kolesnik, Fiz. tverd. tela, 4:1449 (1962).
32. A. G. Tweet, Phys. Rev., 99:1245 (1955).
33. S. A. Kulin and A. D. Kurtz, Acta Met., 2:354 (1954).
34. R. L. Bell and C. A. Hogarth, J. Electron. Control, 3:455 (1957).
35. N. Schwuttke, J. Electrochem. Soc., 106:315 (1959).
36. B. A. Strelkov, Doklady Akad. Nauk SSSR, 125:290 (1959).
37. P. Jacquinot, Rev. Opt., 33:653 (1954).
38. A. S. Toporets, Monochromators. Gostekhizdat (1955).
39. O. D. Dmitrievskii, B. S. Neporent, and V. A. Nikitin, Uspekhi fiz. nauk, 64:447 (1958).
40. W. C. Dash, J. Appl. Phys., 27:1193 (1956).
41. S. G. Gressel and A. A. Powell, Progress in Semiconductors, 2:137 (1957).
42. W. Bardsley, R. L. Bell, and B. W. Strangham, J. Electron. Control, 5:19 (1958).
43. B. M. Vul, V. S. Vavilov, L. S. Smirnov, and G. N. Galkin, Atomnaya energiya, 2:533 (1957).
44. V. S. Vavilov, Uspekhi fiz. nauk, 75:263 (1961).
45. S. M. Ryvkin, Photoelectric Effects in Semiconductors. Fizmatgiz (1963). [English translation: Consultants Bureau, New York (1964).]

# IMPACT IONIZATION OF IMPURITIES IN GERMANIUM
## AT LOW TEMPERATURES †

## É. I. Zavaritskaya

## Introduction

At the beginning of the 20th century, Townsend [1] introduced the concept of impact ionization to explain the appearance and behavior of electric charges in gases. The theory of impact ionization developed by Townsend and his followers has accounted successfully for a wide range of various phenomena observed when an electric current is passed through a gas: the voltage dependence of the current, the influence of the density of the gas and of the interelectrode gap on the breakdown voltage, Paschen's empirical law, etc.

In the late 1920's, A. F. Ioffe [2] suggested that the electrical breakdown of solid dielectrics is also due to impact ionization. Since then, many theoretical investigations have dealt with impact ionization in solid dielectrics, but a convincing experimental proof of impact ionization in a solid has been obtained only recently in an investigation of the breakdown of p−n junctions in semiconductors [3]. In such a junction, the breakdown develops in a thin carrier-depleted layer between the n- and p-type retions in the semiconductor. The depleted layer gets thicker as the reverse voltage is increased, but even in very pure germanium it does not exceed several tens of microns. Therefore, the field intensity at relatively low voltages across a p−n junction reaches several hundreds of kilovolts per centimeter. In such fields, free electrons acquire sufficient energy to transfer, by collision, valence electrons to the conduction band.

The high-field region in a p−n junction is occupied by a space charge and at all temperatures, even the lowest ones, all impurities in this region are ionized [4]. Therefore, the impact ionization in a p−n junction can take place only by the ionization of the atoms of the crystal matrix accompanied by electron transitions from the valence to the conduction band. The minimum energy which must be supplied to an electron which is close to the upper edge of the valence band must be at least about 1.5 times greater than the forbidden band width [5].

Completely different impact ionization conditions obtain in a homogeneous semiconductor at such low temperatures that the energy of thermal vibrations is insufficient for the ionization of impurities. At sufficiently low temperatures, the scattering of carriers by phonons is weak and, in a semiconductor with a fairly low concentration of other scatterers, the mean free path of carriers is quite long [6]. On the other hand, some impurities (for example, elements of groups III and V) have a very low ionization energy in germanium; in particular, this energy is less than the optical phonon energy [7]. Therefore, at low temperatures, very

†Dissertation for the degree of Candidate of Physico-Mathematical Sciences. Scientific directors: B. M. Vul and A. I. Shal'nikov, Corresponding Members of the Academy of Sciences of the USSR.

favorable conditions exist for the impact ionization of impurities, which then occurs in very weak fields, of the order of 1 V/cm [8, 9].

A characteristic feature of the impact ionization of impurities is the participation in the processes of an avalanche of carriers of one sign — either electrons or holes. The low value of the dissipated power, the high thermal conductivity at low temperatures, and the efficient heat removal in liquid helium suppress almost completely the thermal effects, i.e., the thermal generation of carriers, the nonequilibrium distribution of the field due to nonuniform heating, and irreversible structural changes.

The favorable conditions under which this low-temperature "breakdown" occurs provide an opportunity for a comprehensive investigation of the process of impact ionization in a solid [10-19].

CHAPTER I

## Impact Ionization (Review)

### §1. Impact Ionization in Rarefied Gases and in Semiconducting p − n Junctions

Townsend investigated gas discharges and put forward a hypothesis that electrons and ions moving in an electric field can acquire an energy sufficient for the impact ionization of an atom (or a molecule) when these electrons or ions collide inelastically with it. If the electrons liberated in such a collision also acquire the energy necessary for the ionization, an avalanche is produced which increases in intensity during its motion to the anode [1].

According to Townsend, in a path of length dx, n electrons generate dn = $\alpha$dx secondary electrons, where $\alpha$ is the impact ionization coefficient, equal to the number of ionization events which are caused by a charged particle in a path 1 cm long in the direction of the field. If the cathode emits $n_0$ electrons, then the anode receives

$$n = n_0 e^{\alpha h}, \tag{1}$$

where h is the distance between the electrodes. Allowance for the secondary electron emission when the cathode is bombarded with positive ions complicates Eq. (1), which now becomes

$$n = n_0 \frac{e^{\alpha h}}{1 - \gamma (e^{\alpha h} - 1)}, \tag{2}$$

where $\gamma$ is the number of new electrons formed at the cathode.

Townsend suggested the breakdown criterion in the form $\gamma e^{\alpha h} = 1$ when $n \to \infty$.

The dependence of the impact ionization coefficient on the field, $\alpha = f(E)$, is deduced from the standard kinetic theory of gases. The probability of impact ionization is

$$w_i \sim \exp\left(-\frac{\lambda_i}{\lambda}\right) \sim \exp\left[-\left(\frac{\varepsilon_i}{eE\lambda}\right)\right],$$

where $\lambda$ is the mean free path and $\lambda_i$ is the path over which a carrier acquires an energy of the order of the ionization potential of an atom or a molecule, which is $\varepsilon_i = eU_i$.

The impact ionization coefficient, governed by the product of the number of collisions and the probability of ionization, has the form

$$\alpha \simeq \frac{1}{\lambda} \exp\left(-\frac{\varepsilon_i}{Ee\lambda}\right). \tag{3}$$

Since the mean free path in a gas is inversely proportional to the gas pressure P, the relationship (3) results in an expression

$$\alpha \sim P \exp\left(-\frac{\beta P}{E}\right), \tag{4}$$

where $\beta$ is a constant whose value is in good agreement with the experimental data [20].

The process of accumulation of the energy necessary for the ionization by an electron in a gas involves a wide range of interactions with neutral and excited atoms, molecules, and ions. Due to the different mobilities of electrons and ions, a space charge is produced and this charge distorts the field and has a considerable influence on the development of breakdown.

In gases, however, we can establish conditions under which all these processes are much simplified. Thus, the development of impact ionization in a rarefied gas over a small discharge gap in a sufficiently uniform field, i.e., when the recombination and the space charge can be neglected, can be described by simple relationships and is satisfactorily explained by Townsend's elementary theory.

Townsend's remarkable idea has become important in physics by providing a key to the understanding of phenomena taking place in strong electric fields not only in gases but also in solids, both dielectrics and semiconductors.

The concept of impact ionization in a solid as a cause of electrical breakdown was advanced first by A. F. Ioffe in 1928. However, he suggested that the ionization was due to ions and not electrons [2]. The hypothesis of the ionization of a dielectric by electron impact was proposed by Smurov, also in 1928 [21]. These first theories of impact ionization in solids were followed by later theories from von Hippel [22], Fröhlich [23], and Franz [24], who analyzed the problem quantitatively and formulated the breakdown criterion.

The breakdown criterion is deduced from the conditions under which a steady state is no longer possible. Several breakdown criteria have been suggested. According to von Hippel, a steady-state electrical conductivity is impossible if the energy acquired by e a c h electron from the field is greater than the maximum energy losses. The von Hippel criterion predicts breakdown fields higher than the experimental values because it sets quite a stringent requirement for the acceleration of all the electrons.

According to Fröhlich, breakdown occurs when the electrons acquire an energy close to the ionization value. Then, $\bar{\varepsilon} \approx \varepsilon_i$, where $\bar{\varepsilon}$ is the average electron energy. Breakdown fields deduced from this criterion for ionic crystals are several times greater than those found experimentally. Later, it was shown that the Fröhlich criterion is inapplicable to ionic crystals, but it does apply to covalent crystals.

The experimental data obtained in studies of dielectrics are not in conflict with the impact ionization theory, but a convincing experimental proof of the existence of impact ionization in solids has had to wait until studies were made of p—n junctions in very pure semiconductors containing very few imperfections [3].

In investigations of the impact ionization in p—n junctions it has been possible to establish conditions under which the simple Townsend model applies. If the carrier density is low ($\leq 10^{12}$ cm$^{-3}$), the field distribution in a p—n junction is not distorted by the space charge of holes and electrons. Under such conditions, we can neglect the interaction between electrons and holes and the recombination of carriers, since the time $\tau$ for the transit of carriers through a p—n junction is much shorter than the carrier lifetime $\tau^*$. When short pulses are used to investigate p—n junction breakdown, simple relationships are obtained, similar to those for a discharge in a rarefied gas [3].

It has been established experimentally that the breakdown voltage depends on the junction width. As in gases, the breakdown voltage of p—n junctions decreases as the junction thickness increases. The multiplication of carriers is also observed in the breakdown of p—n junctions.

## §2. Impact Ionization in Homogeneous Semiconductors

The first rigorous theory of the behavior of semiconductors in strong electric fields (based on a solution of the transport equation) was developed by Pisarenko in 1938 [25] and then by Davydov and Shmushkevich in 1940 [26]. The theory of impact ionization in semiconductors was taken further by W. Shockley [27], Keldysh [28], and Chuenkov [29].

They have considered the problem of the acquisition of energy by electrons in a strong electric field, allowing for scattering by the acoustical phonons and for the heating of the "electron" gas as the average energy of carriers increases. An analytic expression has been obtained for the distribution function in a strong electric field, and a criterion for the departure from a steady-state electron density has been formulated in a general manner.

The dependence of the average carrier density $\bar{\varepsilon}$ on the electric field intensity is found from the balance of the energy acquired from the field and the energy transferred to the lattice vibrations.

The power acquired by a carrier from the field is

$$A \simeq \left( \frac{e^2 E^2}{m} \tau \right) = \frac{(eE)^2 \, l_{ph}}{\sqrt{2m\bar{\varepsilon}}} \, , \tag{5}$$

where $l_{ph}$ is the mean free path, which is independent of the field in covalent semiconductors, m is the effective carrier mass.

The main energy losses are due to collisions with the acoustical phonons. The power lost in these collisions is

$$B = \frac{1}{\tau_{ph}} \frac{\hbar\omega}{2N + 1} \, , \tag{6}$$

where $1/\tau_{ph}$ is the total number of collisions with the acoustical phonons; $\tau_{ph} = l_{ph}\sqrt{m/2\bar{\varepsilon}}$ is the time interval between two collisions in the case of scattering by the acoustical phonons; $\hbar\omega$ is the energy of the acoustical phonons, where the frequency $\omega$ satisfies the law of conservation of momentum $\hbar\omega/s = 2mv$; $N = 1/[\exp(\hbar\omega/kT) - 1]$ is the number of phonons with an energy $\hbar\omega$; and s is the velocity of sound.

When $(\hbar\omega/kT) \ll 1$, we have $N \approx (kT/\hbar\omega)$, $2N \gg 1$, and

$$\frac{1}{\tau_{ph}} \frac{\hbar\omega}{2N+1} \simeq \frac{1}{\tau_{ph}} \frac{(\hbar\omega)^2}{2kT} = \frac{4m^2v^2s^2}{\tau_{ph}2kT} = \frac{\delta\bar{\varepsilon}}{\tau_{ph}}, \tag{7}$$

where $\delta = (4ms^2/kT)$, $\bar{\varepsilon} = (mV^2/2)$, and T is the temperature; $\delta\bar{\varepsilon}$ is the fraction of the energy lost in each collision with a phonon.

At high temperatures, $\delta \ll 1$, i.e., an electron loses only a small fraction of its energy in each collision. Therefore, in strong fields, electrons may acquire an energy considerably greater than their thermal energy, which is $\frac{3}{2}kT_0$, where $T_0$ is the lattice temperature. As long as the average energy of carriers is not very different from the thermal energy, the carrier mobility $\mu = e\tau/m$ is independent of the field and is equal to $l_{ph}e/\sqrt{3kTm}$.

When the field intensity is increased, the average energy of carriers rises linearly with the field. Equating Eqs. (5) and (6), we find that $\bar{\varepsilon} = eE l_{ph}/\sqrt{\delta}$ and $v \simeq \sqrt{E}$, where v is the velocity of carriers. Electrons in such a state are called "hot." If the electron temperature $T_e$ is defined as the ratio of their kinetic energy to the Boltzmann constant, we obtain the Shockley relationship

$$\frac{T_e}{T_0} = \frac{3\pi}{8} \frac{E}{E_c} \, , \tag{8}$$

where $E_c = 1.5s/\mu_0$, and $\mu_0$ is the mobility of carriers in weak fields.

The electron temperature may reach several thousands of degrees, while the lattice temperature usually changes by only a fraction of a degree.

In very strong fields ($\sim 10^4$ V/cm), electrons become so hot that they can transfer energy to the optical lattice-vibration modes, which absorb the energy more effectively than the acoustical modes. The drift velocity should be independent of the field.

The electron density is found from the balance of the thermal and impact ionization and the recombination

$$G(T) + w_i(E, T)\,n - w_r(E, T)\,n = 0 \qquad (9)$$

or

$$n = \frac{n_0}{1 - \gamma_2}, \qquad (10)$$

where $\gamma_2 = w_i(E, T)/w_r(E, T)$; $n_0 = G(T)/w_r(E, T)$; $G(t)$ is the number of electrons generated per unit time by the thermal ionization process; and $w_i(E, T)$ and $w_r(E, T)$ are the probabilities of impact ionization and of recombination averaged out over the distribution function.

The distribution function is found by solving the transport equation allowing for the processes of impact ionization, recombination, and scattering by the acoustical phonons. It is assumed that in strong electric fields the distribution function can be represented, to within terms of higher orders of smallness with respect to $\delta$, in the same form as for weak fields,

$$f(\varepsilon, \theta) = f_0(\varepsilon) + f_1 \cos\theta, \qquad (11)$$

but now $f_0(\varepsilon)$ is no longer Maxwellian. †

A system of equations for the determination of the symmetric and antisymmetric components of the distribution function is derived by the Davydov method [26]. Since these equations are quite different for $\varepsilon < \varepsilon_i$ and $\varepsilon > \varepsilon_i$, they are solved separately for these two regions and then the two solutions are joined at the point $\varepsilon \approx \varepsilon_i$.

It is found that when $\varepsilon > \varepsilon_i$, the distribution function decreases practically to zero over a very narrow interval $\xi = \varepsilon/\varepsilon_i \leq 0.1$. Thus, the contribution of the $\varepsilon > \varepsilon_i$ region to all the measured quantities (the carrier density, the electrical conductivity, and the average carrier density) is much less than unity. All these quantities can be calculated by averaging out only over the region $\varepsilon < \varepsilon_i$. The function $f(\varepsilon)$ in the region $\varepsilon > \varepsilon_i$ is only required in the determination of the average probability of impact ionization $w_i(E, T)$.

Using this distribution function, expressions are found for the equilibrium number of carriers in a strong field, for the average probability of ionization and recombination, for the critical field and its dependence on temperature, and for the law of interaction between electrons and the lattice.

A detailed analysis of the energy dependence $w_i(\varepsilon)$, carried out by Keldysh, has shown that, near the ionization threshold, $w_i(\varepsilon)$ may be a linear function of $(\varepsilon - \varepsilon_i)$ at low values of the permittivity $\varkappa$, and a quadratic function $w_i(\varepsilon) \approx (\varepsilon - \varepsilon_i)^2$ at high values of $\varkappa$. The value of $\varkappa$ is regarded as high when the binding energy of the Coulomb levels $me^4/\varkappa^2 h^2$ is less than the width of the impact ionization region $\xi\varepsilon_i$; for example, for germanium and silicon, $\varepsilon_i \approx 0.01\text{-}0.04$ eV and $\xi\varepsilon_i \sim 0.1$ eV [28].

---

† Stratton showed in 1957 that, above some critical value of the electron density, collisions between electrons ensure that the Maxwellian form of the electron distribution function $f_0(\varepsilon)$ is retained in strong electric fields. The electron distribution can be described by an electron temperature $T_e$ [30].

The results of such calculations show that, in a strong electric field, the probability of impact ionization $w_i(E, T)$ increases, while the probability of recombination $w_r(E, T)$ decreases. The reason for the decrease of $w_r(E, T)$ lies in the fact that, as the average electron energy increases, the relative number of slow electrons falls and, therefore, the recombination velocity decreases. Because of weaker recombination, the equilibrium number of carriers begins to rise even in fields of about $10^3$ V/cm, i.e., well before impact ionization. For the scattering by the acoustical phonons, the rise of n is proportional to $E^{3/2}$ while for the scattering by the optical phonons, we have $n \propto E^3$.

An analytical expression is available for the field dependence of the impact ionization and of the recombination probability and the breakdown field $E_b$ calculated for $w_i(E_b) = w_r(E_b)$. Thus, a comprehensive and consistent theory is available for impact ionization in the breakdown of homogeneous semiconductors. Unfortunately, experimental investigations of the impact ionization of electrons from the valence bands are very difficult because of the relatively high electrical conductivity of semiconductors.

### § 3. Impact Ionization of Neutral Impurity Atoms

The binding energy of a valence electron of an impurity atom in germanium and silicon is low because of the high permittivity of these materials. Thus, for elements of groups III and V in germanium, we have $\varepsilon_i \approx 10^{-2}$ eV, i.e., $\varepsilon_i/k \approx 100°K$ [31]. Consequently, atoms of group III and V impurities in germanium are thermally ionized at room temperature and even at liquid nitrogen temperature.

When the temperature is lowered sufficiently, the thermal motion energy becomes insufficient for the ionization. When $kT \ll \varepsilon_i$, the density of electrons in the conduction band of n-type germanium decreases exponentially with decrease of temperature

$$n = 2\left(\frac{2\pi mkT}{h^2}\right)^{3/2} \frac{N_d - N_a}{N_a} e^{-\frac{\varepsilon_i}{kT}}, \tag{12}$$

where $N_d$ is the donor concentration and $N_a$ is the acceptor concentration.

A similar relationship can be obtained for the density of holes p in p-type germanium [31].

Near liquid helium temperatures, impurities become almost completely neutral; the resistivity of germanium containing impurities in a concentration far from degeneracy reaches $10^8$-$10^{10}$ $\Omega \cdot$ cm. An impure semiconductor thus becomes a "dielectric" with a very narrow forbidden band ($\sim 0.01$ eV). In addition to the unusually narrow forbidden band, such a "dielectric" is distinguished by two important properties: it can exist only at such low temperatures that the thermal velocity of carriers is low ($v_0 \approx 10^6$ cm/sec) and the mobility of carriers is very high ($\mu_0 \approx 10^5$ cm$^2 \cdot$ V$^{-1} \cdot$ sec$^{-1}$). The low thermal velocities and the high mobility of carriers at helium temperatures result in a considerable increase in the average kinetic energy of carriers even in fields of about 1 V/cm, which in this case should be regarded as strong fields.

An increase in the average carrier energy to values of the same order of magnitude as $\varepsilon_i$ results in the impact ionization of impurity atoms in very weak electric fields. This was established experimentally in 1953.

Sclar, Burstein, Turner, and Davisson [32] investigated n- and p-type germanium immersed in liquid helium and subjected to fields of about 6 V/cm. They found that the electrical conductivity increased so strongly that they regarded the situation as a reproducible "breakdown" which was not accompanied by any structural changes. They also established that this behavior was not observed in samples unprotected from background radiation at room temperature, which caused photoionization of neutral impurities. Hence, Sclar et al. concluded that this "breakdown" was due to the multiplication of carriers by the impact ionization of neutral impurity atoms.

Investigations carried out later by other workers [33-37] and the researches of the present author [10-18] established that the impact ionization of impurities was the only mechanism which could be effective in such very weak fields. It was also established that the impact ionization of neutral impurities was a very special phenomenon, which had the characteristics of a discharge in a rarefied gas, as well as of breakdown in a homogeneous dielectric.

Like the phenomena in a rarefied gas, impact ionization in homogeneous germanium develops at low temperatures in a uniform field when the mean free paths of carriers are long. Although the impact ionization of impurities takes place in a crystal under very favorable conditions, these conditions are much more complex than those under which the Townsend model is valid.

As in dielectrics, the equilibrium density of carriers in germanium at low temperatures is governed by the balance of the generation and recombination processes.

In contrast to dielectrics, the impact ionization of impurities of groups III and V in germanium takes place in very weak electric fields, without a significant evolution of heat, and is not accompanied by structural changes. Since the rapid rise of the current due to the impact ionization of impurities is solely due to a reversible change in the electric strength, the name "low-temperature breakdown" is quite arbitrary.

CHAPTER II

## Experimental Technique

There were many experimental problems to be overcome in the investigation reported here. It was necessary to obtain and measure temperatures between 0.1 and 290°K, to measure high resistivities and low emf's, and to investigate strong currents under isothermal conditions. We shall now consider these problems in detail.

### §1. Production of Temperatures Between 0.1 and 290°K

In accordance with the methods of obtaining these temperatures, the range $0.1 < T < 290$°K can be divided into several regions [38], which are considered below.

a) Temperatures Obtained by Boiling Liquefied Gases. To achieve temperatures in the ranges 77-63, 20.4-14, and 4.2-1.7°K, the investigated samples were placed in liquid nitrogen, hydrogen, and helium at normal and low pressures of saturated vapors of these gases.

b) Intermediate Temperatures (4.2-14, 20-63°K). The range of temperatures between liquid helium and liquid hydrogen was obtained by heating a sample above the liquid helium temperature and then placing it in a vacuum enclosure immersed in liquid helium. The sample was joined to an external bath through a thermal resistance, which ensured a sufficiently rapid cooling of the sample to the bath temperature and an increase in the temperature of the sample by 10-15°K when a small amount of power was provided by a heater. Elementary calculations showed that the time $\tau$ for the establishment of thermal equilibrium between the sample and the bath, as well as the value of the temperature gradient between them, were proportional to the thermal resistance C. The value of C was selected to satisfy these contradictory conditions. Fused quartz was used for the thermal resistance. To establish a temperature drop of $\Delta T \approx 10$°K, we required a power of ~10-100 mW, while the time for the establishment of equilibrium was ~1 min.

In the intermediate temperature range, we used a hermetically sealed chamber consisting of a copper cover and a copper can in good contact with each other or a sealed ampoule with a copper−glass joint, as shown in Fig. 1. Before the experiments were started, the chamber was evacuated with a backing pump to a pressure of ~$10^{-2}$ mm Hg and was shut off with a valve. When the chamber was cooled to liquid helium temperature, the residual gas was practically totally frozen out and a good vacuum was established in the chamber. A thermometer, a sample, and a heater were all enclosed in the chamber. A fused quartz slab was soldered between the sample and a copper cold duct. The quartz also served as the electrical insulation between the circuits serving the sample and the heater; a single-crystal quartz slab was placed between the sample and the thermometer. The temperature gradient between the sample and the thermometer was negligibly small because the thermal conductivity of quartz is very high at low temperatures [39]. To improve the thermal contact, the sample was soldered to the quartz with tin in the case of n-type germanium, and with indium in the case of p-type germanium. A conducting layer on the surface of the quartz was prepared using the

40    É. I. ZAVARITSKAYA

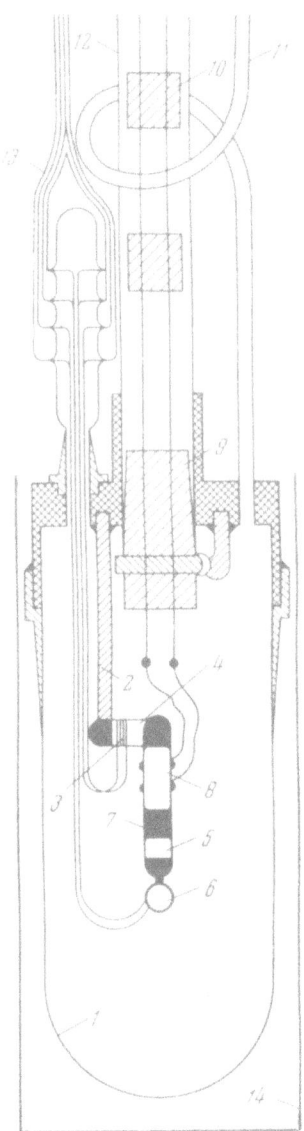

Fig. 1. Device used to obtain "intermediate" temperatures (4.2 < T < 20°K). 1) Glass bulb; 2) copper-rod cold duct; 3) heater; 4) fused quartz; 5) crystalline quartz; 6) thermometer; 7) metal contact; 8) sample; 9) tungsten—quartz seal; 10) Teflon washer; 11) tube for pumping out; 12) tube carrying leads; 13) thermometer and heater leads passing through platinum—glass seal; 14) copper screen.

standard technique. The ground ends of the quartz slabs were coated with a silver paste, which was fired at T ≈ 700°C.

c) Temperature Range Below 1°K. Very low temperatures were produced by the adiabatic demagnetization of a paramagnetic substance [40]. The cooling obtained by this method, due to the magneto-caloric effect, was

$$\left(\frac{\partial T}{\partial H}\right)_{s,p} = \frac{TH}{c_{p,H}}\left(\frac{\partial \chi}{\partial T}\right)_{p,H},\qquad(13)$$

where $\chi$ is the magnetic susceptibility. The main difficulty in the use of this technique was the need for a change, during the experiment, from conditions of good heat exchange between the paramagnetic substance (an alum) and the cooling agent in the magnetization stage to those of thermal insulation during the demagnetization stage.

In our experiments, the residual gas method was employed. This method is based on the strong temperature dependence of the adsorption of gaseous helium by alums.

The chamber contained such an amount of gas which ensured good thermal contact between the alum and the bath down to T ≈ 2°K. On cooling below 1.7°K, the heat exchange conditions became much less favorable (because of the adsorption of the gas). When the alum was magnetized to 10,000 Oe, its temperature rose to 10°K. The gas was desorbed and thus the alum was cooled to a temperature close to the temperature of the helium bath; the gas pressure in the chamber fell considerably. In the demagnetization stage, the temperature of the alum dropped to 0.1°K and the residual traces of the gas were absorbed by the alum. Thus, a high vacuum was established in the apparatus. The spontaneous heating of the device was slight and a temperature between 0.1 and 0.5°K could be maintained for several hours.

The paramagnetic substance used was iron—ammonium alum $NH_4Fe(SO_4)_2 \cdot 12H_2O$. This salt was chosen because iron—ammonium alum has a specific-heat maximum near 0.1°K. Moreover, the properties of this alum have been thoroughly investigated [41]. The temperature dependence of the specific heat of the alum is given in the table below:

| $T°$,K ... | 5 | 2 | 1 | 0.2 | 0.1 | 0,08 |
|---|---|---|---|---|---|---|
| $C, \frac{\text{ergs}}{\text{g} \cdot \text{deg}}$ | $5 \cdot 10^3$ | $2 \cdot 10^3$ | $10^4$ | $2 \cdot 10^4$ | $10^5$ | $9 \cdot 10^4$ |

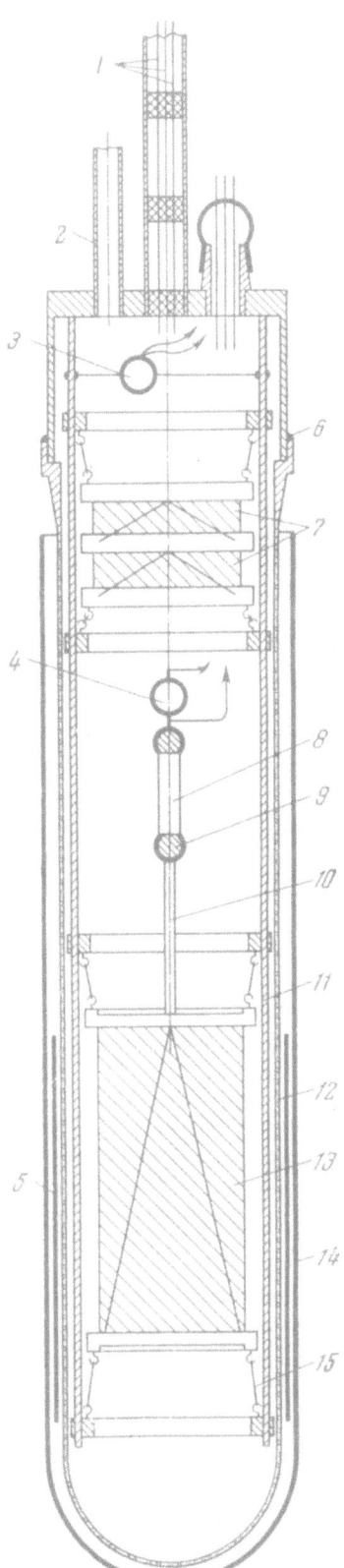

Figure 2 shows the apparatus used in the investigations below 1°K. The evacuated chamber enclosed the principal components: the alum, the sample, and the additional alum pellets used to cool the leads; they were all suspended by thin Capron filaments [42].

At room temperature, the chamber was filled with gaseous helium at 5-7 mm Hg. When the heat-transfer gas was completely absorbed by the alum at T ≈ 2°K, additional gaseous helium was introduced into the chamber in small amounts. The vacuum in the chamber was deduced from the overheating of a carbon resistor (vacuum gauge) when a constant current was passed through it.

§ 2. Measurement of Temperatures Between 0.1 and 290°K

Helium and magnetic thermometers give temperatures which are close to the thermodynamic scale. The state of gaseous helium over a fairly wide range of temperatures is described by the Clapeyron equation

$$pV = R_0 T. \qquad (14)$$

Therefore, to determine the absolute temperature of a body, it is sufficient to measure the helium pressure (p) at this temperature when V = const.

Below 20°K, helium departs considerably from the ideal-gas behavior, and this is allowed for by means of virial coefficients [43].

A gas thermometer can be used only down to 1°K, since the vapor pressure falls sharply at low temperatures. Below 1°K, the ideal law to be used is the Curie–Weiss law

$$\chi = \frac{C}{T - \theta}, \qquad (15)$$

where C = const, θ is the Curie temperature, and χ is the magnetic susceptibility.

The relationship (15) is well satisfied by dilute paramagnetic chromium–potassium alum (θ = 0.004°K) and by iron–ammonium alum (θ = 0.04°K).

We actually used bulk and film thermometers made of various types of carbon and Aquadag and we calibrated them by means of gas and magnetic thermometers.

The characteristic of one of the thermometers used is shown in Fig. 3. Each thermometer had a range of temperatures in which its sensitivity was highest. This was the range in which that thermometer was used.

Fig. 2. Device used to obtain "very low" temperatures (0.1 < T < 1°K). 1) Leads to electrometer circuit; 2) tube for pumping out; 3) vacuum gauge; 4) thermometer; 5) measuring coil; 6) hermetic seal; 7) alum, auxiliary pellet; 8) sample; 9) electrodes of sample; 10) cold duct; 11) supporting rods; 12) glass bulb; 13) alum, main pellet; 14) copper screen; 15) Capron filaments.

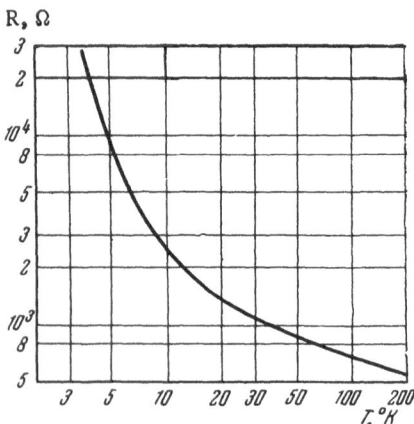

Fig. 3. Characteristic of a thermometer
for intermediate temperatures.

The carbon thermometers were made from standard radio-engineering resistors. Since the thermal conductivity of such resistors is exceptionally low at low temperatures [44], the thickness of the thermometers used was a fraction of a millimeter and their area was 20-30 mm$^2$.

The following method was employed to prepare small thermometers: A radio-engineering resistor was cut across to form a disk which was reduced to the required thickness by grinding. The cut surfaces were coated electrolytically with copper. A copper foil strip was attached electrolytically to the copper coating. The end of the copper foil strip was soldered to a sample, taking care not to overheat the thermometer, which would have resulted in irreversible changes in its properties.

The temperature of the samples immersed in liquid helium was determined from the saturated vapor pressure of helium; we used the tabulated values of $p = f(T)$ approved at the International Conference on Thermometry, June 4, 1949. When a significant amount of power was dissipated in a sample, the temperature was measured with miniature (film) carbon thermometers of the type developed by Shal'nikov and Zavaritskii [45].

§3. Apparatus

Measurements of the resistance of the thermometers and samples (in the case of high conductivities) were carried out using the standard compensation circuit. Dc potentiometers of the KL-48, MS, and PPTV types with galvanometers of about $10^{-7}$ V/division sensitivity were employed. The accuracy of the temperature measurement was $10^{-2}$-$10^{-3}$ deg. The resistance of the carbon vacuum gauge was measured with a portable PP potentiometer. The magnetic susceptibility of the alum was determined with an ac compensation circuit. The "primary" coil was excited with a current of 60-cps frequency from a quartz oscillator of the type used in quartz clocks. The "secondary" coil consisted of two oppositely wound coils connected in series (about $10^4$ turns of copper wire of 30 μ diameter); the alum under investigation was placed in one of these coils. The difference between the emf's induced in the secondary coils was balanced with the ac potentiometer. The readings of this potentiometer were used in each experiment to calibrate a magnetic thermometer in the range 4.2-2°K. The accuracy of the determination of temperature below 1°K was about $10^{-2}$ deg. Since the resistance of the sample changed by more than 10 orders of magnitude when T and E were varied, it was necessary to measure both high and low current densities.

Measurements of high resistivities, the Hall emf, and of the electric-field dependence of the resistance of germanium were carried out using electrometric devices below hydrogen temperature. A voltage from a battery was applied to a germanium sample and to a resistance box with a number of calibrated resistors ranging from $10^{11}$ to $10^3$ Ω, which were used to measure the current flowing through the sample.

The voltage drops across a calibrated resistor and across the potential and Hall contacts of a sample were measured with the following electrometers: an EMU-3 tube electrometer with a voltage sensitivity of 1 mV division and a leakage resistance of about $10^{11}$ Ω, and an SG-1M string electrometer with a lower voltage sensitivity of ~20 mV/division, but with a higher leakage resistance of ~$10^{15}$ Ω. The measurements were carried out by the constant deflection method (the Bronson method) [46], since the current through a sample was several orders of magnitude greater than the leakage current. To reduce the leakage, the whole electrometer was mounted on polystyrene and Teflon insulators. The electrical leads connected to the sample were made of manganin wire, which entered the vacuum chamber through Staybrite tubes and polystyrene washers. To reduce the heat losses along the leads, a quartz plate with a tungsten seal, which was in thermal contact with the external bath, was used as the lower insulating washer.

To provide electrostatic screening, the electrometer was placed in a copper box which was properly grounded.

To prevent heating of the samples by strong currents, the resistance and Hall emf measurements were carried out using short (3 and 100 μsec) pulses. A "long" line was used to form the pulses. The duration of a pulse was approximately an order of magnitude greater than its rise time. A controlled voltage was taken from a load whose impedance was equal to the wave impedance of the line.

The voltages at the potential and current contacts of a sample and across the current-controlling resistance were measured with a voltmeter of the VLI-2 type and a DESO-1 oscillograph. The sample was surrounded by a grounded metal screen. Screened coaxial leads were used for the sample. To minimize the stray coupling between the leads, the system was balanced with respect to the ground [47]. For pulses of 100-μsec duration, the stray signal of the Hall contacts was about 0.1 V when the voltage across the sample was 400 V. However, under these conditions there was a significant heating of the sample. In the case of pulses of 3-μsec duration, the stray signal was an order of magnitude greater, but the change in temperature in fields up to 100 V/cm could be neglected.

## §4. The Preparation of the Samples and Contacts

We investigated several batches of germanium samples doped with elements of groups III and V (Sb, Bi, In) as well as samples doped with zinc. A given batch was a group of samples cut from the same ingot.

In strong electric fields the requirements for the homogeneity of the sample and the quality of the contacts are quite stringent. To reduce the inhomogeneity of the impurity distribution, the samples were cut at right angles to the growth direction. The necessary shape was obtained by grinding or ultrasonic cutting. Irrespective of their shape, the samples could be divided into three groups. Some of the samples, which were intended for the Hall effect measurements, were in the form of slabs whose length ($l$) was not less than four times their width (h). This shape minimized the short-circuiting effect of the current electrodes [48], and the voltage $V_{meas}$ measured between the Hall contacts did not differ from $V_{Hall}$.

The samples prepared from high-resistivity germanium were shaped like dumbbells with thicker ends at the current contacts and projections for the potential probes. Again, the length—width ratio was not less than 4.

In the investigation of the saturation currents, we used samples in the form of thin plane-parallel plates of large area but with small contacts located on opposite faces. The samples were etched in boiling Perhydrol, with a small admixture of NaOH, and then were washed several times in boiling thrice-distilled water and dried.

Tin and indium electrodes of spectroscopic purity were soldered to the faces of a well-etched sample by prolonged heating in vacuum. A graphite matrix of special shape was heated, together with the sample, for 1 hr at 600°K in about $10^{-6}$ mm Hg vacuum. After the attachment of the electrodes, the samples were again etched, washed, and dried.

Large current contacts completely covered the ends of a sample. The potential contacts and the Hall contacts (in the case of the slab-shaped samples) were in the form of small tin or indium spheres whose contact area was 0.2-0.05 mm$^2$ and which were separated by 2-3 mm from one another.

The potential and current electrodes of the dumbbell-shaped samples were of the bulk type and retained ohmic properties at all temperatures.

## §5. Thermal Contact between a Sample and the Liquid Helium

Good thermal contact conditions between a sample and liquid helium are obtained only when the dissipated power is low. Our measurements showed [14] that when the power evolved was greater than a certain critical value, the heat transfer between liquid helium and a solid changed suddenly. The reason for the departure from thermal equilibrium between a sample and a liquid helium bath was a sudden change in the

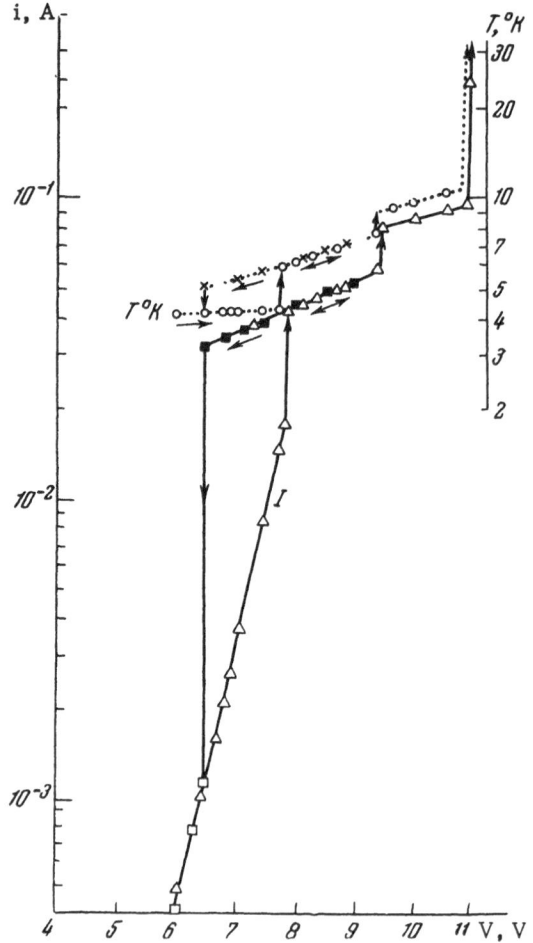

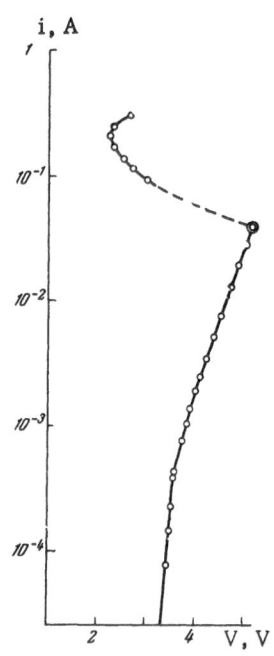

Fig. 4. Current−voltage characteristic (continuous) and temperature dependence of the applied voltage (dashed). The arrows show the direction of variation: → denotes the variation in current and temperature when the voltage was increased; ← denotes the variation in current and temperature when the voltage was reduced.

Fig. 5. Current−voltage characteristic of a sample for a load resistance of 40 Ω. The encircled point shows when the critical power $W_{cr}$ was reached.

TABLE 1

| $\rho$, $\Omega \cdot$ cm (300°K) | Plate dimensions, mm | S, cm² | $W_{cr}$, W | $W_{cr}/S = W_0$, W/cm² | $T$,° K |
|---|---|---|---|---|---|
| 50 | $6 \times 6 \times 0.48$ | 0.84 | 0.39 | 0.465 | 9 |
| 50 | $6 \times 6 \times 1.5$ | 1.0 | 0.44 | 0.44 | 10 |
| 40 | $8 \times 2 \times 2$ | 0.72 | 0.36 | 0.50 | 10 |
| 40 | $4 \times 2.1 \times 1$ | 0.29 | 0.12 | 0.41 | 6 |
|  |  |  | 1.0 | 3.0 | 40—50 |
| 7 | $20 \times 3.8 \times 0.87$ | 1.95 | 2.6 | 1.33 | 20—30 |
| 1 | $3.6 \times 1.25 \times 0.75$ | 0.16 | 0.2 | 1.25 | 20—25 |
| 40 | $25 \times 20 \times 0.4$ | 10.4 | 5 | 0.5 | 10 |

current, the appearance of hysteretic phenomena, and, if a large resistance was connected in series with the sample, the appearance of a falling branch in the current–voltage characteristic.

Figure 4 shows the results of a simultaneous measurement of the current–voltage characteristic and temperature of a sample during two consecutive cycles of voltage variation: from 0 to 9 V and back to zero, and again from 0 to 11 V. It is evident from Fig. 4 that the temperature of the sample differed little from the temperature of the bath (the difference was less than 0.1°K) as long as the power dissipated in the sample was below a certain critical value. When the power was increased above this value, the temperature of the sample suddenly increased; when the voltage was then reduced, the temperature of the sample remained high for several minutes and eventually thermal equilibrium was reestablished between the liquid and the sample. A simultaneous measurement in the current–voltage characteristic showed that the sudden change observed in the temperature when the voltage was increased was accompanied by an irreversible increase in the current. Each temperature discontinuity corresponded to a current discontinuity. When the voltage was reduced, the temperature of the sample remained higher and the current was greater than in the initial run (voltage increasing).

Table 1 gives the results of the measurements of the current–voltage characteristics and of temperature for several samples which showed a correlation between the temperature jump and the transition of the current–voltage characteristic to an irreversible branch. In Table 1, $W_{cr}$ is the power at which the current and temperature discontinuities were observed. The value of $W_{cr}$ varied from sample to sample by a factor of up to 50. The constant parameter for all the samples was the ratio of the power at which the first temperature discontinuity was observed, $W_{cr}^{(1)}$, to the dissipating surface area $S$:

$$W_0 = \frac{W_{cr}^{(1)}}{S} \approx \text{const} = 0.5 \ \text{W/cm}^2.$$

Figure 5 shows the current–voltage characteristic of a sample which was connected in series to a large resistance. A falling branch of the characteristic was observed when the power dissipated reached $W_{cr}$.

Miniature carbon thermometers were used to measure the temperature of a sample in liquid helium. A thermometer 40 $\mu$ thick was placed either between two similar samples through which approximately the same current was passed, or was attached to one of the plate-shaped samples and coated with a layer of cellulose.

The disturbance of the thermal contact between a sample and the liquid helium was probably due to the formation of a gas layer at the boundary of the solid, as observed by Strelkov [49]. The low thermal conductivity of helium ($\lambda \approx 10^{-4}$ W · cm$^{-1}$ · deg$^{-1}$) and the low value of the latent heat of evaporation of helium r suggest that this explanation is probably correct, but the actual mechanism of the sudden appearance of a temperature gradient is not yet clear.

We shall now list the values of the latent heat of evaporation of helium:

| $T°$,K . . . | 1.5 | 2 | 2.19 | 2.5 | 3 | 3.5 | 4 | 4.5 | 4.75 | 5 |
|---|---|---|---|---|---|---|---|---|---|---|
| $r$, cal/deg | 5.3 | 5.5 | 5.41 | 5.5 | 5.6 | 5.52 | 5.17 | 4.37 | 3.63 | 1.9 |

Hence, we see from the above table that the value of r falls sharply when the temperature is increased from 4 to 5°K because of the low value of the critical temperature of helium ($T_{cr} = 5.2°K$).

CHAPTER III

## Results of Measurements and Discussion

We investigated germanium samples with groups III and V impurities whose ionization energies were low. For concentrations less than $5 \cdot 10^{15}$ cm$^{-3}$, they were: $\varepsilon_i = 0.0096$, 0.012, and 0.0112 eV for antimony, bismuth, and indium, respectively [50].

TABLE 2

| Sample No. | Impurity | Batch | Net impurity conc. $N_0$, $cm^{-3}$ | Shape of sample | Length of sample, mm |
|---|---|---|---|---|---|
| 1 | Bismuth | Bi I | $7.5 \cdot 10^{14}$ | Slab | 8 |
| 2 | » | Bi I | $1 \cdot 10^{15}$ | » | 8 |
| 3 | » | Bi I | $2 \cdot 10^{15}$ | » | 8 |
| 4 | » | Bi I | $5 \cdot 10^{15}$ | » | 8 |
| 5 | Antimony | Sb I | $7.0 \cdot 10^{14}$ | » | 8 |
| 6 | » | Sb I | $1 \cdot 10^{15}$ | » | 8 |
| 7 | » | Sb II | $7.5 \cdot 10^{13}$ | » | 10 |
| 8 | » | Sb II | $1.6 \cdot 10^{15}$ | » | 10 |
| 9 | » | Sb II | $2.75 \cdot 10^{15}$ | » | 10 |
| 10 | Indium | In I | $1 \cdot 10^{14}$ | Plate | 0.94 |
| 11 | » | In I | $1.6 \cdot 10^{14}$ | » | 0.45 |
| 12 | » | In I | $1 \cdot 10^{14}$ | Dumbbell | 10 |
| 13 | » | In I | $1.6 \cdot 10^{14}$ | Plate | 0.6 |
| 14 | Antimony | — | $\sim 10^{15}$ | Slab | 10 |
| 15 | Undoped | — | $5 \cdot 10^{13}$ | » | 10 |
| 16 | Antimony | — | $2 \cdot 10^{13}$ | Dumbbell | 10 |

TABLE 3

| Sample No. | $T$, °K | p, $cm^{-3}$ | R, $cm^2/C$ | $\mu$, $cm^2 \cdot V^{-1} \cdot sec^{-1}$ |
|---|---|---|---|---|
| 1 | 4.8 | $1.6 \cdot 10^7$ | $8 \cdot 10^{12}$ | $5 \cdot 10^5$ |
| 7 | 4.2 | $5 \cdot 10^9$ | $1 \cdot 10^{15}$ | $2 \cdot 10^5$ |
| 9 | 6.5 | $5 \cdot 10^6$ | $9 \cdot 10^{10}$ | $1.8 \cdot 10^4$ |
| 12 | 4.2 | $1 \cdot 10^8$ | $2 \cdot 10^{13}$ | $2 \cdot 10^5$ |
| 14 | 4.2 | $8.5 \cdot 10^7$ | $9 \cdot 10^{12}$ | $1 \cdot 10^5$ |

TABLE 4

| $T$, °K | $\rho$, $\Omega \cdot cm$ | R, $cm^2/C$ | p, $cm^{-3}$ | $\mu$, $cm^2 \cdot V^{-1} \cdot sec^{-1}$ |
|---|---|---|---|---|
| 290 | 35 | $9.6 \cdot 10^4$ | $1 \cdot 10^{14}$ | $1.8 \cdot 10^3$ |
| 77 | 2.3 | $6.9 \cdot 10^4$ | $1 \cdot 10^{14}$ | $3 \cdot 10^4$ |
| 20.4 | 1.6 | $1.3 \cdot 10^5$ | $5 \cdot 10^{13}$ | $8 \cdot 10^4$ |
| 14 | 7.15 | $8.15 \cdot 10^5$ | $8 \cdot 10^{12}$ | $1.1 \cdot 10^5$ |
| 4.2 | $10^8$ | $2 \cdot 10^{13}$ | $3 \cdot 10^5$ | $2 \cdot 10^5$ |

The properties of the samples are listed in Table 2. The results of the measurements of the resistivity $\rho$ and of the Hall coefficient R at helium temperatures are given in Table 3. The influence of temperature on the properties of sample No. 12 is illustrated by the results in Table 4. The carrier mobility was calculated from the relationship $u = R/\rho$ and the density was found from the relationship $n = 1/Re$.

A typical dependence of the current density j on the electric field intensity E is given in Fig. 6.

The current−voltage characteristics of n- and p-type germanium exhibited "breakdown" at electric field intensities of 5-7 V/cm.

In fields weaker than the breakdown value, we observed two regions: a weak-field region A in which Ohm's law ($j = \sigma E$) was satisfied and a region of prebreakdown fields B, in which the electrical conductivity $\sigma$ rose monotonically.

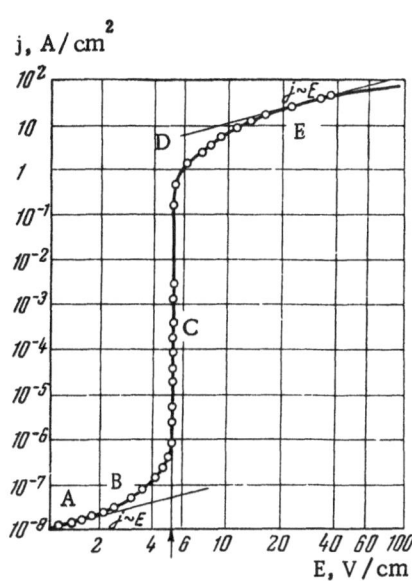

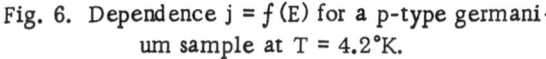

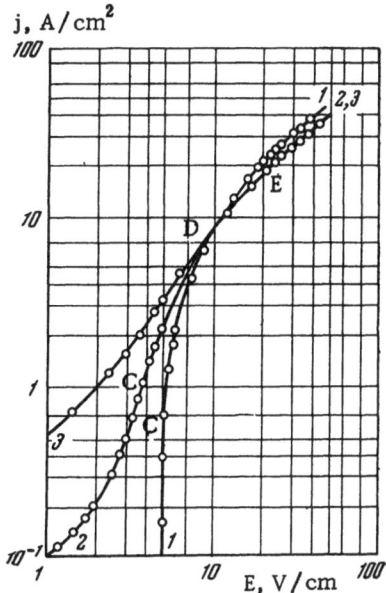

Fig. 6. Dependence $j = f(E)$ for a p-type germanium sample at $T = 4.2°K$.

Fig. 7. Dependence $j = f(E)$ at 4.2°K (1), 14°K (2), and 20.4°K (3). The same sample (No. 12) as in Fig. 6.

Similar dependences $j = f(E)$ were obtained for germanium samples with different impurities and at different temperatures (Fig. 7).

## ELECTRICAL CONDUCTIVITY OF GERMANIUM IN FIELDS LESS THAN THE BREAKDOWN VALUE

### §1. Weak Electric Fields (Region A)

a) Results of Measurements. In weak electric fields, the current density was found to be proportional to the electric field intensity: $j = \sigma E$, where $\sigma$ = const. Ohm's law was satisfied in this region because the carrier density n, the electrical conductivity $\sigma$, and, consequently, the mobility $\mu$ were all independent of the electric field intensity. The values of $\mu$ and n were affected only by temperature and the impurity concentration.

Typical temperature dependences of the mobility of n- and p-type germanium are shown in Fig. 8. Curve 1 represents samples with a net (excess) impurity concentration $N_0 \approx 8 \cdot 10^{13}$ cm$^{-3}$ (which is the difference between the majority and minority impurity concentrations) and a minority (compensating) impurity concentration of $\sim 10^{13}$ cm$^{-3}$. Curve 2 represents a sample with impurity concentrations $N_a - N_d = 1.6 \cdot 10^{14}$ cm$^{-3}$ and $N_d = 1.5 \cdot 10^{13}$ cm$^{-3}$. In the range between room and liquid nitrogen temperatures, the hole mobility $\mu_p$ is less than the electron mobility $\mu_e$. When the temperature is raised, $\mu_p$ and $\mu_e$ increase, but $\mu_p$ rises faster. At liquid nitrogen temperature, the electron and hole mobilities are comparable. Further cooling shows that the temperature dependence of the mobility is governed practically only by the impurity content. The mobility is higher for those samples which have fewer impurities.

A typical dependence of the carrier density on temperature is shown in Fig. 9. At 80 < T < 300°K, the carrier density is constant: n = const $\approx N_d - N_a$. As the temperature is lowered, n decreases and at T < 20°K falls exponentially with $1/T$, in full agreement with Eq. (12). The slope of the $\ln n = f(1/T)$ curve gives, as indicated by Eq. (12), the value of $\varepsilon_i$.

Figure 10 shows the temperature dependence of the resistivity of several samples of germanium doped with antimony and bismuth. The resistivity of all these samples increases as $\exp(1/T)$ between liquid hydrogen and helium temperatures. The ionization energies of these two impurities, calculated from the slopes of the curves $\ln \rho = f(1/T)$, are in good agreement with the published values (mentioned at the beginning of Chapter III) since, in this range of temperatures, the dependence $\rho(T)$ is governed primarily by the dependence $n(T)$.

From the data presented in Fig. 10, it follows that the slope of the $\ln \rho = f(T)$ curves is the same for samples containing different concentrations of the same impurity. Hence, when $N_d - N_a < 5 \cdot 10^{15}$ cm$^{-3}$, the ionization energy is independent of the impurity concentration.

At temperatures lower than 4°K, the experimental dependence $\ln \rho = f(T)$ does not satisfy Eq. (12). This can be seen clearly from Fig. 11, which presents the dependence $\ln \rho = f(1/T)$ between room temperature and $T \approx 0.1$°K.

There are two curves in Fig. 11. One of them represents the resistivity of a sample at liquid hydrogen and helium temperatures and the other at helium and very low temperatures. The results indicate the existence of three activation energies $\varepsilon_{ac}$, corresponding to the different temperature intervals.

The energies $\varepsilon_{ac}$, calculated from the slope of $\ln \rho = f(T)$ for a sample with impurity concentrations $N_d = 2 \cdot 10^{15}$ cm$^{-3}$, $N_a = 2 \cdot 10^{14}$ cm$^{-3}$, have the following values:

| $T$, °K | 5-20 | 1-5 | 0.1-1 |
|---|---|---|---|
| $\varepsilon_{ac}$, eV | ~0.01 | ~$3 \cdot 10^{-3}$ | ~$10^{-5}$ |

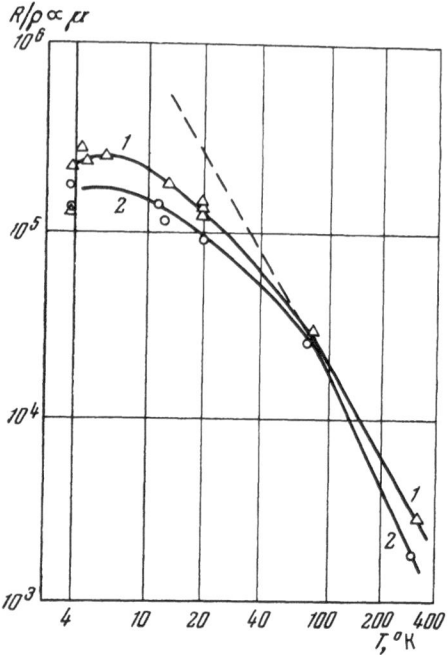

Fig. 8. Temperature dependence of the carrier mobility. 1) Sample of n-type germanium (No.2); 2) sample of p-type germanium (No. 12). Dashed curve represents $\mu \propto T^{3/2}$.

The width† of the weak-field region (A) does not depend strongly on temperature but is much affected by impurities. The width of region A grows with increasing net impurity concentration in a sample. Thus, for example, for the Sb II batch, the width of the weak-field region A is 10% for sample No. 7 with $N_d - N_a = 7.5 \cdot 10^{13}$ cm$^{-3}$, 25% for sample No. 8 with $N_d - N_a = 1.5 \cdot 10^{15}$ cm$^{-3}$, and 80% for sample No. 9 with $N_d - N_a = 3 \cdot 10^{15}$ cm$^{-3}$.

Region A is also affected by an increase in the concentration of the minority impurity in the sample. It follows from the results obtained that the weak-field region A at T = 4.2°K is particularly wide for sample No. 14 prepared from poorly purified germanium (Fig.12).

b) Discussion of Results. In weak electric fields, the electron velocities have a Maxwellian distribution, the average temperature of the electron gas $T_e$ is equal to the lattice temperature $T_0$, the carrier density and mobility are independent of the field, and the relationship between the current and the voltage is given by Ohm's law. The width of the weak-field region depends on the rate at which carriers acquire energy from the electric field; consequently, it depends

---

† The width of regions A and B is expressed as a percentage of the total width of regions A + B, because the breakdown voltage depends on the impurity concentration.

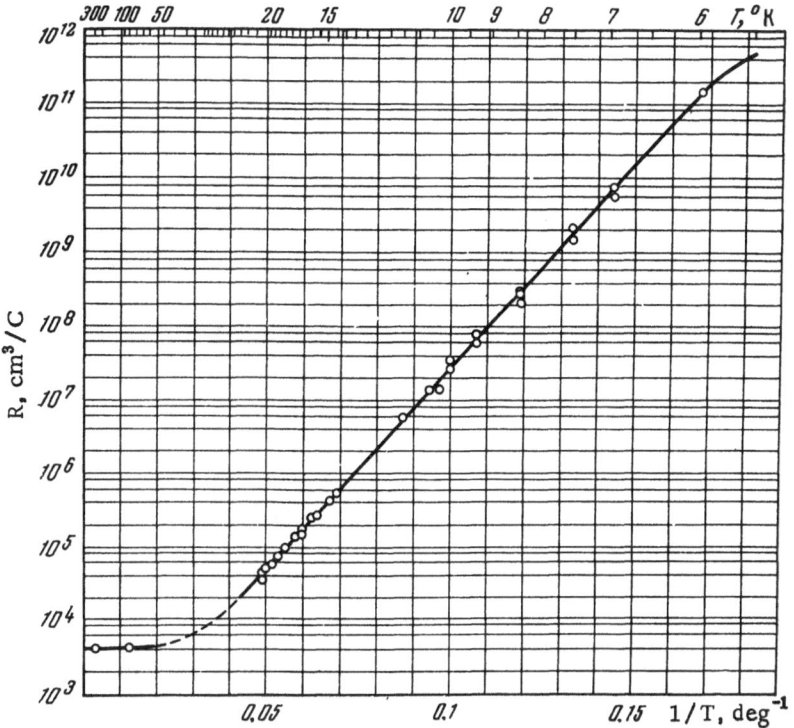

Fig. 9. Temperature dependence of the carrier density (sample No. 9).

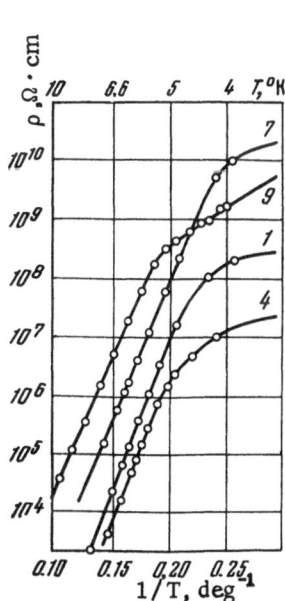

Fig. 10. Temperature dependence of the resistivity. The numbers alongside the curves correspond to the sample numbers.

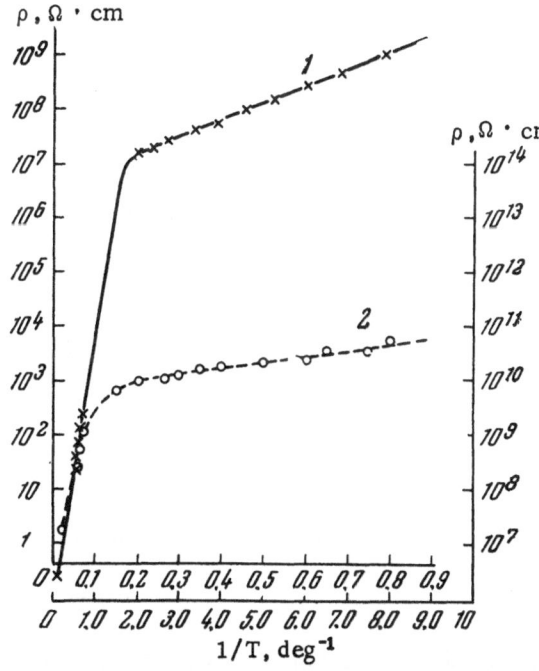

Fig. 11. Temperature dependence of the resistivity of sample No. 14: 1) at hydrogen and helium temperatures (scale on the left); 2) at helium and very low temperatures (scale on the right).

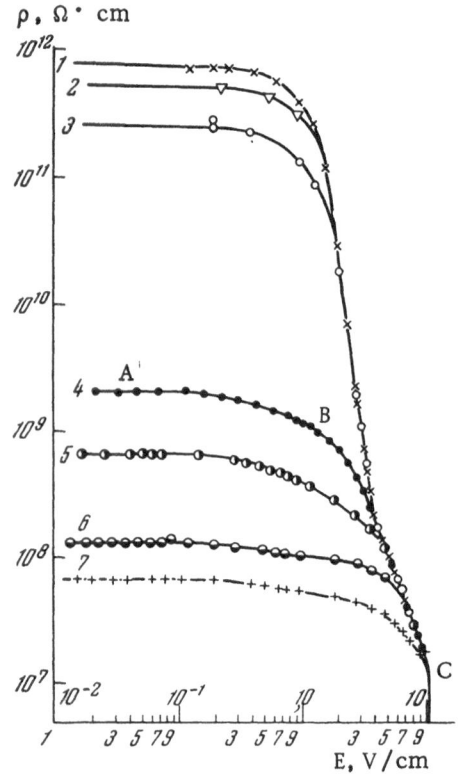

Fig. 12. Dependence of the resistivity on the electric field intensity for sample No. 14 at various temperatures: 1) 0.15; 2) 0.2; 3) 0.4; 4) 1.77; 5) 2.23; 6) 3.12; 7) 4.2°K.

on the carrier mobility. The lower the carrier mobility, the higher are the values of the electric field intensity, which still do not affect the carrier energy distribution.

The results of measurements of $\mu = f(T)$ at low temperatures can be explained by taking into account the scattering by phonons and impurities. Then,

$$\frac{1}{\mu} = \frac{1}{\mu_L} + \frac{1}{\mu_i} + \frac{1}{\mu_N} , \qquad (16)$$

where $\mu_L$ is the mobility associated with the scattering by the acoustical phonons; $\mu_i$ is the mobility associated with the scattering by charged impurities; and $\mu_N$ is the mobility associated with the scattering by neutral impurities.

The temperature dependence of $\mu_L$ is described by the relationship [51]:

$$\mu_L = \frac{\sqrt{8\pi}}{3} \frac{\hbar^4 s^2 \rho^*}{\gamma_\pm^2 m_\pm^{5/2}} (kT)^{-3/2} \qquad (17)$$

or

$$\mu_L = \frac{2.4 \cdot 10^7}{T^{3/2}} . \qquad (18)$$

Here, $\rho^*$ is the average value of the longitudinal elastic constant; s is the velocity of sound; $m_\pm$ is the effective mass of holes and electrons; and $\gamma_\pm$ is the shift of the band edge per unit volume deformation.

The mobility of carriers in the case of scattering by charged impurities can be written in the form [52]:

$$\mu_i = \frac{8\sqrt{2}\, \varkappa^2 (kT)^{3/2}}{\pi^{3/2} \sqrt{m}\, e^3 N_i \ln\left[1 + \left(\frac{\varkappa kT}{e^2 N_i^{1/3}}\right)^2\right]} , \qquad (19)$$

where $\varkappa$ is the permittivity. In a simpler and more convenient form, this expression becomes

$$\mu_i = \text{const}\,\frac{T^{3/2}}{N_i \ln b} , \qquad (20)$$

where

$$b = 1 + \left(\frac{\varkappa kT}{e^2 N_i^{1/3}}\right)^2 .$$

The mobility in the case of scattering by neutral impurities [53] is

$$\mu_N = \frac{me^3}{20\hbar^3 \varkappa N_N} , \qquad (21)$$

where $N_N$ is the concentration of neutral impurities.

If we assume that $\varkappa = 16$, $m = 0.2 m_0$, and if we express $\mu$ in practical units, we find that

$$\mu_N = \frac{1.8 \cdot 10^{20}}{N_N} \frac{\text{cm}^2}{\text{V} \cdot \text{sec}} . \qquad (22)$$

TABLE 5

| Impurity conc., cm$^{-3}$ | $T$, °K | $\mu_N$, cm$^2 \cdot$ V$^{-1} \cdot$sec$^{-1}$ | $\mu_i$, cm$^2 \cdot$ V$^{-1} \cdot$ sec$^{-1}$ |
|---|---|---|---|
| $\sim 10^{13}$ | 2 | $1.8 \cdot 10^7$ | $3.4 \cdot 10^5$ |
|  | 4 | $1.8 \cdot 10^7$ | $6 \cdot 10^5$ |
|  | 8 | $1.8 \cdot 10^7$ | $1.2 \cdot 10^6$ |
| $\sim 10^{14}$ | 2 | $1.8 \cdot 10^6$ | — |
|  | 4 | $1.8 \cdot 10^6$ | — |
|  | 8 | $1.8 \cdot 10^6$ | — |
| $\sim 10^{15}$ | 2 | $1.8 \cdot 10^5$ | — |
|  | 4 | $1.8 \cdot 10^5$ | — |
|  | 8 | $1.8 \cdot 10^5$ | — |

TABLE 6

| Sample No. | Type | Impurity conc., cm$^{-3}$ | $T$, °K | $\mu$, cm$^2 \cdot$ V$^{-1} \cdot$ sec$^{-1}$(meas) | $\mu_N$, cm$^2 \cdot$ V$^{-1} \cdot$ sec$^{-1}$(calc) |
|---|---|---|---|---|---|
| 1 | $n$ | $7.5 \cdot 10^{14}$ | 4.8 | $2.5 \cdot 10^5$ | $2.4 \cdot 10^5$ |
| 5 | $n$ | $7.6 \cdot 10^{14}$ | 4.2 | $2 \cdot 10^5$ | $2.4 \cdot 10^5$ |
| 7 | $n$ | $7.5 \cdot 10^{13}$ | 4.2 | $2 \cdot 10^5$ | $2.4 \cdot 10^6$ |
| 9 | $n$ | $2.75 \cdot 10^{15}$ | 6.5 | $1.76 \cdot 10^4$ | $6.5 \cdot 10^4$ |
|  |  |  | 7 | $2.17 \cdot 10^4$ | $6.5 \cdot 10^4$ |
|  |  |  | 7.7 | $2.37 \cdot 10^4$ | $6.5 \cdot 10^4$ |
|  |  |  | 8.7 | $2.1 \cdot 10^4$ | $6.5 \cdot 10^4$ |
| 12 | $p$ | $\sim 10^{14}$ | 4.2 | $2 \cdot 10^5$ | $1.8 \cdot 10^6$ |
| 13 | $p$ | $1.6 \cdot 10^{14}$ | 4.2 | $1.5 \cdot 10^5$ | $1.1 \cdot 10^6$ |
| 14 | $n$ | $\sim 10^{15}$ | 4.2 | $\sim 10^5$ | $1.8 \cdot 10^5$ |

At $T > 80$°K, we can ignore the scattering by impurities for impurity concentrations $N < 10^{15}$ cm$^{-3}$.

It is evident from the data presented in Fig. 8 that the electron mobility is given by $\mu_L \propto T^{-1.6}$, which is in good agreement with the relationship applicable to the scattering by the acoustical phonons. In the same range of temperatures, the mobility of holes decreases more strongly ($\mu \propto T^{-2.3}$), which may be due to the additional scattering associated with the valence band degeneracy.

At $T < 80$°K (cf. Fig. 8), we must allow for the scattering by impurities.

The majority of impurities are ionized down to 10-15°K and, therefore, the mobility is governed primarily by the scattering on charged impurities. At helium temperatures, carriers are scattered both by charged and neutral impurities. The majority impurity is practically completely neutral, and $2N_a$ compensated donor—acceptor centers remain charged right down to the lowest temperatures.

To compare the intensity of the scattering by neutral and charged impurities, Table 5 lists the results of the calculation of $\mu_i$ and $\mu_N$ using Eqs. (20) and (22).

Table 6 gives the experimental values of the mobility for several germanium samples.

It is evident from these data that the order of magnitude of the carrier mobility at helium temperatures can be estimated using Eq. (22). This does not mean that the scattering by charged impurities can simply be neglected. A small reduction in the mobility is observed at $T < 4$°K, which may in fact be due to the scattering

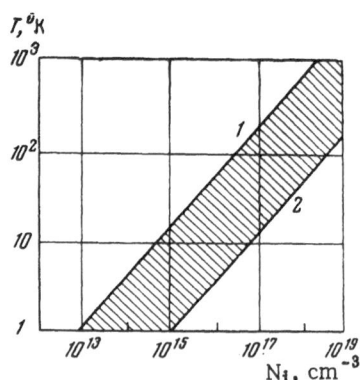

Fig. 13. Range of applicability of Conwell's (1)
and Sclar's (2) formulas.

by charged impurities. Unfortunately, the low-temperature value of $\mu_i$ of the majority of our samples cannot be represented in terms of elementary functions. This is because the range of applicability of Eq. (20) is limited by the range of applicability of the Born approximation, which is given by the condition $\mathbf{k}a \ll 1$, where $\mathbf{k}$ is the wave number and $a$ is the range of action of the scattering potential.

In the opposite case, $\mathbf{k}a > 1$, the expression for the mobility has been obtained by Sclar [53], who has suggested a scattering potential in the form of a quadratic barrier

$$V_r = \begin{cases} \pm \dfrac{e^2}{\varkappa a} & \text{when } r \leqslant a, \\ 0 & \text{when } r > a \end{cases}$$

(where r is the coordinate). It is found that in such an analysis the following parameter is important:

$$\alpha_0 a = \left( \mathbf{k}^{-2}a^2 - \frac{2mV_r}{h^2}\,a^2 \right)^{1/2} .$$

If $\alpha_0 a \to (2n+1)\pi/2$, then $\mu \approx v$ and the scattering is of the resonance type. If

$$\frac{\tanh \alpha_0 a}{\alpha_0 a} \to 1, \text{ then } \mu \sim \frac{1}{v^5} \to \infty.$$

This type of scattering corresponds to the Ramsauer effect in semiconductors.

For the majority of attracting and repulsing centers, we have

$$\mu \simeq \frac{\text{const}}{v} \simeq \text{const} \sqrt{\frac{m}{kT}} . \tag{23}$$

Figure 13 shows graphically the range of applicability of the formulas considered here. It is evident from this figure that the Conwell and Brooks—Herring formulas [54] should be used at fairly high temperatures and low impurity concentrations. The range of applicability of the Sclar formulas extends further in the direction of high concentrations. In the shaded region, which represents the condition $ka \approx 1$, the mobility cannot be described by a simple formula. Unfortunately, the mobility of most of the samples investigated lay in this region.

c) Temperature Dependence of the Carrier Density. It is evident from the results presented in Figs. 10 and 11 that there is a deviation from Eq. (12) at helium temperatures. Several mechanisms have been proposed to explain the anomalous conductivity observed at helium temperatures. Hung et al. and Fritzsche et al. [55] have suggested that the anomalous conductivity is associated with an impurity band formed by the overlap of the wave functions of the donor and acceptor states. They have assumed that the properties of electrons in the impurity band are identical with their properties in the conduction band and that the density of electrons in the impurity band is approximately equal to the net impurity concentration; the mobility, governed by the impurity band width, depends on the net impurity concentration [55].

It has been established that an impurity band is formed in germanium at a donor concentration of $\sim 10^{16}$ cm$^{-3}$ or an acceptor concentration of $\sim 5 \cdot 10^{16}$ cm$^{-3}$. The activation energy determined on the basis of this model does not contradict the results of measurements of $n = f(T)$ at $T < 4°K$.

To explain similar results obtained for samples with impurity concentrations of $\sim 10^{14}$-$10^{15}$ cm$^{-3}$, Mott and Conwell [56] have proposed a new conduction mechanism associated with the presence of a minority im-

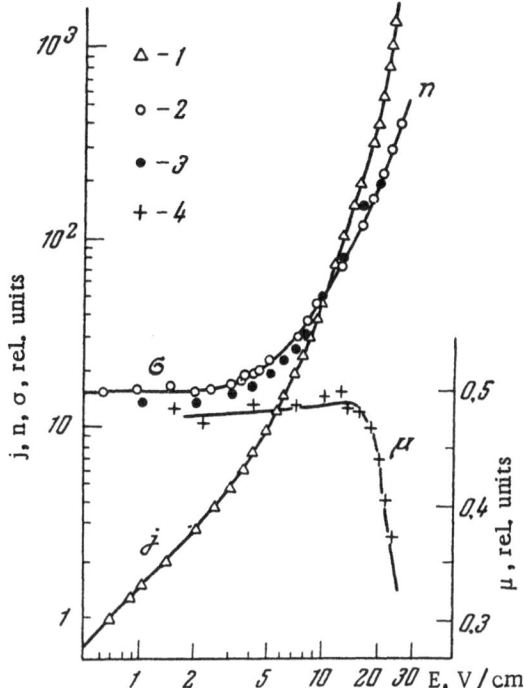

Fig. 14. Field dependence of the current j (1), electrical conductivity σ (2), carrier density n (3), and mobility μ (4). Sample No. 7 at T = 5.7°K.

purity in a sample. Thus, n-type germanium always contains compensated acceptors which are equal in number to positively charged donors. The latter act as "holes" or "donor" defects in an array of neutral donors. Since in n-type germanium the majority impurity is of the donor type, each acceptor is surrounded by g donors, one of which is ionized. A tunnel transition of a bound electron from a neutral donor to a charged donor is equivalent to the motion of a hole and thus contributes to the electrical conductivity.

The results of the measurements of $\ln \rho = f(T)$ at helium temperatures, presented in Fig. 11, can be accounted for, at least as far as the order of magnitude is concerned, by the Mott and Conwell mechanism if we assume that $\varepsilon_{ac} = 3 \cdot 10^{-4}$ eV, where $\varepsilon_{ac}$ is the activation energy of a donor defect, and $\mu \approx 3 \cdot 10^{-4}$ cm$^2$ · V$^{-1}$ · sec$^{-1}$.

However, the experimental results show that the mobility of most of the samples at helium temperatures is very high and so is the magnetoresistance (Chapter IV).

The absence of magnetoresistance, indicating low carrier mobility, has been observed only for samples with impurity concentrations $\geq 5 \cdot 10^{15}$ cm$^{-3}$. It is possible that an impurity band may form at these concentrations. On the basis of these considerations, we may assume that the more likely cause of the additional electrical conductivity observed for samples with impurity concentrations less than $10^{15}$ cm$^{-3}$ is the thermal radiation absorbed from the hotter environment.

It is known that the electrical conductivity of germanium may be altered by very weak thermal radiation ($\sim 10^{-12}$ W) of wavelengths up to 100 μ [57]. The radiation reaching a sample from the parts of a cryostat which are at higher temperatures may, in spite of a complex system of blackened screens within the chamber, cause a considerable increase in the carrier density in the conduction band. The contribution of the stray photoconductivity to the electrical conductivity of germanium is particularly important at very low temperatures. The experimentally determined activation energies are $\sim 10^{-3}$-$10^{-4}$ eV at helium temperatures and $\sim 10^{-5}$ eV at very low temperatures; these energies may be exceeded by the radiation of bodies kept at temperatures of 11 and 0.4°K, respectively.

§ 2. Prebreakdown Fields (Region B)

a) Results of Measurements. From the data presented in Fig. 14, it follows that the electrical conductivity rises smoothly in fields which are an order of magnitude smaller than the breakdown value. The results of measurements of R and μ show that the carrier mobility is practically independent of the field right up to $E_b$, and that the rise in the electrical conductivity in prebreakdown fields is due to a monotonic rise of the carrier density (Fig. 14).

b) Discussion of Results. The power received by a carrier from the field, $A = e\mu E^2$, where μ is the mobility governed by the overall scattering (by the optical and acoustical phonons and by the charged and neutral impurities), is balanced, under steady-state conditions, by the power lost in collisions with the acoustical phonons:

$$B \approx \frac{1}{\tau_{ph}} \overline{\delta \varepsilon}.$$

If the scattering by neutral impurities is the dominant process, then the mobility is independent of the field: $\mu_N(E) = \mu_N(0) = \text{const}$.

In the case of scattering by charged impurity atoms, the mean free path increases with the carrier energy and, as demonstrated by Conwell, the ratio $l/v$ remains approximately constant up to fields $E_*$; when $E > E_*$, $l$ increases. The value of

$$E_* \simeq \left(\frac{N_a}{n_*}\right)^{1/2} \frac{\sqrt{2\delta\epsilon_i}}{el_{ph}} \tag{24}$$

can be found from the condition A = B when $\mu = \mu_i$ and $n_* = (\pi a_0^2 l_{ph})^{-1}$, where $a_0$ is the radius of an impurity atom, $l_{ph}$ is the mean free path in the scattering by phonons, and $N_a$ is the minority impurity concentration in n-type germanium.

If $\sqrt{N_a/n_*} \simeq 0.1$, then $E_* \simeq 4$ V/cm. Thus, when carriers are scattered by charged or neutral impurities, the mobility $\mu$ is practically independent of the field up to fields close to $E_b \approx 5$ V/cm, as observed experimentally (Fig. 14).

c) Field Dependence of the Carrier Density. We may assume that there are two causes of the monotonic increase in the equilibrium carrier density in strong electric fields: the thermal ionization, which is assisted by an external field [58], and the reduction in the rate of recombination in strong electric fields [59].

The Frenkel mechanism, based on the classical thermal ionization, assisted by an external field, is represented by the following expression for the electrical conductivity:

$$\sigma(E) = \sigma_0 \exp \frac{\Delta W}{kT} , \tag{25}$$

where

$$\Delta W = 2e \sqrt{\frac{eE}{\varkappa}} .$$

At T = 4.2°K, we have $\Delta W/2kT \approx 0.3\sqrt{E}$. Hence, it follows that the Frenkel mechanism can cause a slight increase in the carrier density even in a field of 1 V/cm; in fields of 5-6 V/cm, the carrier density should be doubled.

The experimental results presented in Fig. 14 show, however, that the carrier density increases more rapidly than predicted by Eq. (25). Therefore, the Frenkel mechanism can account for the increase in the carrier density only at the very beginning of region B, in fields of about 1 V/cm (Fig. 14).

Weakening of the recombination in strong electric fields. The carrier density is governed by the balance between the thermal generation and recombination processes

$$w_{\scriptscriptstyle T}(N_d - N_a) - w_r(TE)N_a n = 0, \quad n = \frac{w_{\scriptscriptstyle T}(N_d - N_a)}{w_r(TE)} , \tag{26}$$

where $w_T$ is the probability of thermal ionization and $w_r$ is a quantity proportional to the probability of simple electron capture.

The carrier density will increase if $w_r(TE)$ decreases when the field intensity is increased. At $T > \hbar\omega_{opt}/k$, the recombination of hot electrons becomes weaker because simple recombination is no longer possible while multiphonon processes have a low probability [28].

At low temperatures, this mechanism of increase in the carrier density is inoperative because $\epsilon_i < \hbar\omega_{opt}$ and the recombination of "hot" electrons requires the emission of one phonon.

$\sigma, \Omega^{-1} \cdot cm^{-1}$

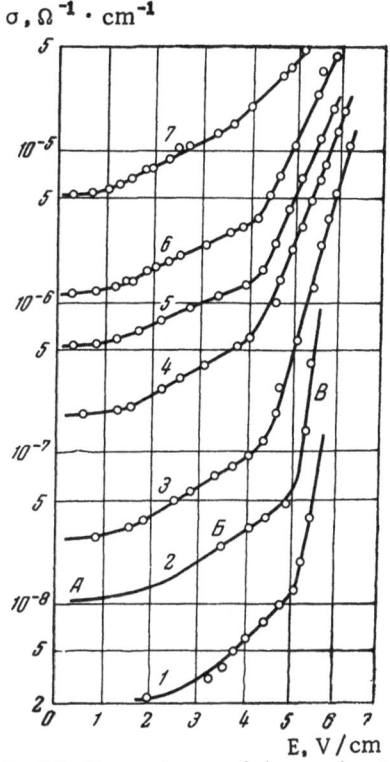

E, V/cm

Fig. 15. Dependence of the conductivity $\sigma$ on the field intensity for sample No. 2 at various temperatures: 1) 4.5; 2) 4.65; 3) 4.8; 4) 5.2; 5) 5.4; 6) 5.75; 7) 6.35°K.

At $T \approx 4.2°K$, the dependence $w_r f(\bar{\varepsilon})$ can be found only by a detailed analysis of the recombination mechanism.

The problem of carrier recombination at low temperatures has been considered recently by Lax [59]. To account for the experimentally observed "giant" recombination cross sections of Coulomb centers at helium temperatures, Lax has proposed carrier "capture" by an excited level followed by a cascade of one-phonon transitions in which some of the captured carriers go over to the ground state.

The probability of capture by the ground state becomes sufficiently high if the binding energy is $e^2/\varkappa r \sim kT$.

Since, at low temperatures, the capture may take place at orbitals of increasing radii r, the recombination cross section Q should increase strongly when the temperature and the electron energy are reduced.

The cascade theory yields the following results:

$$Q = \frac{4^5}{6} \left( \frac{C}{\gamma^4} \right) \left[ \ln \frac{\gamma}{1,75 \, \delta} + \frac{\delta}{\gamma} \right], \tag{27}$$

where

$$C = \left( \frac{\pi}{12} \right) \frac{Ze^2}{\frac{1}{12} mc^2} (l_{ph}\gamma)^{-1} = const;$$

$\gamma = kT/\frac{1}{2}mc^2$ is a dimensionless temperature; $l_{ph}\gamma \approx 1.84 \cdot 10^{-2}$ is independent of T; and Z is the nuclear charge.

At $T = 4.2°K$, $\gamma = 20$, $\delta = 6$, $C \approx 7 \cdot 10^{-9}$ cm$^2$ and the theory predicts, in agreement with experiment, that $Q \approx 3 \cdot 10^{-12}$ cm$^2$. Moreover, $Q(T) \propto T^{7/2}$.

In strong electric fields, the effective electron temperature rise reduces the recombination cross section because of a reduction in the radius of the orbital by which the electron is captured. In the case of interest to us, when the main energy losses are due to collisions with the acoustical phonons, we have

$$Q(x) = \frac{4^6}{6} \frac{C}{\gamma^4} \frac{F(x)}{x}, \tag{28}$$

where $x = \varepsilon/kT$, $\gamma = kT/\frac{1}{2}mc^2$, and $F(x) \propto x^{-1} - x^{-1.5}$ in a wide range of values of x. Hence,

$$Q(x) \simeq \frac{3 \cdot 10^{12}}{x^{(2-2.5)}}, \tag{29}$$

and the recombination probability is

$$w_r = Qv \simeq \frac{7 \cdot 10^6}{x^{(1.5-2)}}, \tag{30}$$

because $v = \sqrt{2\bar{\varepsilon}/m} = 2.5 \cdot 10^6 \sqrt{x}$.

It is evident from Eq. (30) that the probability of carrier recombination at helium temperatures decreases as $E^{1.5} - E^2$, which is in reasonable agreement with experiment.

CHAPTER IV

# Breakdown

## §1.  Results of Measurements

At helium temperatures, in fields of the order of 10 V/cm, the current and carrier density increase rapidly, as indicated by the data presented in Fig. 6.

The drift mobility of carriers remains practically constant, equal to $10^6$ cm/sec. The value of $E_b$ is independent of the geometry of the sample and of the surface treatment to which it is subjected.

When the temperature is increased, the dependence of the carrier density on the field becomes smoother (Fig. 15). Between 4 and 7°K, the slope of the $dj/dE = f(E)$ curve decreases by a factor of several tens. When the temperature is increased, the change in the carrier density due to the application of the field becomes small compared with the change due to thermal generation, and at $T > 7°K$, this increase can no longer be called "breakdown." Therefore, in considering the relationships governing the "low-temperature breakdown," we shall restrict ourselves to temperatures below 7°K. The breakdown field varies weakly with increasing temperature, as indicated by the data presented in Fig. 16a.

An interesting feature of the "low-temperature breakdown" in a sample is the dependence of $E_b$ on the carrier density, on the carrier mobility, and on the magnetic field intensity.

The results of the measurements of $E_b = f(N_d - N_a)$ are presented in Fig. 16b and in Table 7. The dependence $E_b = f(N_d - N_a)$ is described by a family of similar curves, each of which represents samples of one batch. For samples prepared from the same ingot (i.e., samples of the same batch), the value of $E_b$ is independent of the net impurity concentration if $(N_d - N_a) < 10^{15}$ cm$^{-3}$, but $E_b$ increases linearly with this concentration if $(N_d - N_a) > 10^{15}$ cm$^{-3}$.

These data do not support Burstein and Sclar's conclusion that the breakdown voltage is determined solely by the net impurity [8]. In fact, $E_b$ depends not only on the net impurity concentration, but also on the concentration of the minority impurity. In our measurements, $E_b$ increased by a factor of 2-3 when the minority impurity concentration was increased by a factor of 10. This could be used, in particular, for quality control of the degree of compensation of impurities in germanium. For minority impurity concentrations less than $10^{13}$ cm$^{-3}$, $E_b \approx$ 5-7 V/cm for n- and p-type germanium. The dependence of $E_b$ on the total impurity concentration is presented in Fig. 16c and the dependence of $E_b$ on the carrier mobility is shown in Fig. 16d. These results show that samples with different impurity contents have different breakdown voltages, but $E_b \times \mu \approx$ const $\approx 10^6$ cm/sec.

The value of $E_b$ is affected considerably by a magnetic field. The results presented in Chapter V show that

$$E_b(H)\,\mu(H) = E_b(0)\,\mu(0) = \text{const.} \tag{31}$$

We may assume that, in general, the dependence of $E_b$ on various factors ($N_d$, $N_a$, T, H) is due to the influence of these factors on the carrier mobility.

## §2.  Breakdown Mechanism

The breakdown at helium temperatures in very weak electric fields may be associated with the ionization of neutral donors or acceptors whose electron binding energy is very small (only 0.01 eV). Since the electrostatic ionization of impurities would have required fields about one hundred times stronger than the experimentally observed values [60], we shall consider in detail the impact ionization mechanism.

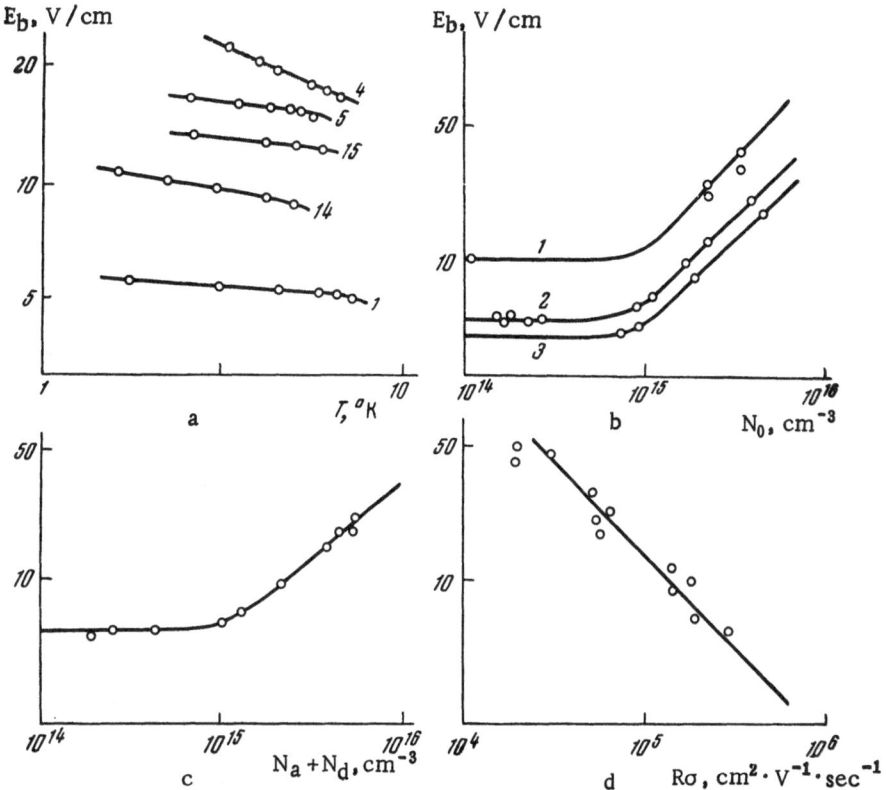

Fig. 16. Dependence of the breakdown field $E_b$ on: a) temperature (the numbers alongside the curves correspond to the sample numbers); b) net impurity concentration $N_0 = N_d - N_a$ (1 — batch Sb II; 2 — batch Sb I; 3 — batch Bi I); c) total impurity concentration $N = N_d + N_a$ (points represent different samples); d) carrier mobility $\mu \propto R\sigma$ (the points represent different samples).

TABLE 7

| Sample No. | $\rho$, $\Omega \cdot$ cm | | $E_b$, V/cm | Sample No. | $\rho$, $\Omega \cdot$ cm | | $E_b$, V/cm |
|---|---|---|---|---|---|---|---|
| | $T = 300°$ K | $T = 4,2°$ K | | | $T = 300°$ K | $T = 4,2°$ K | |
| 1 | 2.5 | $10^8$ | 4.8 | 9 | 0.65 | $10^9$ | 55—70 |
| 2 | 1.8 | $10^8$ | 4.8 | 10 | 35 | $10^8$ | 4.8 |
| 3 | 0.9 | $10^9$ | 10 | 11 | 22 | $10^8$ | 4.8 |
| 4 | 0.35 | $10^7$ | 20 | 12 | 35 | $10^8$ | 4.8 |
| 5 | 2.5 | $1,2 \cdot 10^8$ | 6.4 | 13 | 22 | $10^8$ | 4.8 |
| 6 | 1.8 | $1,4 \cdot 10^8$ | 7.5 | 14 | 1 | $8,5 \cdot 10^7$ | 9 |
| 7 | 25 | $5 \cdot 10^9$ | 14—16 | 15 | 45 | $1,2 \cdot 10^9$ | 13 |
| 8 | 1.5 | $10^{11}$ | 38—41 | 16 | 52 | $10^9$ | 7 |

TABLE 8

| Sample No. | $\varepsilon_i$, eV | $T$,°K | $\mu$, cm$^2 \cdot$V$^{-1} \cdot$sec$^{-1}$(exp.) | $E_b$, V/cm (exp.) | $\xi$, calc. |
|---|---|---|---|---|---|
| 1 | 0.012 | 4.8 | $2 \cdot 10^5$ | 4.7 | 0.3 |
| | | 5.25 | | 4.7 | 0.34 |
| 12 | 0.011 | 4.2 | $2 \cdot 10^5$ | 5 | 0.34 |
| 5 | 0.0097 | 5 | $1.5 \cdot 10^5$ | 5.5 | 0.2 |

The Fröhlich breakdown criterion can be used for covalent crystals. According to this criterion, the breakdown occurs when the energy of the fastest carriers is close to the ionization energy $\bar{\varepsilon} = \xi \varepsilon_i$, where $\xi < 1$. Then, the breakdown field is given by the relationship

$$E_b \sqrt{\mu \mu_{ph}} = s \sqrt{\frac{\xi \varepsilon_i}{2kT}} \ . \tag{32}$$

In this case, the scattering by the acoustical phonons governs not only the energy losses but also the carrier mobility, and the condition (32) yields

$$E_b \mu_{ph} = s \sqrt{\frac{\xi \varepsilon_i}{2kT}} \ , \tag{33}$$

from which it follows that, if T = const,

$$E_b \mu = \text{const,} \tag{33'}$$

in agreement with the experimental results presented in Fig. 16d.

The experimental values and the values calculated using Eq. (32) are compared in Table 8.

The last column of Table 8 lists the values of $\xi$ for which the calculated values of $E_b$ agree with the experimental values. It is evident from these data that, if $\xi$ = 0.2-0.35, the experimental results can be described by the elementary formula (33).

Although allowance for all the scattering processes leads to Eq. (32), which differs from the empirical dependence $E_b \mu \approx$ const, the latter dependence is a convincing proof of the impact ionization mechanism, since the rate of acquisition of energy by carriers from an electric field is proportional to their mobility.

Let us consider the processes which govern the equilibrium carrier density, allowing for the thermal and impact ionization and recombination, including bimolecular and triple (impact) $w_A$ processes:

$$w_t (N_d - N_a - n) + w_i n (N_d - N_a - n) - w_r n (N_a + n) - w_A n^2 (N_a + n) = 0. \tag{34}$$

At low temperatures, in weak electric fields, we have

$$n < N_a < N_d - N_a. \tag{35}$$

When the field intensity is increased, n increases due to the impact ionization. However, in spite of the rapid rise of the current, the condition (35) is still satisfied because, in weak fields, n $\approx 10^4$-$10^6$ cm$^{-3}$.

If we consider only the range n $\ll N_a < N_d - N_a$, then the impact recombination can be neglected. Then

$$n = \frac{w_t (N_d - N_a)}{w_r N_a - w_i (N_d - N_a)} = \frac{n_0 (E, T)}{1 - \gamma}, \tag{36}$$

where $n_0 = w_t(N_d - N_a)/w_r \times N_a$ is the carrier density in fields for which the impact ionization can be

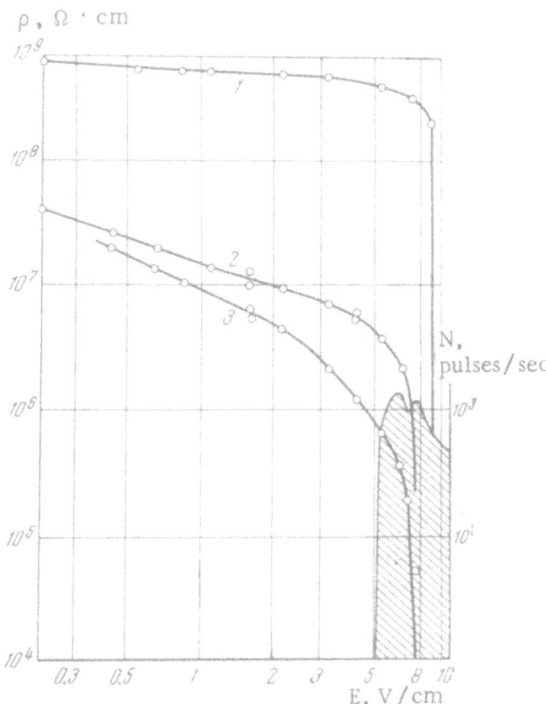

Fig. 17. Dependence of the resistivity of sample No. 14 on the field intensity E. 1) In the absence of background radiation; 2,3) for different intensities of background radiation. The dashed region represents fields in which "noise" pulses were observed. The curve gives the dependence of the number of "noise" pulses N = $f$(E) for the higher background radiation intensity (3).

neglected, and

$$\gamma = \frac{w_i N_0}{w_r N_a.}$$

The rapid rise of the carrier density begins when $\gamma \to 1$ or

$$w_i N_0 \to w_r N_a. \qquad (37)$$

The theoretical problem of finding $w_i/w_r = f(E)$ at helium temperatures has been considered by Chuenkov. By assuming that $w_r$ is a very strong function of the energy, he has found

$$\frac{w_i}{w_r} = C_1 \left(\frac{E}{E_0^*}\right)^4 \exp\left(-\frac{E_0^2}{E^2}\right), \qquad (38)$$

where $E_0^*$ and $E_0$ are characteristic fields whose values are close to the breakdown value and $C_1$ is a constant.

The ratio $(w_i/w_r) \to 1$ when $E \to E_b$. The numerical result [29] for germanium, obtained using Eq. (38), gives $E_b \approx 6$ V/cm, in agreement with the experimental results. The relationships represented by Eqs. (35) and (36) are not valid at higher temperatures because we cannot neglect the term $w_t \propto \exp(-\varepsilon_i/kT)$. The rise in $w_t$ results in an increase in the denominator in Eq.(36) and a smaller value of the ratio $n/n_0$.

Consequently (Fig. 15), the dependence $\sigma = f(E)$ already loses the sudden breakdown features at $T \approx 6$-$7°K$.

## §3. Noise

The impact ionization process is known to be accompanied by noise [34]. Therefore, we carried out additional experiments to determine the nonstationary nature of the electrical conductivity in fields close to $E_b$. The measurements were made using a circuit in which a variable voltage was supplied from a battery to a germanium sample and to a calibrated resistance, the signal from which was fed to an oscillograph and a D-class amplifier with a pulse counter. To reduce the resistance of the sample in the prebreakdown region, we used steady-state illumination (this illumination was simply the thermal background radiation).

The experimental results showed that, in fields close to $E_b$, the passage of a current through a sample was accompanied by current pulses (which were recorded by the counter and displayed on the oscillograph screen). The fluctuations of the current at breakdown were 3-4 times larger than the average current.

Figure 17 shows the results of the simultaneous measurements of the resistivity $\rho = f(E)$ and of the number of pulses per unit time N = $f$(E) for a certain intensity of the background radiation. It follows from Fig. 17 that pulses appeared almost suddenly in a field close to $E_b$. When $E \to E_b$, the number of pulses and their average amplitude increased strongly. When $E \neq E_b$, we observed (on the oscillograph screen) pulses with very different amplitudes U, for which $\overline{U} \ll U_{max}$. When $E \approx E_b$, the average pulse amplitude approached the maximum amplitude.

Nonstationary processes in fields $E \to E_b$ may be associated with the generation—recombination noise which accompanies the process of impact ionization.

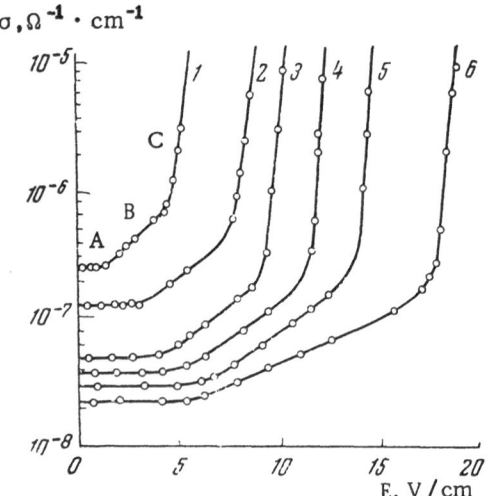

$\sigma, \Omega^{-1} \cdot cm^{-1}$

Fig. 18. Field dependence of the conductivity $\sigma = f(E)$ at T = 5.2°K for lightly doped germanium subjected to magnetic fields H of the following intensities: 1) 0; 2) 600; 3) 1100; 4) 1700; 5) 2650; 6) 4400 Oe.

CHAPTER V

# Influence of a Magnetic Field on the Electrical Conductivity of Germanium in Prebreakdown and Breakdown Fields

## §1. Results of Measurements

To obtain additional information on the carrier scattering mechanism at low temperatures, we investigated the electrical conductivity of germanium subjected simultaneously to strong electric and magnetic fields.†

An investigation of $\sigma = f(E)$ showed that the application of a magnetic field did not alter the nature of the curves but did result in an increase in the resistivity, a broadening of the regions A and B, and an increase in the breakdown field intensity $E_b = f(H)$ (Fig. 18).

Heavily doped samples, with impurity concentrations higher than $3 \cdot 10^{15}$ cm$^{-3}$, behaved differently at helium temperatures. The electrical conductivity of such samples (Fig. 19) in weak electric fields was independent of the magnetic field intensity, but in prebreakdown fields the usual dependence was observed. When the temperature was increased, this anomaly in the behavior of heavily doped samples disappeared, while the dependence $\sigma(E, H)$ (for samples having a range of impurity concentrations) remained the same, like that shown in Fig. 18.

The ratios $\rho(H)/\rho(0)$ and $E_b(H)/E_b(0)$ at a fixed value of the magnetic field (Fig. 20) differed widely for samples having different impurity concentrations. Higher values of $\rho(H)$ and $E_b(H)$ were obtained for samples with a lower impurity content. When the temperature was lowered, the increase in the resistivity became larger.

The general features of the dependences presented in Fig. 20 are easily discernible. The resistivity and the breakdown field have similar dependences on the magnetic field intensity

$$\frac{\rho(H)}{\rho(0)} = K \frac{E_b(H)}{E_b(0)}, \tag{39}$$

where K is a constant.

For germanium samples with N < $10^{15}$ cm$^{-3}$ at T ⩾ 70°K and N > $10^{15}$ cm$^{-3}$ at T ≥ 4.2°K, the ratios $\rho(H)/\rho(0)$ and $E_b(H)/E_b(0)$ not only depended similarly on the magnetic field but even had equal absolute values (Fig. 20).

The values of $\rho(H)$ and $E_b(H)$ depended linearly on the magnetic field intensity for fields H > 2 kOe. In weak fields, H < 1 kOe, $\rho(H)$, and $E_b(H)$ depended quadratically on the magnetic field. For clarity, these ratios are presented in Fig. 21 as a function of the square of the magnetic field intensity. We can see that in weak fields the experimental points fit a straight line.

_____

† Strong electric fields at helium temperatures are those ~1 V/cm. Strong magnetic fields, defined by the inequality $\mu H > 1$, represent about 1 kOe at T = 4.2°K.

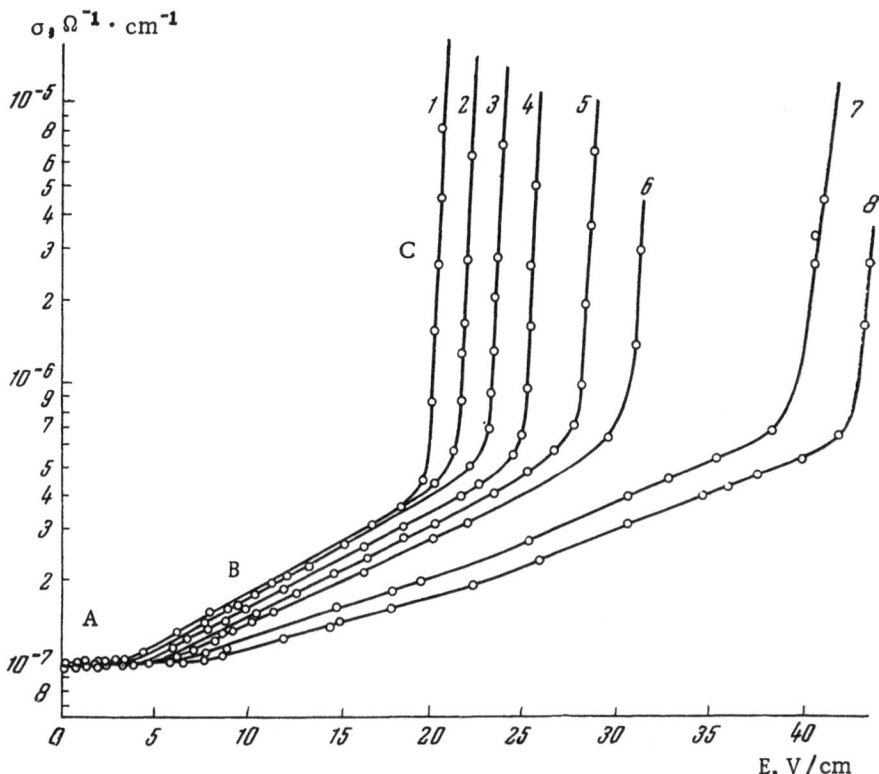

Fig. 19. Field dependence of the conductivity $\sigma = f(E)$ at T = 4.2°K for heavily doped germanium subjected to magnetic fields H of the following intensities: 1) 0; 2) 240; 3) 580; 4) 870; 5) 1525; 6) 2180; 7) 5800; 8) 7000 Oe.

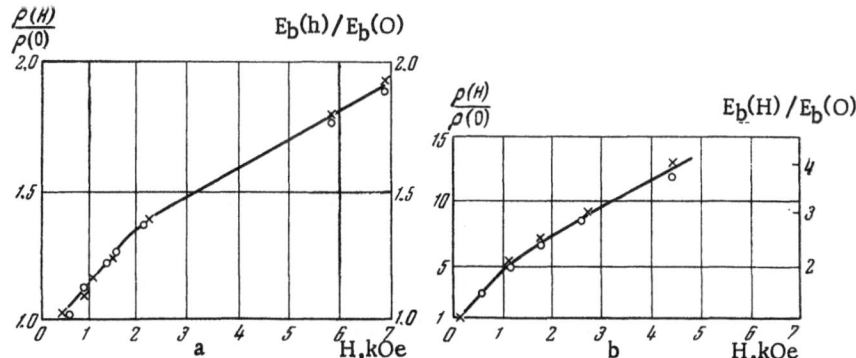

Fig. 20. Magnetic field dependence of the ratios $E_b(H)/E_b(0)$ (points) and $\rho(H)/\rho(0)$ (crosses) in heavily (a) and lightly (b) doped samples.

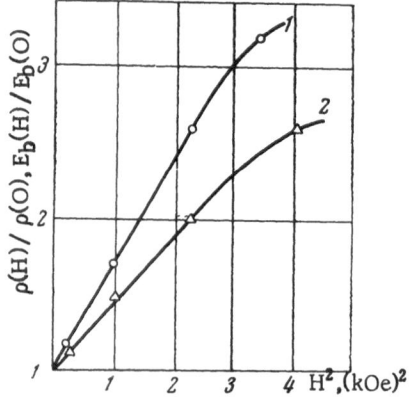

Fig. 21. Dependences $\rho(H)/\rho(0) = f(H^2)$ (curve 1) and $E_b(H)/E_b(0) = f(H^2)$ (curve 2).

TABLE 9

| j | H | $\left[\dfrac{\rho(H)}{\rho}\right]_{sat}$ | $(\mathscr{K} \approx 20)$ |
|---|---|---|---|
| $\mathbf{j}\perp\mathbf{H}$ | [111] | $\dfrac{(\mathscr{K}+2)\;(5\mathscr{K}+4)}{3(8\mathscr{K}+1)} = 2.7$ | |
| [100] | [010] | $\dfrac{32\;(2\mathscr{K}+1)}{27\pi\;(\mathscr{K}+2)} = 1.5$ | |
| [110] | [1$\bar{1}$0] | $\dfrac{32\;(2\mathscr{K}+1)\;(\mathscr{K}+2)}{81\pi\;\mathscr{K}} = 5.6$ | |

## §2.  Weak Electric Fields

If all the carriers have the same velocity, then the Hall field completely compensates the Lorentz force, and the application of a magnetic field does not bend the carrier trajectories. However, carriers have a range of velocities and the Hall field compensates only the Lorentz force of carriers traveling at an average velocity. Consequently, the trajectories of most of the carriers in magnetic fields are curved; this reduces the mobility and increases the resistivity.

The problem of finding the ratio $(\Delta\rho/\rho) = [\rho(H) - \rho(0)]/\rho = f(H)$ in its general form has not yet been solved [61].

To carry out a quantitative calculation of $\Delta\rho/\rho$, it is necessary to make certain assumptions about the form of the energy surface and about the momentum dependence of the relaxation time. The dependence $\Delta\rho/\rho = f(H)$ has been calculated theoretically, allowing for the complex energy band structure of germanium, for two extreme cases only: very weak and very strong magnetic fields [62]. In weak magnetic fields, $\Delta\rho/\rho \propto H^2$. In strong magnetic fields, the theory predicts the saturation of $\Delta\rho(H)/\rho$. The limiting value of $\rho(H)/\rho$ depends on the orientation of $\mathbf{j}$ and $\mathbf{H}$ with respect to the principal crystallographic axes. $(\Delta\rho/\rho)$ are given for ellipsoids whose principal axes are directed along [111]; the results are taken from [62]. The following quantities are used in Table 9: $\mathscr{K}$ is the anisotropy coefficient, defined by $\mathscr{K} = m_L/m_T$, where $m_L$ and $m_T$ are, respectively, the longitudinal and transverse components of the effective mass. For germanium, $\mathscr{K} \approx 20$ [6].

The results of the measurement of $\rho(H)/\rho(0)$ at helium temperatures (Fig. 20) cannot be described completely by this theory.

In weak magnetic fields, the resistivity is proportional to the square of the field intensity, in agreement with the theory. However, in strong magnetic fields, the resistivity in a transverse field, $\rho(H)/\rho(0)$, is a linear function of the field intensity, which does not agree with the theory.

The results of the measurement of $\rho(H)/\rho(0)$ in strong magnetic fields can be divided into two groups:

The first comprises the results for $\rho(H)/\rho(0)$ at $T \gtrsim 7°K$, which can still be accounted for by the theory. In this case, $\rho(H)/\rho(0)$ at $H \approx 7$ kOe does not differ greatly from the theoretically predicted value, which is 2.7 (for H ∥ [111] and $\mathscr{K} = 20$). However, the tendency to saturation is not observed in fields up to $H \approx 10$ kOe.

The second group of results relates to the behavior of $\rho(H)/\rho(0)$ at helium temperatures. In weak electric fields, $\Delta\rho(H)/\rho = 0$ for samples with an impurity content higher than $3 \cdot 10^{15}$ cm$^{-3}$ in magnetic fields up to $H \approx 10$ kOe (Fig. 19).

On the other hand, the change in the resistivity in a magnetic field is extremely large for samples with impurity contents less than $10^{15}$ cm$^{-3}$. Thus, for sample No. 1 in a field H = 4.5 kOe, we find that $\Delta\rho/\rho$ = 4, 6, and 14 at temperatures of 7, 6, and 5°K, respectively.

The experimentally observed constancy of the resistivity of heavily doped samples in a magnetic field (region A, Fig. 19) is in agreement with the hypothesis of a narrow impurity band with a low carrier mobility $\mu = 10^{-2}$-$10^{-4}$ cm$^2 \cdot$ V$^{-1} \cdot$ sec$^{-1}$ [55, 63].

Since $\Delta\rho/\rho \propto (\mu H)^2/c^2$, it is not surprising that $\Delta\rho/\rho \to 0$ and cannot be measured.

A strong increase in the resistivity in a magnetic field observed for weakly doped samples at T = 4-5°K cannot be explained within the framework of the classical theory of Abeles and Shibuya [62]. This increase in the resistivity is not due to a reduction in the carrier density in strong magnetic fields resulting from the Zeeman splitting of an n-fold degenerate level, because the magnetic fields in our experiments were not sufficiently strong to significantly alter the impurity energy level $\varepsilon_i$.

The position of this energy level in a magnetic field

$$\varepsilon_i(H) = \varepsilon_i(0) + \delta^* H, \tag{40}$$

where $\delta^* = eh/mc$ is a quantity which is proportional to the Bohr magneton, changes by only 1-2% in H $\approx$ 2 · $10^3$ Oe. Obviously, such a change in the ionization potential cannot greatly alter the carrier density.

The hypotheses about the structure of the energy surface in the Brillouin zone, on which the theory of Abeles and Shibuya [62] is based, have been confirmed by the diamagnetic resonance results and can be regarded as definitely proved. Therefore, disagreement between this theory and the results of the measurements of $\rho(H)/\rho$ at T = 4°K indicates that the purely classical treatment is insufficient in the case of the motion of carriers in strong magnetic fields [64].

## §3. Strong Electric Fields

In our analysis of the results in this range of fields, we shall use only approximate estimates because the theory has not yet been developed.

The increase in the carrier density in prebreakdown fields is due to the weakening of the recombination of the "hot" carriers. The heating of the electron gas is obviously different in magnetic fields of different intensity. The stronger the magnetic field, the greater is the reduction in the mobility. Therefore, to obtain significant heating of the electron gas, high-intensity fields are required. This broadens region B and increases the breakdown voltage (Figs. 18 and 19).

In weak electric fields at low temperatures, the electrical conductivity may be governed by several mechanisms (in particular, by the impurity band conduction). In prebreakdown fields, the electrical conductivity increases solely because of the contribution of the conduction band. This allows us to conclude that, irrespective of the mechanism of conduction in weak electric fields, the electrical conductivity in strong electric fields is governed by the direct influence of the conduction band.

When the elementary theory of the dependence of $\rho$ on H is applicable, we have the following equalities:

$$\frac{\rho(H)}{\rho(0)} = \frac{\mu(0)}{\mu(H)} \quad \text{and} \quad \frac{E_b(H)}{E_b(0)} = \frac{\rho(H)}{\rho(0)}$$

(see Fig. 20). Then

$$E_b(H)\mu(H) = E_b(0)\mu(0) = \text{const} \simeq 10^6 \text{ cm/sec}.$$

Note. The similarity of the dependences $\rho(H)$ and $E_b(H)$ is observed only in the compensation measurements, which eliminate the influence of the contacts. When two electrodes are used, the breakdown fields are overestimated, and in weak magnetic fields the quadratic dependence of $\rho(H)$ is absent.

CHAPTER VI

# Electrical Conductivity of Germanium
# in Fields Higher than the Breakdown Value

In the "breakdown" region, the carrier density in the conduction band increases rapidly, reaching several percent of the net impurity concentration. Further increase in the field intensity produces a monotonic rise in the current.

We shall now consider the phenomena responsible for the rise of the current in electric fields considerably higher than the breakdown value.

## §1.  Results of Measurements

Figure 7 shows the dependence $j = f(E)$ obtained at various temperatures. It follows from this figure that the dependence of the current on the voltage is stronger at lower temperatures when the thermal ionization is less significant than the impact ionization. When E becomes much greater than $E_b$, $dj/dE$ decreases and the density of the current depends weakly on temperature. In the range $4E_b < E < 7E_b$, the current density rises linearly with increasing field intensity, in the same way as in weak fields. When $E > 7E_b$, the current density increases sublinearly as the field intensity is increased.

The power evolved in a sample in the dc measurements is small compared with the "critical" value of the power ($0.5$ W/cm$^2$) at which the normal thermal contact with liquid helium is disturbed [14]. Under these conditions, we could assume that the temperature of a sample was constant and equal to $4.2°$K. To ensure similar conditions in the determination of the Hall effect and of the current–voltage characteristics in stronger fields, we carried out the measurements using 3-μsec pulses. The heating of a sample could then be neglected.

These measurements gave values of R very close to $1/e(N_a - N_d)$ in fields $E \geq 20$ V/cm for $j \geq 20$ A/cm$^2$, from which it followed that in fields $E \geq 20$ V/cm the majority (acceptor) impurity was practically completely ionized.

Figure 22 shows the dependence of the drift velocity v on the field intensity E, calculated from the current–voltage characteristics in fields $20 \leq E \leq 400$ V/cm at temperatures of 77 and $4.2°$K. It is clear from Fig. 22 that the drift velocity at both temperatures at first increases linearly with the field intensity in fields $E > 20$ V/cm; the linear rise is followed by a slower rise in stronger fields.

The observed dependence of the drift velocity is due to the application of the field and is not a thermal effect, but it should be mentioned that, in the case of high values of jE, heat exchange between a sample and the liquid helium surrounding it becomes difficult. This follows from the results of measurements of the current–voltage characteristic at $4.2°$K using 100-μsec pulses; these results are represented by the dashed curve in Fig. 22. It follows from this figure that in strong electric fields, the j–E characteristic lies approximately in the middle between characteristics obtained at helium and nitrogen temperatures. A calculation carried out assuming adiabatic conditions gives about $50°$K as the temperature of a sample under 400 V/cm.

In fields $E \leq E_b$, the j–E characteristics and the Hall coefficient were measured using dc. The mobility in fields $E < E_b$ calculated from these measurements was about $2 \cdot 10^5$ cm$^2 \cdot$ V$^{-1} \cdot$ sec$^{-1}$, and it did not change much up to $E = E_b = 5$ V/cm, since the drift velocity was about $10^6$ cm/sec at breakdown [8]. We had found earlier that the drift velocity remained constant in fields up to $E = 1.2E_b$. In the present investigation, we established that $v \approx 10^6$ cm/sec, even for $E = 4E_b$. Taking cognizance of these results, we may assume that the drift velocity of carriers changes little over the whole range of fields in which the carrier density increases.

In fields close to $E_b$, the hole density p increase approximately as $\exp(E - E_b)$. In the region $E > 4E_b$, $p \approx$ const. The field dependence of the hole density in the region $E_b < E < 4E_b$ can be determined from the j–E curve shown in Fig. 23, remembering that, in this range of fields, $v =$ const.

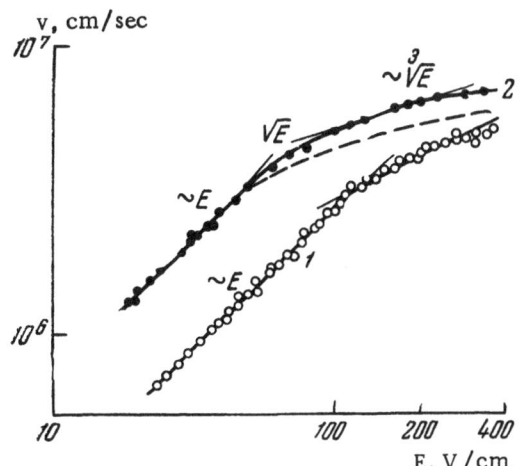

Fig. 22. Dependence of the carrier drift velocity on the electric field at temperatures of: 1) 77; 2) 4.2°K.

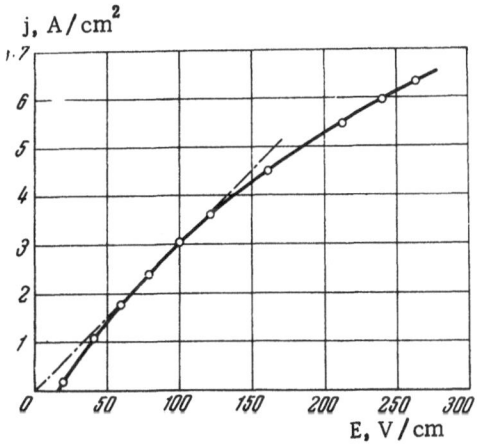

Fig. 23. Dependence $j = f(E)$ for p-type germanium in strong electric fields.

Remarks on the Results of the Measurements. In the calculation of the hole density from the measurements of the Hall emf, we met a difficulty associated with the influence of the magnetic field (used in these measurements) on the experimental results.

It is known [65] that in a weak field

$$R_{H \to 0} = \frac{3\pi}{8ep_1} \frac{1 + \left(\frac{p_2}{p_1}\right)\left(\frac{\mu_2}{\mu_1}\right)^2}{\left[1 + \left(\frac{p_2}{p_1}\right)\left(\frac{\mu_2}{\mu_1}\right)\right]^2} , \qquad (41)$$

where the subscript "1" refers to the normal holes and "2" to the light holes. The simple relationship $p = 1/Re$, where $p = p_1 + p_2$, is valid for p-type germanium only in strong magnetic fields [65] defined by the inequality $\beta > 25$, where

$$\beta = \frac{9\pi}{16}\left(\frac{\mu H}{s}\right)^2 , \qquad (42)$$

where s is the velocity of sound.

At helium temperatures, the condition (42) is satisfied by fields of about 1 kOe. To satisfy the condition (42) at nitrogen temperatures, much stronger fields are required. However, an actual analysis carried out by Willardson and Beer [65] for germanium with an impurity concentration $N_a - N_d \approx 10^{14}$ cm$^{-3}$ has shown that, at T = 77°K, R = $f$(H) has a minimum close to $1/pe$ in a field H $\approx$ 2000 Oe. However, magnetic fields of the order of 1 kOe affect considerably the carrier relaxation time, and this increases the value of the breakdown field. Consequently, impact ionization is affected by a magnetic field.

The results obtained from the measurements of the Hall effect show that in the presence of a magnetic field, impurities are almost completely ionized in a field E $\geq$ 20 V/cm. This applies even more for H = 0. On the other hand, the curve presented in Fig. 23 shows that, in the region E < 20 V/cm and H = 0, the current density falls away sublinearly as E is decreased. This may simply be due to a reduction in the carrier density, since in this range the mobility can only increase, because of weaker scattering by ionized atoms.

Thus, a magnetic field up to 2000 Oe does not greatly affect the value of the electric field $E_M$ at which the impurities are almost completely ionized; with or without such a field, we have $E_M \approx$ 20 V/cm.

§2. Elementary Theory and Discussion of Results

a) Dependence of the Average Carrier Energy on an Electric Field E. The power adquired by carrier from the field is

$$Q \approx (eE)\left(\frac{eE}{2m}\tau\right) = \frac{(eE)^2\tau}{2m} , \qquad (43)$$

where $1/\tau$ is the total number of collisions in 1 sec.

At low temperatures, the carriers are scattered by the acoustical phonons and by charged and neutral impurities. When $E > E_b$, the scattering by neutral impurity atoms can be neglected and $\tau^{-1} = \tau_{ph}^{-1} + \tau_i^{-1}$, where $\tau_{ph}$ is the average time interval between collisions with the acoustical phonons:

$$\tau_{ph} = \frac{l_{ph}}{v} = \sqrt{\frac{m}{2}}\, l_{ph}\,(\bar{\varepsilon})^{-1/2} . \qquad (44)$$

Here, $l_{ph}$ is the mean free path (in covalent conductors, this path is independent of the field).

The average time between collisions with charged impurities is found from the relationship

$$\tau_i^{-1} \simeq [p(\bar{\varepsilon}) + N_d]\,\pi a_0^2\,\sqrt{\frac{2\bar{\varepsilon}}{m}}\left(\frac{\varepsilon_i}{\bar{\varepsilon}}\right)^2 , \qquad (45)$$

where $\bar{\varepsilon}$ is the average carrier energy; $\varepsilon_i$ is the impurity ionization energy; $p(\bar{\varepsilon})$ is the carrier density; and $\pi a_0^2$ is the geometrical cross section of an acceptor in p-type germanium.

When $p(\bar{\varepsilon}) > N_d$,

$$\tau \simeq \sqrt{\frac{m}{2}}\,\frac{l_{ph}}{\sqrt{\bar{\varepsilon}}}\,\frac{1}{1 + \left(\frac{\varepsilon_i}{\bar{\varepsilon}}\right)^2\left(\frac{p(\varepsilon)}{\bar{p_0}}\right)} , \qquad (46)$$

where $p_0^{-1} = \pi a_0^2\, l_{ph}$.

Since the main energy losses are due to the collisions of holes with the acoustical phonons, the power lost by a carrier is

$$A = \frac{1}{\tau_{ph}}\frac{\hbar\omega}{2N+1} , \qquad (47)$$

where $N = 1/[\exp(\hbar\omega/kT) - 1]$, and the phonon energy is $\hbar\omega = mvs$.

If $\hbar\omega/kT \ll 1$, then

$$A_{ph} \simeq \frac{1}{2}\frac{\delta\bar{\varepsilon}}{\tau_{ph}} , \qquad (48)$$

where $\delta = 2ms^2/kT$ is the fraction of the total energy lost in one collision with a phonon [25]. The condition $\hbar\omega/kT \ll 1$ ceases to be valid at helium temperatures in fields $E > E_b$. However, the difference between the approximate formula (48) and the exact formula (47) is small as long as the average temperature of the electron gas does not exceed several tens of degrees Kelvin.

Under steady-state conditions, $Q = A$ and

$$\left(\frac{e l_{ph} E}{\sqrt{2\delta\varepsilon_i}}\right)^2 = \left(\frac{\bar{\varepsilon}}{\varepsilon_i}\right)^2 + \frac{p(\bar{\varepsilon})}{p_0} . \qquad (49)$$

It follows from this formula that if $p(\bar{\varepsilon})/p_0 \ll (\bar{\varepsilon}/\varepsilon_i)^2$, then $\bar{\varepsilon} \approx a'E$, where $a' = e\, l_{ph}/\sqrt{2\delta}$. In the range of fields where the scattering takes place mainly on ionized impurity atoms, the average energy increases more slowly:

$$\bar{\varepsilon} \simeq \sqrt{(a')^2 E^2 - \frac{p(\bar{\varepsilon})}{p_0}} . \qquad (50)$$

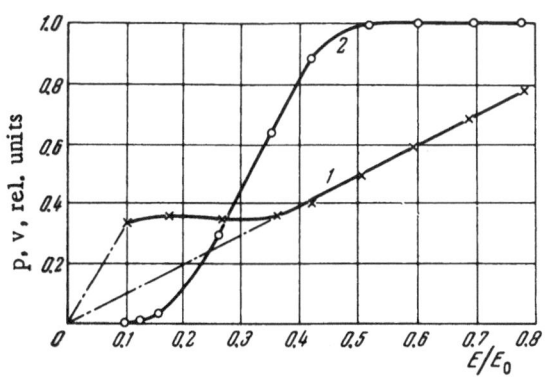

Fig. 24.  Calculated dependences $v = f(E)$ (curve 1) and $p = f(E)$ (curve 2); $E_0$ = const.

b) Dependence of the Carrier Density on the Average Carrier Energy.

At equilibrium, the number of holes generated is equal to the number of holes which recombine:

$$w_\tau [N_a - N_d - p(\bar{\varepsilon})]$$
$$+ w_i p(\varepsilon)[N_a - N_d - p(\bar{\varepsilon})]$$
$$= w_r p(\bar{\varepsilon})[N_d + p(\bar{\varepsilon})], \qquad (51)$$

where $w_t$ is the probability of thermal ionization, $w_i$ is the probability of impact ionization, and $w_r$ is the probability of recombination.

In fields $E > E_b$, we can neglect the thermal generation as well as the value of $N_d$ compared with $p(\bar{\varepsilon})$; then

$$p(\bar{\varepsilon}) \simeq \frac{N_a - N_d}{1 + \dfrac{w_r(\bar{\varepsilon})}{w_i(\bar{\varepsilon})}} . \qquad (52)$$

In deriving (52), we have neglected the impact recombination and assumed that the recombination probability $w_r$ is a weak function of the energy: $w_r(\bar{\varepsilon}) \approx$ const.  Bearing in mind that

$$w_i(\bar{\varepsilon}) = \frac{2}{\sqrt{\pi}} w_{i0} \left( \frac{\bar{\varepsilon}}{\varepsilon_i} \right)^{1/2} e^{-\frac{\varepsilon_i}{\bar{\varepsilon}}}, \qquad (53)$$

where $w_{i0} \simeq \pi a_0^2 \sqrt{2\varepsilon_i / m}$, we find

$$p(\bar{\varepsilon}) \simeq \frac{N_a - N_d}{1 + \beta \left( \dfrac{\bar{\varepsilon}}{\varepsilon_i} \right)^{-1/2} \exp\left( \dfrac{\varepsilon_i}{\bar{\varepsilon}} \right)} , \qquad (54)$$

where

$$\beta = \frac{\sqrt{\pi}}{2} \frac{w_r}{w_{i0}} .$$

It follows from Eq. (49) that, in the range of fields where $(\bar{\varepsilon}/\varepsilon_i)^2 < p(\bar{\varepsilon})/p_0$, the hole density p increases as the square of the field E.  When the field intensity is increased further, $\bar{\varepsilon} \to \varepsilon_i$ and $p \to (N_a - N_d)$, which follows from Eq. (54) and agrees with the experimental data.

c) Dependence of the Carrier Density and Drift Velocity on the Electric Field Intensity.  Dependence of the carrier density and drift velocity on the average carrier energy for $0.1\varepsilon_i < \varepsilon < \varepsilon_i$ is obtained from Eqs. (46) and (54).  The relationship between the average carrier energy and the electric field intensity is found from Eqs. (49) and (54):

$$\xi^2 = y^2 + \frac{\alpha}{1 + \beta y^{-1/2} \exp(1/y)} , \qquad (55)$$

where

$$\xi = \frac{e l_{ph}}{\sqrt{2\delta \varepsilon_i}} E', \quad y = \frac{\bar{\varepsilon}}{\varepsilon_i}, \quad \alpha = \frac{N_a - N_d}{p_0}, \quad \beta = \frac{\sqrt{\pi}}{2} \frac{w_r}{w_{i0}} .$$

The numerical solution of Eq. (55) for given values of the parameters $\alpha$ and $\beta$ yields $\bar{\varepsilon} = f(E)$.  Using Eqs. (46) and (54), we find $v = f(E)$ and $p = f(E)$.

Figure 24 presents the results of a numerical calculation of the electric-field dependence of the hole density and drift velocity for $\alpha = 0.1$ and $\beta = 0.03$; these values of the parameters were selected to obtain the

TABLE 10

| Sample No.† | Main impurity | $N_0$, cm$^{-3}$ | $q$, % | $N_d$, cm$^{-3}$ | $N_a$, cm$^{-3}$ | $d$, cm |
|---|---|---|---|---|---|---|
| 1 | Bi | $1.4 \cdot 10^{14}$ | 10 | $1.5 \cdot 10^{14}$ | $10^{13}$ | 1 |
| 2 | Sb | $2.1 \cdot 10^{14}$ | 10 | $2.3 \cdot 10^{14}$ | $2 \cdot 10^{13}$ | 1 |
| 3 | Sb | $4.2 \cdot 10^{14}$ | 10 | $4.6 \cdot 10^{14}$ | $4 \cdot 10^{13}$ | 1 |
| 4 | Sb | $4.2 \cdot 10^{14}$ | 10 | $4.6 \cdot 10^{14}$ | $4 \cdot 10^{13}$ | 1 |
| 5 | Sb | $1.6 \cdot 10^{15}$ | 60 | $3.7 \cdot 10^{15}$ | $2.1 \cdot 10^{15}$ | 1 |
| 6 | Bi | $1.0 \cdot 10^{15}$ | 10 | $1.1 \cdot 10^{15}$ | $10^{14}$ | 1 |
| 7 | Bi | $5.5 \cdot 10^{15}$ | 10 | $6 \cdot 10^{15}$ | $5 \cdot 10^{14}$ | 1 |
| 8 | In | $1.4 \cdot 10^{14}$ | 10 | $1.5 \cdot 10^{13}$ | $1.5 \cdot 10^{14}$ | 1 |
| 9 | In | $1.4 \cdot 10^{14}$ | 10 | $1.5 \cdot 10^{13}$ | $1.5 \cdot 10^{14}$ | 1 |
| 10 | In | $6.5 \cdot 10^{14}$ | 80 | $3 \cdot 10^{15}$ | $3.65 \cdot 10^{15}$ | 1 |
| 11 | In | $1.4 \cdot 10^{14}$ | 10 | $1.5 \cdot 10^{14}$ | $10^{13}$ | $5.5 \cdot 10^{-3}$ |
| 12 | In | $6.5 \cdot 10^{14}$ | 80 | $3 \cdot 10^{15}$ | $3.65 \cdot 10^{15}$ | $2 \cdot 10^{-3}$ |
| 13 | In | $6.5 \cdot 10^{14}$ | 80 | $3 \cdot 10^{15}$ | $3.65 \cdot 10^{15}$ | $1 \cdot 10^{-3}$ |
| 14 | In | $6.5 \cdot 10^{14}$ | 80 | $3 \cdot 10^{15}$ | $3.65 \cdot 10^{15}$ | $6 \cdot 10^{-4}$ |
| 15 | In | $6.5 \cdot 10^{14}$ | 80 | $3 \cdot 10^{15}$ | $3.65 \cdot 10^{15}$ | $3 \cdot 10^{-4}$ |
| 16 | In | $6.5 \cdot 10^{14}$ | 80 | $3 \cdot 10^{15}$ | $3.65 \cdot 10^{15}$ | $2 \cdot 10^{-4}$ |
| 17 | Sb | $1.4 \cdot 10^{14}$ | 10 | $1.5 \cdot 10^{14}$ | $10^{13}$ | $5 \cdot 10^{-4}$ |

†Numbering of samples (in order of increasing impurity concentration) is different from that used in Table 2.

TABLE 11

| Sample No. | $q$, % | $E_b$, V/cm | $E_m$, V/cm | $E_1$, V/cm | $E_2$, V/cm | $j_i$, A/cm$^2$ | $E_1/E_3$ | $v = (j_1/eN_0) \cdot 10^{-6}$, cm/sec |
|---|---|---|---|---|---|---|---|---|
| 1 | 10 | 4.5 | — | — | — | — | — | — |
| 2 | 10 | 5.0 | — | 20 | 65 | 60 | 4 | 1.94 |
| 3 | 10 | 5.2 | — | 20 | 80 | 120 | 4 | 1.9 |
| 4 | 10 | 5.5 | — | 22 | 80 | 120 | 4 | 1.9 |
| 5 | 60 | 23 | 20 | 50 | 100 | 450 | 2.5 | 1.76 |
| 6 | 10 | 7.5 | — | — | — | — | — | — |
| 7 | 10 | 20 | — | 80 | 100 | 1400 | 4 | 1.6 |
| 8 | 10 | 4.8 | — | 20 | 50 | 35 | 4 | 1.45 |
| 9 | 10 | 5.0 | — | 20 | 50 | 35 | 4 | 1.45 |
| 10 | 80 | 24 | 15 | 25 | 50 | 150 | 1.5 | 1.5 |

best agreement with the experimental data. We find, however, that the results of the calculations are very sensitive to the value of the parameter $\alpha = N_0/\pi a_0^2 l_{ph}$; when $\alpha \neq 0.1$, the dependence $v(E)$ differs considerably from the linear law. Moreover, we obtain $\alpha = 0.1$ for $N_0 \approx 10^{14}$ cm$^{-3}$ if we make the unlikely assumption that $\pi a_0^2 = 3 \cdot 10^{-12}$ cm$^2$, which is an order of magnitude greater than the geometrical cross section of an impurity atom.

In view of this, it was of interest to extend the investigations and to study the dependence $j = f(E)$ in fields greater than the breakdown value, using both n- and p-type germanium with various impurity concentrations.

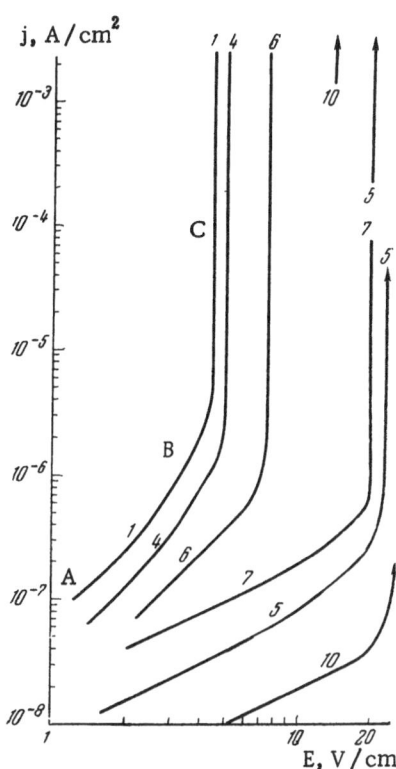

Fig. 25. j−E characteristics of samples with various concentrations of impurities; $E \leq E_b$. The numbers alongside the curves are sample numbers.

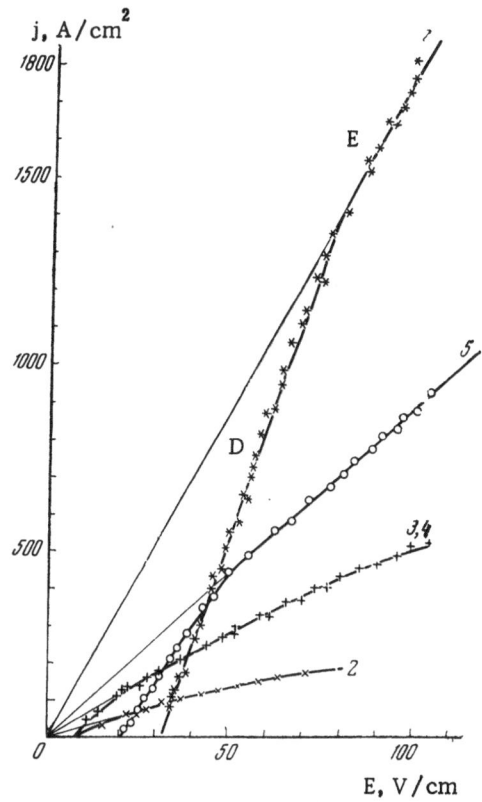

Fig. 26. j−E characteristics of n-type germanium for $E > E_b$. The numbers alongside the curves are sample numbers.

## §3. Results of Control Measurements

We determined the j−E characteristics of germanium doped with antimony, bismuth, and indium in concentrations ranging from $10^{14}$ to $5 \cdot 10^{15}$ cm$^{-3}$; the measurements were made at liquid helium temperature using pulses of 3-$\mu$sec duration. We also investigated the dependence of the various quantities on the dimensions of the breakdown "gap," using p-type germanium samples with thicknesses down to several microns.

The properties of the samples used are given in Table 10, where the following quantities are tabulated: the thickness of the breakdown gap d; the uncompensated impurity concentration $N_0$; and the degree of impurity compensation q, calculated by the Lee method [66] from the temperature dependence of the Hall coefficient using the relationship

$$\frac{1}{q} = \frac{N_d}{N_a} = \left(\frac{N_d - N_a}{n_1} - 1\right)^2. \tag{56}$$

Here, $n_1$ is the carrier density at the temperature defined by the point of intersection of the extrapolated dependences $n \approx N_0$ and $n \propto \exp(-\varepsilon_i/kT)$, where $\varepsilon_i$ is the impurity ionization energy and T is the absolute temperature.

Figure 25 shows the j−E characteristics of samples in fields close to the breakdown value. We can see from this figure that the j−E characteristics of samples with different impurity contents differ in the pre-breakdown and breakdown fields. The compensated samples have two critical fields: $E_b$, which is the field at which breakdown begins, and $E_m$, which is the field at which breakdown can still be maintained [67]. The higher the degree of impurity compensation, the greater is the difference between the fields $E_b$ and $E_m$.

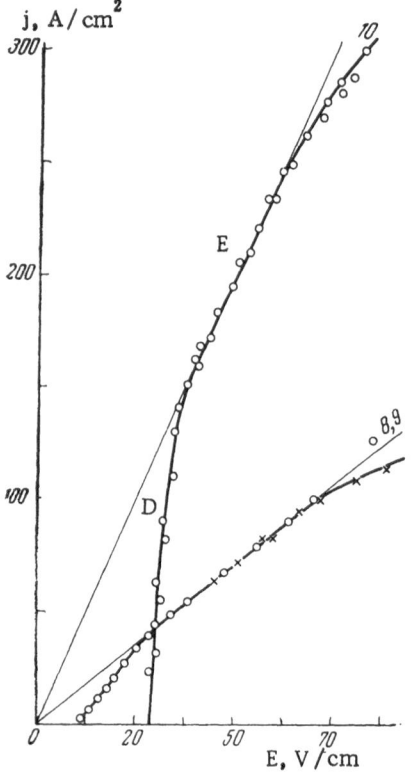
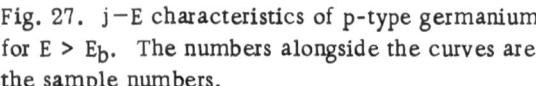

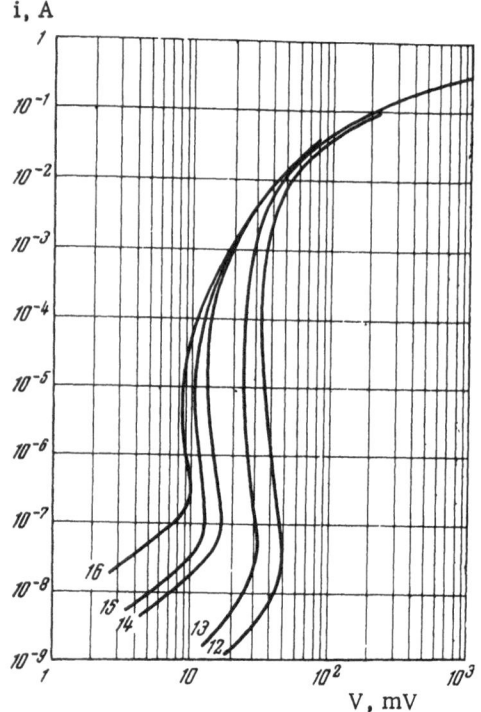

Fig. 27. j—E characteristics of p-type germanium
for E > $E_b$.  The numbers alongside the curves are
the sample numbers.

Fig. 28.  Current—voltage characteristics of thin
samples.  The numbers alongside the curves are the
sample numbers.

Figure 16c shows that the breakdown fields for all samples with $(N_d + N_a) \leq 10^{15}$ cm$^{-3}$ are about 5 V/cm.
When the impurity concentration is increased further, the breakdown field $E_b$ rises almost linearly with $(N_d + N_a)$.

The results of the measurements of the dependence of the current density on the electric field intensity
E > $E_b$ are given in Figs. 26 and 27 for n- and p-type germanium.

It follows from these results that the j—E characteristics of all the investigated samples have, in very
strong electric fields, a region of direct proportionality between the current and the voltage similar to Ohm's
law in weak fields.  The width of this region is particularly large for n-type germanium.  Thus, for samples
Nos. 3 and 4, it extends from 20 to 80 V/cm, i.e., from $4E_b$ to $16E_b$.  The width of this region for p-type
germanium (Fig. 27) is about half that for n-type germanium.

The results of the measurements in strong electric fields are summarized in Table 11.

In this table, $E_1$ and $j_1$ represent the field intensity and current density corresponding to the beginning of
the region of direct proportionality between the current and the voltage; $E_2$ represents the field corresponding
to the end of this region; and v is the carrier velocity calculated on the assumption that in a field $E_1$ the carrier
density is equal to $N_0$ and the current density is $j_1 = eN_0v$.

From the results presented in Figs. 26 and 27 and in Table 11, it is clear that the current—voltage characteristics of weakly compensated n- and p-type germanium samples have a rectilinear region which begins
at $E_1 \approx 4E_b$.  The values of the field $E_1$ for compensated samples agree with the values obtained for uncompensated samples with the same majority impurity concentration.  For $(N_d + N_a) < 10^{15}$ cm$^{-3}$, $E_1 \approx 20$ V/cm
for samples compensated to the extent ranging from 10 to 80% (in the same range, the breakdown fields change
by a factor of 5).

The dependence of the current on the voltage in fields higher than the breakdown value is governed mainly by the majority impurity and obeys relationships which are common for all samples; this can be seen particularly clearly if the results of the measurements are presented in the form $j/j_1 = f(E/E_1)$. Then, all the curves shown in Figs. 26 and 27 (except that for sample No. 10) merge into one curve.

We shall now consider the results obtained for thin samples, whose voltage−current characteristics are shown in Fig. 28. It follows from this figure that, in very strong electric fields, the current−voltage characteristics of samples of various thicknesses almost merge into one curve which, in contrast to the curve for thick samples, has a very narrow region in which the current is directly proportional to the voltage. The current−voltage characteristics of the samples for which $V_b \approx V_i$, where $V_i$ is the ionization potential of the impurity, differ considerably from the characteristics obtained for thick samples.

The $i/i_0 = f(V/V_0)$ characteristics of the thinnest samples are a family of curves, which is in contrast to the common curve obtained for thick samples and for samples with $V_b > 3V_i$. The calculated value of the current density for very thin samples is reached at $V/V_0 \approx 20$; the direct proportionality between the current and the voltage is completely absent.

## §4. Discussion of Results

From the balance of the power acquired by a carrier in the field [Eq. (43)]

$$Q = \frac{(eE)^2}{2m} \tau(\bar{\varepsilon})$$

and power lost to the acoustical phonons $\beta \simeq \delta \bar{\varepsilon}/2\tau_{ph}(\bar{\varepsilon})$, it follows that:

1) the average carrier energy increases linearly with the field,

$$\bar{\varepsilon} = \frac{elE}{\sqrt{2\delta}}, \tag{57}$$

if the phonon collisions are dominant and the time between each collision is $\tau(\varepsilon) \approx \tau_{ph}(\varepsilon)$;

2) the average carrier energy increases more slowly [Eq. (50)],

$$\bar{\varepsilon} = \sqrt{\frac{e^2 l_{ph}^2 E^2}{2\delta} - \frac{N_i}{p_0}},$$

if the additional scattering by charged impurities, whose concentration is $N_i$, is sufficiently strong; then,

$$\tau^{-1}(\varepsilon) = \tau_{ph}^{-1} + \tau_i^{-1}, \quad p_0 = (\pi a_0^2 l_{ph})^{-1},$$

where

$$\tau_i^{-1} \simeq N_i \pi a_0^2 \sqrt{\frac{2\bar{\varepsilon}}{m}} \left(\frac{\varepsilon_i}{\varepsilon}\right)^2;$$

3) in compensated samples there are additional energy losses due to the excitation of complexes such as ionized hydrogen molecules.

Hence we see that the average carrier energy increases in the field in different ways, depending on the scattering mechanism and energy losses. If we assume that the impact ionization of impurities takes place when the average carrier energy reaches the value $\bar{\varepsilon} = const \approx 0.2\varepsilon_i$ [68], it follows that in samples with different impurity concentrations the energy necessary for the ionization can be acquired in different electric fields and this results in a dependence of the breakdown voltage on the impurity concentration.

For samples with low impurity concentrations, such that we can neglect the scattering by impurities, it follows from Eq. (57) and from the condition $\bar{\varepsilon} \approx 0.2\varepsilon_i$, that

$$E_b \simeq \frac{\sqrt{2\delta} \cdot 0.2\varepsilon_i}{el_{ph}} \approx 5 \text{ V/cm} \tag{58}$$

on the assumption that $\delta = 0.3$, $l_{ph} = 5 \cdot 10^{-4}$ cm.

From the results of the measurements presented in Fig. 16c, it is evident that the breakdown field remains almost constant up to impurity concentrations $N \leq 10^{15}$ cm$^{-3}$. At $T = 4.2°K$, the scattering of electrons by phonons in strong electric fields predominates over the impurity scattering right up to these impurity concentrations. Therefore, it is doubtful that at $N_i \approx 10^{14}$ cm$^{-3}$ there would be the combined scattering by phonons and charged impurities which would have resulted in an anomalously large value of $\pi a_0^2$, the cross section of impurity atoms.

We shall now consider the results obtained in fields higher than the breakdown value, assuming that the impact ionization process is balanced by the bimolecular and triple (or impact) recombination

$$w_i(N_0 - n) = w_r(N_a + n) + w_A n(N_a + n), \tag{59}$$

where $w_A$ is a coefficient proportional to the impact recombination probability. The carrier density in such fields is very high and its field dependence is described by fairly simple relationships in two limiting cases:

1) a low degree of impurity compensation, when $n \gg N_a$,

$$n = \frac{1}{2w_A}\left[\sqrt{4w_iN_0 - (w_r + w_i)^2} - (w_r + w_i)\right]; \tag{60}$$

2) a high degree of impurity compensation, when

$$n = \frac{N_0 - N_a \dfrac{w_r}{w_i}}{1 + N_a \dfrac{w_A}{w_i}}. \tag{61}$$

In those cases when the impact recombination is unimportant,

$$n = \frac{N_0}{1 + \dfrac{w_r}{w_i}} - \frac{N_a}{1 + \dfrac{w_i}{w_r}} \tag{62}$$

and

$$n = \frac{N_0}{1 + \dfrac{w_r}{w_i}} \text{ when } n \gg N_a, \tag{63}$$

$$n = N_0 - \frac{w_r}{w_i}N_a \text{ when } n \ll N_a. \tag{64}$$

Since the drift velocity of carriers $v = el\,E/\sqrt{2m\varepsilon}$ depends weakly on the field, we can assume, in the first approximation, that $j \propto n$. We shall compare the results of the measurements of $j/j_1$ and the calculations of $n/N_0$.

The results of the measurements on samples whose net impurity concentrations varied by a factor of 50 were described by a single dependence $j/j_1 = f(E/E_1)$ for $q \leq 50\%$. It follows from Eqs. (60)-(64) that this can be true only if the impact recombination is unimportant compared with simple recombination in samples in which the carrier density ranges up to $10^{15}$ cm$^{-3}$. Since $nw_A < w_r$ right up to $n \approx 10^{15}$ cm$^{-3}$, we can estimate the upper limit of the value of $w_A$ for $E \geq 4E_b$, taking, from [69], the value of $w_r$ for thermal electrons at $T = 4.2°K$ and using the dependence $w_r = f(E)$ [59]; this limit is found to be $w_A \leq 10^{-23}$. A similar value of $w_A$ is obtained from an analysis of the results for strongly compensated samples. It follows from Fig. 27 and Table 11 that the current density in sample No. 10 reaches a value comparable with $j = eN_0v$. But Eq. (61) shows that this result is possible only if $N_a(w_A/w_i) \ll 1$ right up to $N_a = 5 \cdot 10^{15}$ cm$^{-3}$. Since, in very strong electric fields, $w_i \leq 10^{-6}$ [72], it follows that $w_A \ll 2 \cdot 10^{-22}$.

Thus, a comparison of the results of the measurements and of the calculations by means of Eq. (63) shows that Eq. (63) is in qualitative agreement with experiment.

It is interesting to compare quantitatively the results of the calculations and of the measurements. From the dependence $n/N_0 = f(\bar{\varepsilon})$, calculated from Eq. (63) on the assumption that [13]:

$$w_r \simeq 10^{-8} \left(\frac{\varepsilon_i}{\bar{\varepsilon}}\right)^2 \text{and } w_i \simeq 10^{-6} \left(\frac{\bar{\varepsilon}}{\varepsilon}\right)^{1/2} \exp\left[-\left(\frac{\varepsilon_i}{\bar{\varepsilon}}\right)\right],$$

it follows that the carrier density reaches a value close to $N_0$ at $\bar{\varepsilon} \approx 0.8\varepsilon_i$. This corresponds to a field $E \approx 4E_b$ if $\bar{\varepsilon} \approx E$ and $E_b \approx 0.2\varepsilon_i$.

The carrier density can also be estimated from the current density in strong electric fields, given in Table 11. Representing the current $j_1$ in a field $E_1 = 4E_b$ in the form $j_1 = eN_0v$, we obtain $v \approx 2 \cdot 10^6$ cm/sec for n-type germanium and $v \approx 1.5 \cdot 10^6$ cm/sec for p-type germanium. Since $v \approx 1 \cdot 10^6$ cm/sec for $E = E_b$ and the carrier drift velocity increases with the field not faster than $\propto \sqrt{E}$, we easily find that the values of v listed in Table 11 are close to the carrier drift velocity and, consequently, in agreement with the calculations, the value of n is close to $N_0$ in fields $E > 4E_b$.

Comparison of the results of the calculations and of the measurements shows that the increase in the carrier density up to $n \approx N_0$ takes place in fields $E < 4E_b = E_1$, while in fields $E > E_1$ the rise in the current is mainly due to an increase in the drift velocity.

## §5. Field Dependence of Carrier Drift Velocity

The region of direct proportionality between the current and the voltage discovered in fields $E > 4E_b$, where the carrier density is approximately constant and close to $N_0$, is especially interesting. We have shown that this linear dependence in uncompensated samples may be explained by the combined effect of the scattering by phonons and charged impurities if $N_0/p_0 \approx 0.1$. Since $p_0 = $ const, the theory given in [13] can explain the results of the measurements only at one fixed net impurity concentration, i.e., at $N_0 \approx 10^{15}$ cm$^{-3}$, since $\pi a_0^2 = 3 \cdot 10^{-3}$ cm$^2$, $l_{ph} = 3 \cdot 10^{-4}$ cm at $T = 10°K$.

However, additional experiments have established that in fields several times greater than the breakdown value, the direct proportionality between the current and the voltage is observed in all n- and p-type germanium samples having net impurity concentrations ranging from $10^{14}$ to $5 \cdot 10^{15}$ cm$^{-3}$. Obviously, there is a more general mechanism which results in a proportionality of j and E. One such mechanism has been described recently by Paranjape [70]. Paranjape pointed out that, since the phonon–phonon interaction is very weak at low temperatures, high-energy phonons emitted by hot electrons may considerably alter the electron energy distribution function when the electron density exceeds $10^{14}$ cm$^{-3}$.

The field dependence of the carrier mobility is then described by the relationship

$$\mu(E) = C_2^{-\frac{3}{4-\beta}} E^{-\frac{2(1-\beta)}{4-\beta}},$$ (65)

where $C_2$ is a constant which depends on the impurity concentration and $\beta$ is a constant which depends on the phonon scattering mechanism.

Since $\beta \approx 1$ for phonons of energies higher than kT [71], it follows from Eq. (65) that the carrier mobility is independent of the field in those fields in which the carrier density is greater than $10^{14}$ cm$^{-3}$.

The region of direct proportionality between j and E is limited on the high-field side. The value of the critical field $E_2$, which represents the upper limit of this region, depends, according to Paranjape's theory, on the effective carrier mass. Therefore, for n-type germanium this region is about twice as wide as that for the p-type material.

If a sample is very small, then excess phonons may leave a crystal without interacting with carriers. Then Eq. (65) becomes invalid and the region of direct proportionality between j and E should disappear. The results presented in Figs. 26-28 show that the conclusions which follows from Paranjape's theory agree well with the experimental results. Thus, the width of the region $j \propto E$ is greater for n-type germanium samples, and the region narrows when the thickness d of the discharge gap is reduced, disappearing almost completely when $d < 10 \, \mu$.

CHAPTER VII

# Electrical Discharge in Very Thin Germanium Layers at Low Temperatures

It seemed interesting to investigate the process of impact ionization in samples so thin that their thickness is $l < l_\omega$, where $l_\omega$ is the mean free path when energy is transferred to phonons. It is known that $l_\omega \approx (l_{ph}/\sqrt{\delta})$, where $\delta \approx (2ms^2/kT)$ is the fraction of the energy transferred by an electron to a phonon in a collision; m is the effective carrier mass; s is the velocity of sound; k is Boltzmann's constant; and T is the absolute temperature.

In germanium, at helium temperatures, $l_{ph} \approx 10 \, \mu$ and $l_\omega \approx 20 \, \mu$.

We carried out measurements on germanium samples of thicknesses right down to $2-3 \, \mu$. To the best of our knowledge, such investigations had not until now been carried out.

## §1. Samples and Measurement Method

The investigations were conducted on two batches of samples prepared from gallium-doped germanium.

The first batch was prepared from an ingot of weakly compensated germanium with an acceptor concentration $N_a \approx 1.4 \cdot 10^{14} \, cm^{-3}$ and a donor concentration $N_d \approx 1.5 \cdot 10^{13} \, cm^{-3}$; $N_a - N_d = 1.25 \cdot 10^{14} \, cm^{-3}$. The degree of compensation of this ingot was $q = N_d/N_a \approx 10\%$. A second batch of samples was prepared from strongly compensated germanium with $q \approx 80\%$, $N_a \approx 3.6 \cdot 10^{15} \, cm^{-3}$, $N_d \approx 3.0 \cdot 10^{15} \, cm^{-3}$, $N_a - N_d = 6 \cdot 10^{14} \, cm^{-3}$. The donor and acceptor concentrations were determined from the temperature dependence of the Hall coefficient; the results of the measurements of this coefficient are presented in Fig. 29 for weakly and strongly compensated germanium. The degree of compensation was determined as in [70].

Plates were cut from ingots at right angles to the crystal growth axis <111>; they were ground down to the required thickness and etched for one minute in boiling Perhydrol with an admixture of an alkali. This treatment removed a germanium layer about $10 \, \mu$ thick. We etched separately the lumps of indium used to make the electrodes. The indium electrodes were fused into the samples whose thickness was greater than $20 \, \mu$. The thinner samples were prepared by a method used for surface-barrier transistors. This operation was carried out at relatively low temperatures and the calculations showed that the diffusion of indium into germanium during the deposition of electrodes could, in practice, be neglected.

The thickness of the samples subjected to breakdown was determined in various ways: with a microscope, after removal of indium electrodes; from the capacitance at helium temperature; and from the resistivity at room temperature. Different measurement methods gave similar results.

The breakdown voltage was measured using dc either by gradually increasing the voltage applied or by the application of $10^{-6}$ sec pulses with a front rise time of $10^{-7}$ sec.

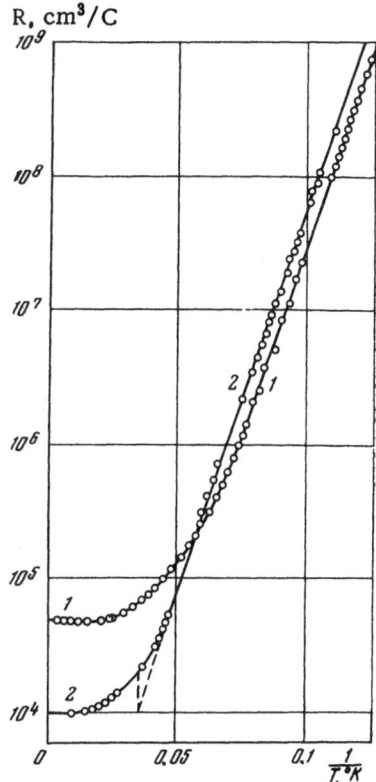

Fig. 29. Temperature dependence of the Hall coefficient: 1) weakly compensated germanium with $q \approx 10\%$; 2) strongly compensated germanium with $q \approx 80\%$.

All the necessary measures were taken to suppress stray emf's. To reduce these strays, all the measurements were carried out in a screened room. Some of the measurements were carried out inside a double electrostatic screen with a special grounding circuit. The leads were made of copper. Control experiments showed that the value of the thermoelectric emf in the circuit was less than $10^{-4}$ V.

During the measurements, the samples were immersed in liquid helium, and since the power evolved by them was less than the critical value (at which overheating took place), we could assume that the temperature of the samples was the same as that of the surrounding medium and, therefore, equal to 4.2°K [14].

Some measurements were also carried out while pumping helium vapor.

## § 2. Results of Measurements

The measurements showed that the i—V characteristics of thin and thick samples were similar. For comparison, the results of the measurements carried out on weakly compensated germanium are presented in Fig. 30 for samples whose thickness d = 4 μ, and in Fig. 31 for those with d = 11 mm. Similar results for heavily compensated germanium are given in Fig. 32 (d = 5 μ) and Fig. 33 (d = 11 mm).

From the results obtained it follows that samples with different degrees of impurity compensation had different current—voltage characteristics. Weakly compensated samples exhibited a strong rise in the current ("breakdown") in fields $E_b \approx 5$ V/cm, while strongly compensated samples exhibited a "breakdown" in much higher fields and this breakdown was accompanied by a falling current—voltage characteristic. The breakdown of compensated material had two characteristic fields: $E_m$ and $E_b$.

For samples whose degree of compensation is $q \approx 80\%$, the ratio of these critical fields $E_b/E_m \approx 1.7$.

The results of the measurements of the breakdown voltage as a function of the sample thickness are summarized in Fig. 34 for weakly compensated germanium and in Fig. 35 for strongly compensated germanium. Figure 35 includes the values of the voltages at which breakdown begins (curve 1) and of voltages at which breakdown is still maintained (curve 2).

It follows from the results presented in Fig. 34 that the breakdown field of thin, weakly compensated samples and of similar thick samples remained practically constant (5 V/cm) up to a thickness of 20 μ, but below this the breakdown voltage decreased to about 10 mV, which was close to the ionization potential of neutral atoms of the uncompensated impurity. For samples whose thickness was less than 20 μ, the breakdown voltage was practically constant and equal to the ionization potential $V_i$. The thinnest samples were 2-3 μ thick, as deduced from the measurements of the resistivity and geometrical dimensions.

We measured the breakdown voltage of 24 pure germanium samples about 3 μ thick. For 20 of these samples, the dc breakdown voltage was 10-11 mV, but for four of them the voltage was less than the ionization potential by 2-4 mV. When the temperature was reduced to 1.8°K, the breakdown voltage of these four samples increased, approaching the value of $V_i$. Therefore, the strong rise in the current could not be explained

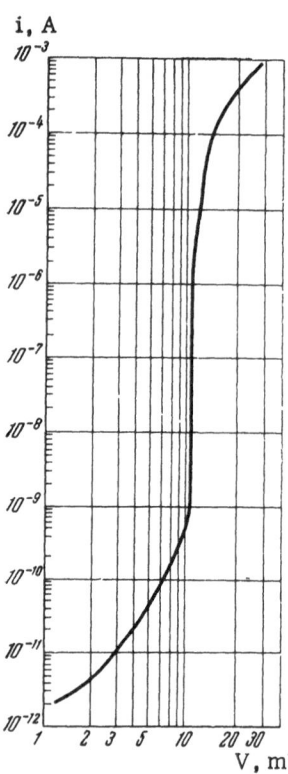

Fig. 30. Current−voltage characteristic
of a thin sample of weakly compensated
germanium (d = 4 μ, S = 1.8 · 10⁻⁴ cm²).

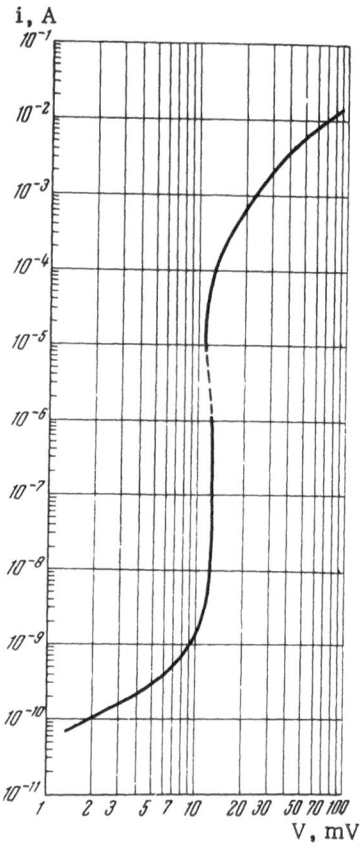

Fig. 32. Current−voltage characteristic
of a thin sample of strongly compensated
germanium (d = 5 μ, S = 1.8 · 10⁻⁴ cm²).

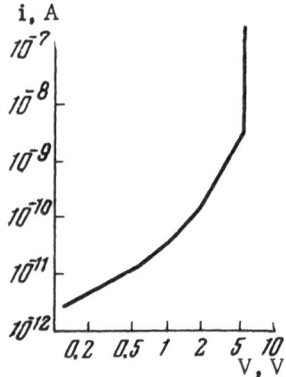

Fig. 31. Current−voltage characteristic
of a thick sample of weakly compensated
germanium (d = 11 mm, S = 4 · 10⁻² cm).

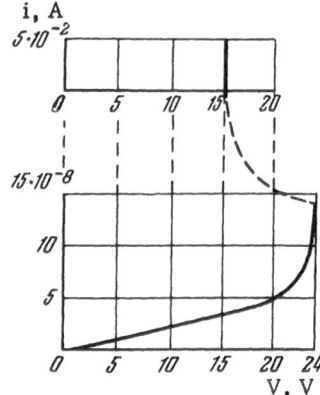

Fig. 33. Current−voltage characteristic
of a thick sample of strongly compensated germanium (d = 11 mm, S = 4 ·
10⁻² cm²).

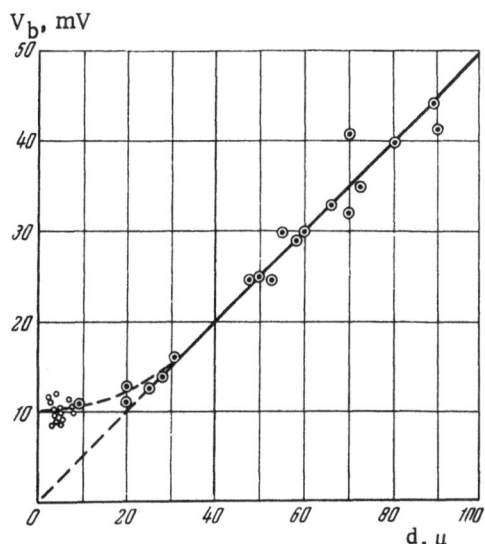

Fig. 34. Dependence of the breakdown voltage $V_b$ on the sample thickness d for weakly compensated germanium (points). The continuous curve represents the relationship $E_b = $ const $= 5$ V/cm, which is valid for thick samples.

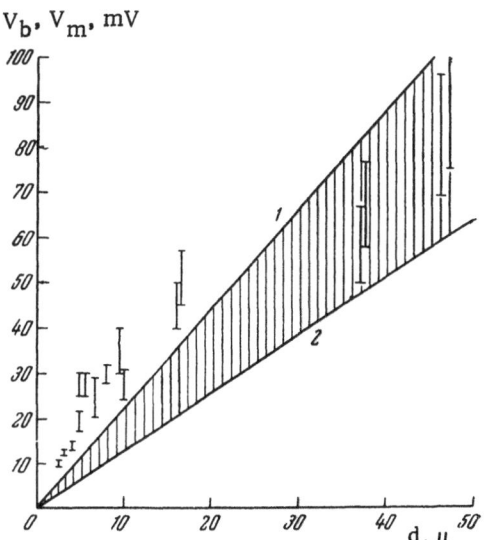

Fig. 35. Dependence of the breakdown voltages $V_b$ and $V_m$ on the sample thickness d for strongly compensated germanium. The continuous lines represent the relationships $E_b = $ const $= 22$ V/cm (1) and $E_m = $ const $= 15$ V/cm (2), which are valid for thick compensated samples. The vertical marks represent the results obtained for thin samples: the top of each mark corresponds to $E_b$; the bottom to $E_m$.

by the tunnel effect on the assumption that for some reason the thickness was sufficiently low for this effect to take place. Moreover, the strong rise in the current under these conditions could not be due to injection because of the lowering of the potential barrier by the applied voltage. The recombination of injected holes at immobile ionized acceptors would have produced a space charge preventing further injection.

It is known [72] that the ionization energy of impurities in germanium may be reduced by several millivolts by a uniform pressure $P \approx 50$ kg/mm². Such high pressures could not have appeared at the germanium—indium contacts because of the high plasticity of indium. However, in some cases, a sample of germanium could have been in direct contact with a nickel wire and not with a plastic layer of indium. In such cases, the mechanical stresses present would have been sufficient to reduce the ionization potential to the values obtained in our measurements.

In strongly compensated thick samples, the breakdown began at $E_b = 22$ V/cm and was maintained by $E_m = 15$ V/cm. When $d < 40\ \mu$, the fields $E_m$ and $E_b$ began to increase (Fig. 35) well before the breakdown voltage fell to $V_i$.

The breakdown voltage of these samples decreased to $V_i$ at a thickness of about $2\ \mu$, and there was not even a single case of a strongly compensated sample with a breakdown voltage less than the ionization potential.

The mean free path in samples with a high impurity concentration $N_0$ was governed mainly by the scattering on ionized centers: $l = l_i$. Here, $l_i \approx 7.86 \cdot 10^{15}(\varepsilon^2/N_i \ln \gamma_0)$, where $\gamma_0 = 1 + 1.23 \cdot 10^{16}(\varepsilon^2/N_i^{2/3})$. When $\varepsilon_i = eV_i$, $N_i = 2N_d = 6 \cdot 10^{15}$ cm⁻³, this mean free path is $l_i \approx 1\mu$.

If we assumed that the scattering by ionized centers was purely elastic, we found that the energy of holes depends only on the potential drop across the path traveled by holes. Therefore, samples of thickness $l \approx l_{ph}$ should also have a breakdown voltage $V_b \approx V_i$. In fact, this applied when $l = l_i$ and was evidently due to the fact that the scattering by ionized centers was not purely elastic [67].

## §3. Breakdown in Thin Layers

Breakdown in very thin samples, manifested by a sudden rise of the current by 3-4 orders of magnitude, occurs at a voltage equal to the ionization potential of the uncompensated impurity atoms. Since the field intensity in such samples does not exceed several tens of volts per centimeter, the sudden rise in the current cannot be explained by the tunnel effect.

The impact ionization in samples whose thicknesses lie in the range $1 < l \leq 10 \, \mu$ can be described as follows:

In weakly compensated germanium with an uncompensated impurity concentration $N_0 < 10^{15} \, cm^{-3}$, the mean free path at helium temperature is governed by the scattering on the acoustical phonons, and, in the thinnest samples, this path is greater than the breakdown gap. Then, many holes travel the whole path without collision and acquire an energy eV, where V is the applied voltage. When $V = V_i$, holes should travel the whole gap between the electrodes to acquire an energy sufficient for the ionization process. Initially, the impact ionization region lies only near the cathode and a sample consists of two regions of different electrical conductivities. In one region (I) there is no impact ionization and the carrier density is low, while in the other region (II) the carrier density increases under the action of impact ionization.

The voltage V applied to the sample is distributed between these two regions in inverse proportion to the electrical conductivities of these regions. Practically all the voltage is then concentrated in the first region between the anode and the boundary with the second region. Holes acquire the energy necessary for the ionization at the boundary between these two regions. Thus, impact ionization is possible not only near the cathode but also at some distance away from it in the interior of a sample. This makes the second region grow at the expense of the first, and the boundary between the two regions travels from the cathode to the anode until the voltage is concentrated practically completely in a layer near the anode.

These phenomena, well known from gas discharges, have been recently considered in connection with the low-temperature breakdown in germanium [73]. However, in gas discharges, electrons are knocked out from the cathode by secondary processes, whose efficiency is low: they are knocked out by positive ions or photons resulting from recombination. In solids, holes are easily extracted, due to the ohmic contacts, from the anode by a negative space charge and, under steady-state conditions, they replace exactly the number of holes lost from the discharge gap.

At any given moment, the same current of holes is passing through every cross section of the sample:

$$j_{\mathrm{I}} = j_{\mathrm{II}} = p_{\mathrm{I}} e v_{\mathrm{I}} = p_{\mathrm{II}} e v_{\mathrm{II}},$$

where the subscript "I" refers to the first region and the subscript "II" to the second region. When the boundary between the regions moves from the cathode to the anode, the field intensity $E_{\mathrm{I}} = V/d_{\mathrm{I}}$ rises. This increases the drift velocity $v_i$ as well as the carrier density $p_{\mathrm{I}}$ due to the reduction in the rate of recombination in strong electric fields [59]. Then

$$p_{\mathrm{I}} = \frac{p_0}{w_r(E)/w_r(0)},$$

where $p_0$ is the equilibrium density of holes in weak electric fields, governed by the thermal generation and recombination at T = 4.2°K.

However, since the drift velocity of carriers in germanium in a strong electric field tends to saturation [74] and does not exceed $v \leq 10^7 \, cm/sec$, while the increase in the hole density due to the reduction in the rate of recombination is small compared with that resulting from an avalanche, the process of impact ionization in the second region and the associated rise in the carrier density are limited by those current densities which can be passed through the first region under steady-state conditions.

Calculation of the maximum current density which can be reached under conditions such that a sample consists of two regions can be carried out making the following assumptions:

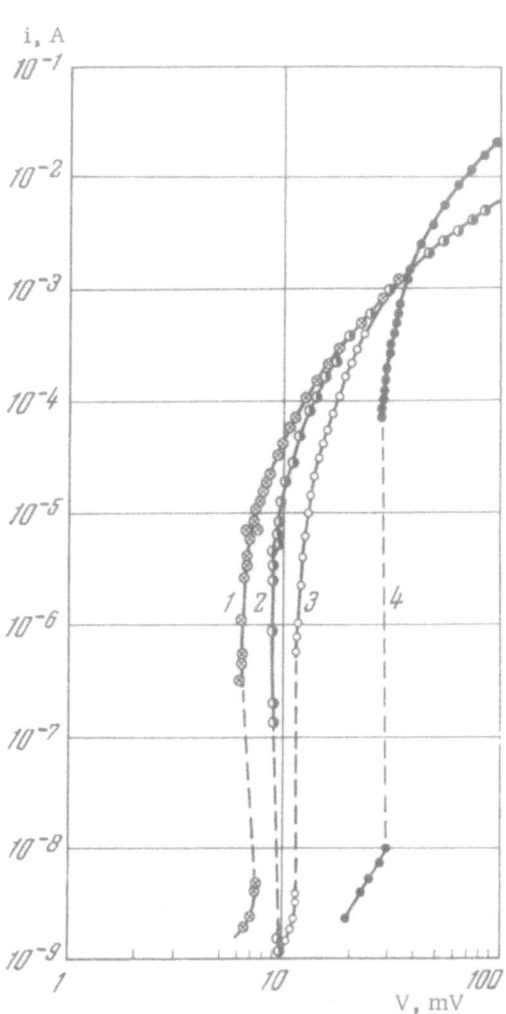

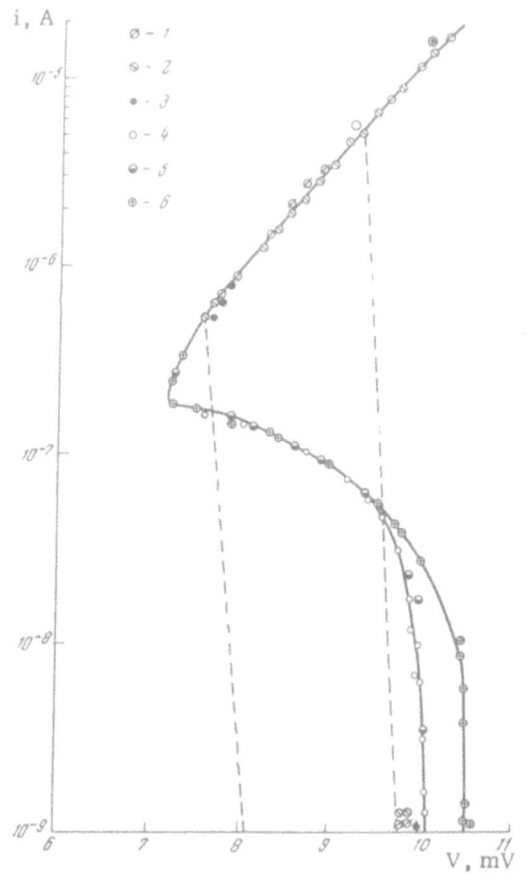

Fig. 36. Current–voltage characteristics of samples ~5 μ thick. 1, 2, 3) Weakly compensated samples, with $E_b$ governed by the effect of stresses; 4) strongly compensated sample.

Fig. 37. Current–voltage characteristic of a thin sample connected in series to various resistances: 1) $10^2$; 2) $10^3$; 3) $5 \cdot 10^3$; 4) $10^4$; 5) $2 \cdot 10^4$; 6) $4 \cdot 10^4$ Ω. The dashed line on the right represents the current "jump" when breakdown begins, the line on the left shows the "jump" when breakdown stops (for a series resistance of $10^2$ Ω). When the series resistance is increased, the distance between these lines decreases; when $R \geq 10^4$ Ω, the current–voltage characteristic becomes a continuous S-shaped curve.

1. The rise in the density, due to a reduction in the recombination rate in a strong electric field, is described by $p(E) \approx p_0 E^2$, in accordance with [59]. This formula has been confirmed by the results of measurements in the range of fields $0.5 < E \leq 10$ V/cm. The question of the applicability of this formula in stronger fields remains open.

2. The motion of the boundary between the two regions continues until the field intensity in the first region exceeds a certain critical value $E_{cr}$. The concentration of the applied voltage in the region of the anode cannot proceed without limit because of the increase in the electrical conductivity due to electrostatic ionization.

In these calculations, we assumed [75] that $E_{cr} = V_i/d_{cr} \approx 3 \cdot 10^2$ V/cm. The calculations showed that in germanium samples with an electrode area $S = 2 \cdot 10^{-4}$ cm$^2$ and a carrier density in weak electric fields $n_0 = 2 \cdot 10^{14}$ cm$^{-3}$ at $T = 4.2°$K, the maximum value of the current is $i \leq 10^{-7}$ A, while at $V = V_b$ the current suddenly increases to $i \approx 10^{-5}$ A, as indicated by the experimental results presented in Fig. 36. In an analysis of

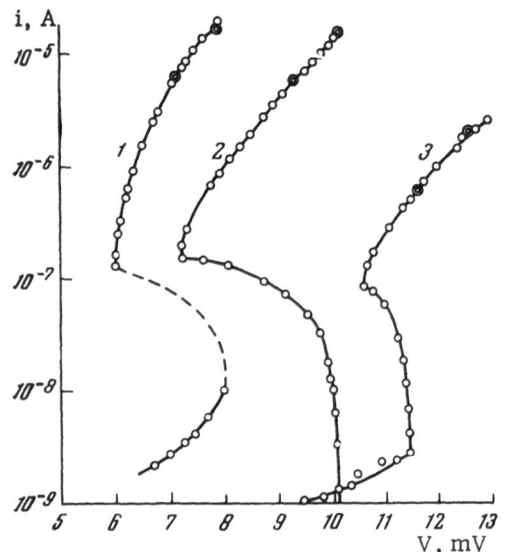

Fig. 38. Current−voltage characteristics of weakly compensated samples ∼5 μ thick for a load resistance R ≈ $10^4$ Ω. The numbers alongside the curves correspond to different samples (see Fig. 36).

the breakdown in thin samples for i > $10^{-7}$ A, we cannot restrict ourselves to the impact ionization of impurities, but we must allow for the electrostatic ionization.

Thus, when a breakdown voltage $V_b$ ≈ 10 mV is applied to a sample whose thickness is d ≈ 5 μ, this voltage is at first distributed over the whole sample. The field in the sample is uniform and equal to 20 V per cm; the equilibrium density is p(E) ≈ 5 · $10^2 p_0$. After a certain time from the beginning of the impact ionization process, almost the whole of the voltage becomes concentrated in region $d_{cr}$, whose dimensions are governed by the critical field intensity $E_{cr} = V/d_{cr}$, and are, therefore, 3 · $10^{-5}$ cm if $E_{cr}$ = 300 V/cm. This means that there are few carriers in the first region, which is about 0.3 μ thick, but the carrier drift velocity in this region is $v_I$ ≈ $10^7$ cm/sec, while in the second region (which is about 5 μ wide), we have v < $10^6$ cm/sec.

When electrostatic ionization takes place, the first region ceases to limit the rise of the current in the second region and, moreover, the two regions exchange their roles. Now the current is limited by the low drift velocities of carriers in the second region. The current−voltage characteristic of thin samples will, of course, be flat in the region V > $V_i$, since the "excess voltage" V − $V_i$ will be concentrated in the second region where it increases the drift velocity, which is a weak function of the field.

The data presented in Fig. 36 show that the region where the current rises suddenly at V = $V_b$ is indeed replaced by a region of monotonic and even slow rise of i = ƒ(V). This can be seen very clearly by comparing the i−V characteristics of thin and thick samples.

Before we compare in detail the calculated and measured values, we must point out that the impact ionization takes place under different conditions which vary from the moment of the appearance of the impact ionization near the cathode to the moment when the ionization region extends right up to the anode; this is because the impact ionization probability is a function of the field and the field increases by a factor of several tens as the boundary between the two regions moves from the cathode to the anode. Therefore, under steady-state conditions, we may expect that the maintenance of the ionization would require a smaller voltage than that necessary at the beginning of the ionization and that the current−voltage characteristic would have a negative resistance region right up to the limiting current, which is obtained when the first region has shrunk to its minimum size.

Further rise in the current (i > $10^{-7}$ A) is possible only by stepping up the voltage across a sample, since, when the electrostatic ionization appears, it is no longer possible for the potential to become redistributed between the various parts of the sample.

To compare the results of the calculations and the measurements, we repeated the experiments on the thin samples under different conditions.

To suppress the current "jump," we connected a sample in series with calibrated radio-engineering resistors of various values. Figures 37 and 38 show the dependence i = ƒ(V − iR) obtained in such measurements; V is the applied voltage and (V − iR) is the voltage across the sample.

TABLE 12

| Impurity | Ionization energy, eV | |
|---|---|---|
| | $(\varepsilon_i)_1$ | $(\varepsilon_1)_2$ |
| Zn | 0.029 | 0.095 |
| Cu | 0.04 | 0.30 |
| Pt | 0.04 | 0.58 |
| Au | 0.05 | 0.15 |

Our results show that the dependence of the current on the voltage for a sample connected in series with a resistance $R \approx 10^4$ $\Omega$ is described by a complex but smooth curve, in complete agreement with the results of the analysis presented in this section.

It is evident from Fig. 38 that, up to $i \approx 10^{-7}$ A, the current increases along the negative resistance region of the current—voltage characteristic, while for $i > 10^{-7}$ A, the dependence of the current on the voltage is described by a curve with a positive slope.

The resistance of the samples at the point of inflection of the $i-V$ curve is approximately constant and equal to $r_{min} \approx 10^4$ $\Omega$. Therefore, it is not surprising that to obtain an S-shaped $i-V$ characteristic, we need a resistance $R \geq 10^4$ $\Omega$. The value of the resistance of a sample $r = V_{min}/i = \rho d_I / S$ depends on the resistivity and the dimensions of the first region. Comparison of the results of the calculations and the measurements shows that $d_I \approx 3 \cdot 10^{-5}$ cm and, consequently, $E_{cr} = 10^{-2}/(3 \cdot 10^{-5}) \approx 300$ V/cm, in full agreement with the theory [75] and the assumptions on which our calculations are based.

TABLE 13

| Material | Impurities, $N_0 < 10^{15} cm^{-3}$ | $\varepsilon_i$, eV | E , V/cm | Phonon energy, eV† |
|---|---|---|---|---|
| $n$-Ge | Sb, P, As | 0.0096—0.0127 | ~5 | $\varepsilon_{opt} = 0.037$ |
| $p$-Ge | Al, B, Ga, Zn | 0.0102—0.0112 | ~5 | |
| $p$-Ge | Zn | 0.03 | >100 | $TO = 0.034$ |
| | Cu | 0.04 | ~1000 | $TA = 0.010$ |
| | Cd | 0.05 | not observed | $LO = 0.029$ |
| $n$-Si | Li, Sb, As | 0.033—0,049 | » » | $\varepsilon_{opt} = 0.063$ |
| | P | 0.044 | >100 | |
| $p$-Si | B | 0.046 | >100 | $TA = 0.02$ |
| | Al | 0.057 | not observed | $TO = 0.06$ |
| | Ga | 0.065 | » » | $LA = 0.04$ |
| | | | | $LO = 0.05$ |

† Phonon modes: TA — transverse acoustical; TO — transverse optical; LA — longitudinal acoustical; LO — longitudinal optical.

TABLE 14

| $T$,° K | $\mu \cdot 10^{-4}$, cm$^2 \cdot$ V$^{-1} \cdot$ sec$^{-1}$ | $E_b$, V/cm | $E_b\mu \cdot 10^{-6}$, cm/sec |
|---|---|---|---|
| 11,6 | 5.5 | 31.5 | 1.76 |
| 14 | 8.5 | 22 | 1.9 |

APPENDIX

## Electrical Conductivity of Zinc-Doped Germanium

In addition to the impact ionization of group III and V impurities, it seemed interesting to also investigate the ionization of those impurities which give rise to nonhydrogen-like acceptor levels in the forbidden band of germanium. The ionization energies of several such impurities are listed in Table 12.

It follows from Table 12 that these impurities are capable of introducing into the forbidden band of germanium several levels with ionization energies considerably higher than the ionization energies of the elements of groups III and V.

Investigations carried out at low temperatures on germanium doped with zinc, copper, and cadmium showed that the breakdown in such samples was observed in fields 100-1000 times greater than the fields necessary to ionize impurities of groups III and V [8].

The high values of $E_b$ observed for germanium with deep impurity levels were ascribed by Lampert, Herman, and Steele [7] to large energy losses from hot electrons by the emission of optical and acoustical phonons. They suggested that if carriers could lose energy by the emission of optical phonons before their energy reached $\varepsilon_i$, breakdown in weak electric fields would become impossible [7]. Since $l_{opt}$ is practically independent of temperature, it follows that if $\varepsilon_i > \hbar\omega_{opt}$, the value of $E_b$ can be very high and even comparable with $E_b$ for p−n junctions.

The data on the phonon energy $\hbar\omega_{opt}$ on $\varepsilon_i$ and on $E_b$ are given in Table 13. It is evident from this table that the ionization energies of impurities with deep levels usually exceed the energy $\hbar\omega_{opt}$. Thus, the breakdown fields are high, in agreement with the theory. Zinc is the exception to the general rule, and, therefore, the "breakdown" in zinc-doped germanium is of special interest.

It is known that the diffusion coefficient of zinc in germanium is of the same order as the coefficients of the elements of groups III and V, and that zinc can be used as a substitutional impurity. When zinc, which belongs to group II of the periodic system, replaces a germanium atom, two unsatisfied valence bonds are obtained and zinc may become a doubly charged acceptor. When it captures one electron, which is equivalent to the ionization of a hole, its energy level lies 0.029 eV from the valence band edge, and when it captures another electron the level becomes even deeper. At low temperatures, zinc acceptor impurities are practically completely neutral.

The heating of the electron gas in a strong electric field may result in the impact ionization of this impurity.

According to the model proposed in [7], breakdown in germanium doped with zinc, whose ionization energy is $\varepsilon_i < \hbar\omega_{opt}$, should be observed in fairly weak fields. Burstein and Sclar [8] have shown that the breakdown in zinc-doped germanium at helium temperatures takes place in fields of about 400 V/cm, which are considerably higher than $E_b$ for group III and V impurities. The problem of the cause of these high values of the breakdown field of zinc-doped germanium has not yet been solved. The high values of $E_b$ may be associated with a lower carrier mobility or with a lower probability of ionization of zinc atoms.

Because of this, we investigated the behavior of zinc-doped germanium in strong electric fields over a wide range of temperatures in order to establish a correlation between $E_b = f(T)$ and $\mu = f(T)$. Our measurements showed that the resistivity of such samples increased as $\exp(-\varepsilon_i/kT)$, where $\varepsilon_i \approx 0.03$ eV. At T $\approx$ 12°K, the resistance of such samples reaches about $10^{11} \Omega$. The $\ln\rho = \text{const} \cdot \exp(-\varepsilon_i/kT)$ curve had no regions with an activation energy less than 0.029 eV. Hence, we concluded that the concentration of zinc in these samples was considerably higher than the concentration of residual impurities, and that the results obtained could be ascribed to zinc.

The results of the measurements of the mobility gave the following values:

| $T°$, K | 11.6 | 14 | 20 |
|---|---|---|---|
| $\mu$, cm$^2 \cdot$ V$^{-1} \cdot$ sec$^{-1}$ | $5.5 \cdot 10^4$ | $8.5 \cdot 10^4$ | $\sim 10^5$ |

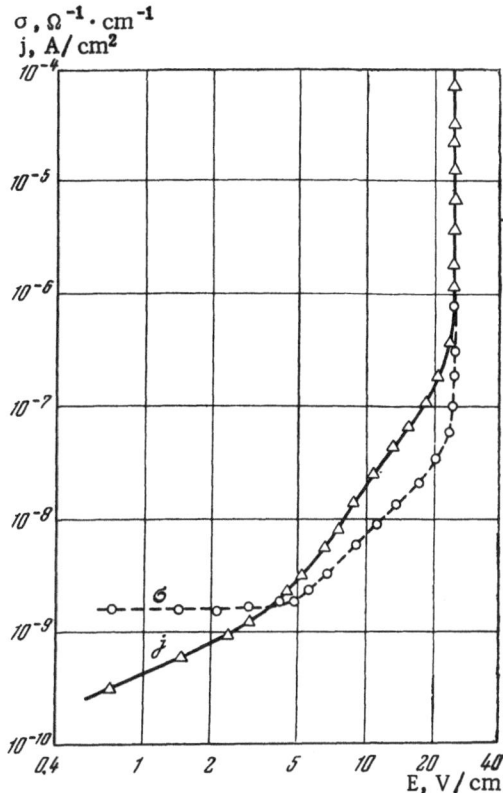

Fig. 39. Dependences $\sigma = f(E)$ and $j = f(E)$ for a zinc-doped germanium sample at T = 11.6°K.

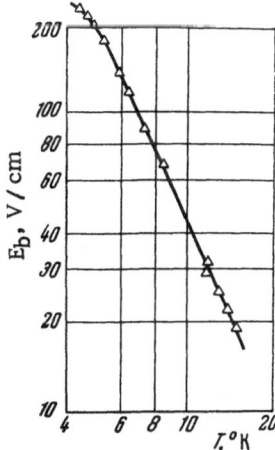

Fig. 40. Temperature dependence of the breakdown field of zinc-doped germanium.

The mobility was calculated on the assumption that the Hall factor was r = 1, which was only correct as far as the order of magnitude was concerned.

In view of the high carrier mobility and the low thermal velocities, we could expect heating of the electron gas and the impact ionization of zinc even in relatively weak fields.

The dependence of the current density and electrical conductivity on the electric field intensity for zinc-doped germanium is shown in Fig. 39. It follows from this figure that the electrical conductivity $\sigma$, considered as a function of the electric field E, has several regions, as in the case of germanium doped with group III and V impurities.

In weak fields (region A), Ohm's law is satisfied: $j = \sigma E$, where $\sigma$ = const. In the prebreakdown region B, the electrical conductivity rises monotonically as the field intensity is increased and $\sigma \propto E^2$. In breakdown fields (region C), the electrical conductivity increases very rapidly. The conductivity of zinc-doped germanium in each of the regions A, B, C can be explained by processes similar to those described in Chapters I and II.

The rise in the electrical conductivity in prebreakdown fields may be associated with an increase in the carrier density due to weaker recombination in strong electric fields. The observed dependence $\sigma \propto E^2$ is described well by the relationship (30), deduced from the cascade theory [59].

The "breakdown" observed in fields of 20-30 V per cm is evidently due to the impact ionization of zinc atoms. In the "breakdown" of zinc-doped germanium — as in the case of germanium doped with group III and V impurities — the field $E_b$ is correlated with the carrier mobility $\mu$ (Table 14).

From the results presented in Table 14, it follows that in the case of zinc-doped germanium

$$E_b \mu \simeq \text{const} \sim 2 \cdot 10^6 \text{ cm/sec.} \quad (66)$$

The relationship (66) can be used to estimate the carrier mobility in that range of temperatures in which it is difficult to measure the mobility — for example, at T < 10°K.

The temperature dependence of the breakdown field of zinc-doped germanium is given in Fig. 40. It follows from this figure that the breakdown field depends very strongly on temperature ($E_b \propto 1/T^2$).

If we assume that the relationship $E_b\mu \propto T^{-1/2}$ is valid for zinc-doped germanium, it follows from the experimental dependence $E_b \propto 1/T^2$ that $\mu \propto T^{+3/2}$, and that the scattering by charged impurities predominates over the scattering by neutral impurities.

It follows from Fig. 40 that the experimentally observed breakdown fields at $T \approx 5°K$ reach 200 V/cm. It then follows from Eq. (66) that $\mu = 10^4 \ cm^2 \cdot V^{-1} \cdot sec^{-1}$. A similar value of the mobility follows from Conwell and Weisskopf's formula [52] for a charged impurity concentration $2N_i \approx 10^{15} \ cm^{-3}$, i.e., for a 20% compensation of a sample.

The data presented in Table 13 show that the results of the measurements of $\mu = f(T)$ in the region $T > 10°K$ suggest scattering by charged impurities. The scattering by neutral impurities can be neglected. This could be due to the strong localization of the wave function of the zinc atom compared with the localization of hydrogen-like atoms. The geometrical cross section of the zinc atom, $S = \pi a_0^2$ (where $a_0 = h^2/\varkappa m Z e^2$, $Z$ is the nuclear charge), is one-fourth of the cross section of a hydrogen-like atom. Thus, the majority of the results obtained for zinc-doped germanium samples can be described qualitatively by analogy with the processes observed in strong electric fields in germanium doped with group III and V impurities.

## Literature Cited

1.  J. S. Townsend, Nature, 62: 340 (1900); Phil. Mag., 1:198 (1901).
2.  A. F. Ioffe, Uspekhi fiz. nauk, 8:141 (1928).
3.  K. McKay, Phys. Rev., 91:1079 (1953); 94:877 (1954); B. M. Vul and A. P. Shotov, Zhur. tekh. fiz., 27:2189 (1957); A. P. Shotov, Zhur. tekh. fiz., 28:465 (1958).
4.  B. M. Vul and É. I. Zavaritskaya, Zhur. éksp. i teor. fiz., 38:10 (1960).
5.  V. A. Chuenkov, Fiz. tverd. tela (Sbornik), 2:200 (1959).
6.  G. Dresselhaus, A. F. Kip, and C. Kittel, Phys. Rev., 98:368 (1955).
7.  M. A. Lampert, F. Herman, and M. C. Steele, Phys. Rev. Letters, 2:394 (1959).
8.  N. Sclar and E. Burstein, J. Phys. Chem. Solids, 2(1):1 (1957).
9.  S. H. Koenig and G. R. Gunther-Mohr, J. Phys. Chem. Solids, 2(4):268 (1957).
10. É. I. Abaulina-Zavaritskaya, Zhur. éksp. i teor. fiz., 30:1158 (1956).
11. É. I. Abaulina-Zavaritskaya, Zhur. éksp. i teor. fiz., 36:1342 (1959).
12. B. M. Vul and É. I. Zavaritskaya, Proc. Internat. Conf. Semiconductor Phys. (Prague, 1960).
13. B. M. Vul, É. I. Zavaritskaya, and L. V. Keldysh, Doklady Akad. Nauk SSSR, 135:1361 (1960).
14. É. I. Zavaritskaya, Fiz. tverd. tela, 2:3009 (1960).
15. É. I. Zavaritskaya, Fiz. tverd. tela, 3:1884 (1961).
16. B. M. Vul, É. I. Zavaritskaya, and I. V. Davydova, Proc. Internat. Conf. Semiconductor Phys. (Exeter, 1962).
17. B. M. Vul, É. I. Zavaritskaya, and I. V. Davydova, Fiz. tverd. tela, 5:1107 (1963).
18. É. I. Zavaritskaya, Fiz. tverd. tela, 6:3545 (1964).
19. É. I. Zavaritskaya, Fiz. tverd. tela, 7:2459 (1965).
20. L. B. Loeb, Fundamental Processes of Electrical Discharge in Gases, 1939. [Russian translation, GITTL, Moscow-Leningrad (1950).]
21. A. A. Smurov, Vestn. éksp. i teoret. elektrotekhniki, 1:239 (1928).
22. A. von Hippel, Ergebn. exper. Naturwissenschaft, 14:104 (1935).
23. H. Fröhlich, Proc. Roy. Soc. (London), A160:230 (1937); 188:521 (1947).
24. W. Franz, Z. Physik, 113:607 (1939).
25. N. L. Pisarenko, Izv. Akad. Nauk SSSR, Ser. Fiz., Nos. 5-6:631 (1938).
26. B. I. Davydov and I. I. Shmushkevich, Uspekhi fiz. nauk, 24:21 (1940); Zhur. éksp. i teor. fiz., 10:1043 (1940).
27. W. Shockley, Bell System Tech. J., 30:990 (1951).
28. L. V. Keldysh, Zhur. éksp. i teor. fiz., 37:713 (1959).

29.   V. A. Chuenkov, Fiz. tverd. tela, 2:799 (1960).
30.   R. Stratton, Proc. Roy. Soc. (London), A242:355 (1957).
31.   W. Shockley, Electrons and Holes in Semiconductors, 1950. [Russian translation, IL (1953)].
32.   N. Sclar, E. Burstein, W. J. Turner, and J. W. Davisson, Phys. Rev., 91:215 (1953); 92:858 (1953).
33.   E. J. Ryder, I. M. Ross, and D. A. Kleiman, Phys. Rev., 95:1342 (1954).
34.   G. Lautz, Proc. Internat. Confer. Semiconductor Phys. (Prague, 1960).
35.   G. Ascarelly and S. Brown, Phys. Rev., 120:1615 (1960).
36.   S. H. Koenig, Phys. Rev., 110:986 (1958).
37.   J. Yamashita, J. Phys. Soc. Japan, 16:720 (1961).
38.   Low-Temperature Physics (Collection, ed. A. I. Shal'nikov) [Russian translation, IL (1959)].
39.   R. Berman, Proc. Roy. Soc. (London), A208:90 (1951); G. K. White and S. B. Woods, Phys. Rev., 103:569 (1956).
40.   N. Kurti and F. E. Simon, Proc. Roy. Soc. (London), A149:161 (1935); Phil. Mag., 26: 849 (1935).
41.   H. B. G. Casimir, W. J. de Haas, and D. de Klerk, Physica, 6:241 (1939).
42.   N. V. Zavaritskii, Zhur. éksp. i teor. fiz., 33:1085 (1957).
43.   W. H. Keesom, Helium, 1942. [Russian translation, IL (1949)].
44.   N. V. Zavaritskii and A. G. Zel'dovich, Zhur. tekh. fiz., 26:2032 (1956).
45.   N. V. Zavaritskii and A. I. Shal'nikov, Pribory i tekh. éksp., No. 1:189 (1961).
46.   Experimental Techniques (Collection, ed. A. F. Ioffe). Moscow-Leningrad (1929).
47.   R. F. Broom and A. C. Rose-Innes, Proc. Phys. Soc. (London), B69:1269 (1956).
48.   I. Isenberg, B. R. Russell, and R. F. Greene, Rev. Sci. Instr., 19:685 (1948).
49.   P. G. Strelkov, Zhur. éksp. i teor. fiz., 10:1225 (1940).
50.   J. A. Burton, Physica, 20:845 (1954).
51.   M. J. Morin, Phys. Rev., 93:62 (1954).
52.   E. M. Conwell and V. F. Weisskopf, Phys. Rev., 77:388 (1950).
53.   G. Erginsoy, Phys. Rev., 79:1013 (1950); N. Sclar, Phys. Rev., 104:1559 (1956).
54.   H. Brocks, Phys. Rev., 83:879 (1951).
55.   C. S. Hung and J. R. Gliesman, Phys. Rev., 96:1126 (1954); H. Fritzsche, K. Lark-Horovitz, Physica, 20:834 (1954).
56.   E. M. Conwell, Phys. Rev., 103:51 (1956); N. F. Mott, Ottawa Conf. Electron Transport, September, 1956.
57.   S. H. Koenig, Phys. Rev. Letters, No. 4:170 (1960).
58.   Ya. I. Frenkel', Zh. éksp. i teor. fiz., 8:1292 (1938).
59.   M. Lax, Phys. Rev., 119:1502 (1960).
60.   K. B. McAfee, E. J. Ryder, W. Shockley, and M. Sparks, Phys. Rev., 83:650 (1951).
61.   M. Glicksman, Progr. Semiconductors, 3:1 (1958).
62.   B. Abeles and S. Meiboom, Phys. Rev., 95:31 (1954); M. Shibuya, Phys. Rev., 95:1385 (1954).
63.   W. Baltenspenger, Phil. Mag., 44:1355 (1953); F. Stern and R. M. Talley, Phys. Rev., 100:1638 (1955).
64.   G. Lautz and W. Ruppel, Z. Naturforsch., 10a:521 (1955).
65.   R. K. Willardson and A. C. Beer, Phys. Rev., 110:1286 (1958).
66.   P. A. Lee, Brit. J. Appl. Phys., 8:340 (1957).
67.   A. L. McWhorther and R. H. Rediker, Proc. Internat. Conf. Semiconductor Phys. (Prague, 1960).
68.   J. Yamashita, J. Phys. Soc. Japan, 16:720 (1961).
69.   S. H. Koenig, Phys. Rev., 110:988 (1958).
70.   V. V. Paranjape, Proc. Phys. Soc. (London), 80:971 (1962).
71.   P. G. Klemens, Solid State Physics, 7:1 (1958).
72.   J. J. Hall, Proc. Internat. Conf. Semiconductor Phys. (Exeter, 1962).
73.   S. H. Koenig, Proc. Internat. Conf. Semiconductor Phys. (Brussels, 1958).
74.   R. D. Larabee, J. Appl. Phys., 30:857 (1958).
75.   L. V. Keldysh, Zhur. éksp. i teor. fiz., 33:994 (1957).

# INVESTIGATION OF THE INFRARED ABSORPTION SPECTRUM
# OF NEUTRON-IRRADIATED SILICON

## É. N. Lotkova

## Introduction

A departure from the periodicity of the crystal lattice (i.e., the presence of defects) in silicon, can, as in other semiconducting crystals, result in considerable changes in its electrical and optical properties (the electrical conductivity, the transparency, etc.). The properties of a semiconducting crystal are usually altered by the introduction of impurity atoms of certain substances into the lattice of the crystal, i.e., by doping. The irradiation of semiconducting crystals with high-energy particles (neutrons, deuterons, etc.) produces various types of radiation defect which can also alter the properties of the crystals. Knowing the actual effect of radiation on a semiconducting crystal and carrying out the irradiation under controlled conditions (temperature, dose), we can alter the properties of the crystal in a required manner. It follows that investigations of the effect of high-energy radiations on semiconducting crystals are of great interest. They are also of importance in relation to the use of semiconducting devices in situations where they are subjected to radiation or where they are used directly as radiation detectors.

Systematic investigations of the effects of radiation on semiconductors, in particular on germanium and silicon, began soon after the Second World War in connection with the development of semiconductor technology, nuclear physics and reactors, and various elementary particle accelerators. The first brief communications on this subject appeared in 1948-49 outside the USSR [1, 2]. The investigations reported in these communications were concerned with the conductivity of germanium and silicon single crystals after irradiation with deuterons, neutrons, and $\alpha$ particles. It was established that the electrical properties of germanium single crystals changed continuously during irradiation. The conductivity of p-type germanium increased as the radiation dose was increased, while the conductivity of n-type germanium first decreased, reached a minimum, and then increased smoothly; this was accompanied by a change in the type of conduction from n-type to p-type. Irradiation always reduced the conductivity of n- and p-type silicon. When silicon crystals were heated, their initial conductivity was reestablished. The nature of the reestablishment of the conductivity showed that irradiated silicon had traps with deep local levels in the forbidden band which captured free electrons and holes; therefore, irradiation reduced the conductivity. The exact positions of the local levels in the forbidden band were not determined.

In spite of the fact that the investigations of irradiated silicon and germanium were begun almost at the same time, initially the volume of published results on silicon was considerably less than that on germanium. In particular, this was due to the fact that (based on electrical measurements) the investigations of irradiated silicon by the standard methods used to study semiconducting crystals met with considerable difficulties in the preparation of ohmic contacts [3]. Therefore, it was found that optical methods, based on investigations of the absorption spectra, were more convenient for investigating irradiated silicon.

The reduction in the conductivity of silicon as a result of irradiation is due to the decrease in the number of free carriers. There should be a corresponding reduction in the absorption of light by free carriers, which is proportional to the free-carrier density. It is, in fact, observed in silicon in the infrared range, beginning from about $1 \mu$.

If the forbidden band contains local levels at 0.1-0.6 eV from the edges of the conduction or valence bands, the infrared absorption spectrum may have absorption bands at $\lambda = 1$-$10 \mu$ which are associated with electron transitions from these local levels to the conduction band or from the valence band to these levels. Therefore, investigations of the infrared absorption spectra of irradiated silicon can yield information on the positions of the local levels associated with radiation defects. This was why the present author investigated the infrared absorption spectra of silicon irradiated with neutrons. It was known (from the brief communications of Lark-Horovitz et al. [4, 5] and from a paper by Vavilov et al. [6]) that irradiation with deuterons or neutrons increased the transparency of silicon beyond the fundamental absorption edge (in the infrared region, beginning from $1 \mu$), shifted the absorption edge toward longer wavelengths, and produced a new narrow absorption band near $1.8 \mu$. Irradiation considerably reduced the resistivity of n- and p-type samples. The origins of the $1.8$-$\mu$ band and of the shift of the fundamental absorption edge were not determined. Later, in 1959, when the present author's investigation was under way, Fan and Ramdas published a detailed paper [7] on the infrared absorption spectrum and photoconductivity of silicon irradiated with neutrons and electrons. They reported that the infrared spectrum of irradiated silicon had several absorption bands, which could be ascribed to definite local levels in the forbidden band. At the same time, many other papers were published which reported investigations of the positions of the local levels in the forbidden band of irradiated silicon by various nonoptical methods, and a model of radiation defects associated with these levels was proposed.

These investigations made it possible to compare the results of the present author with the data obtained independently by different methods.

The present paper describes, in three chapters, the results of an investigation of the infrared absorption spectrum (in the region $1$-$14 \mu$) of silicon single crystals irradiated with fast neutrons in a reactor. The first chapter considers the fundamentals of the current theories of the effect of neutron radiation on a semiconductor and reviews the main papers on neutron-irradiated silicon. The second chapter describes the method used in the present investigation. The third chapter presents the results and the discussion

## CHAPTER I

## Effects of Radiations on a Semiconducting Crystal. Review of the Main Papers on Radiation Defects in Silicon

### §1. Types of Radiation Defect

Irradiation of a crystalline solid with high-energy particles produces various types of defects of the crystal lattice. These radiation defects can be in the form of vacancies, interstitial atoms, impurity atoms, and whole disordered regions (known as "thermal" or "displacement spikes"). The actual nature of the defects depends both on the type of incident radiation and on the kind of solid; this is discussed in detail in several reviews [8-10].

The simplest radiation defects which can form in the lattice of a solid due to the collisions of high-energy particles with the lattice atoms are vacancies and interstitial atoms. Usually, the energy transferred to a lattice atom in this kind of collision is such that the knocked-out atom can form new vacancies in subsequent collisions. Thus, the initial collision can result in a cascade of collisions which produce vacancies and interstitial atoms.

Impurity atoms are usually formed when crystals are irradiated with slow neutrons (~10 eV) produced in nuclear reactions. The process of impurity formation depends on the cross section for neutron capture and is observed mainly when heavy elements are irradiated.

Thermal (displacement) spikes are regions consisting of several thousand atoms each, which melt for a very short time and then solidify. Such regions can be produced by the motion of very high energy particles or as a result of high-amplitude vibrations of atoms which have suffered collisions but have not been knocked out and can thus transfer a large amount of energy to neighboring atoms. This process ir particularly important in heavy metals.

When charged particles or gamma rays pass through a solid, intense ionization and excitation of electrons may take place, which may result in the breaking of bonds, the formation of free radicals, etc. These ionization effects are typical of insulators, dielectrics, glasses, etc.

Thus, the interaction of high-energy particles with a solid is a complex process and the theory of the interaction is not yet fully developed. Most calculations are concerned with the initial damage, i.e., the number of initially knocked-out atoms, since it is assumed that the displacement of atoms is the main initial effect.

A theory developed by Seitz, Kinchin, Pease, et al. [10-14], shows that if charged particles of velocity greater than a certain minimum value are used, most of the energy of the bombarding particle is used up in the excitation of electrons and only about a thousandth part of the energy is transmitted to the nuclei. In the case of bombardment with neutrons, the greater part of the energy of the bombarding particle is used up in the displacement of atoms. We shall be most interested in neutron irradiation, and, therefore, we shall consider it in more detail.

## §2. Displacement of Atoms by Neutron Bombardment

Since a neutron carries no charge, it produced radiation defects only by its direct interaction with the nuclei of the bombarded crystal. The most likely type of interaction is an elastic collision in which the fast neutron transfers some of its energy to the nucleus. It is assumed that an atom is always displaced when it receives energy greater than a certain threshold value $\varepsilon_d$, and is never displaced if it receives a smaller amount. In the case of neutron bombardment, the energy transfer to a lattice atom may vary from zero for glancing collisions to a maximum value $\varepsilon_m$ which is transmitted in a head-on collision and is found from the law of collision of elastic spheres:

$$\varepsilon_m = \frac{4M_1 M_2}{M_1 + M_2} \varepsilon \; . \tag{1}$$

Here, $M_1$ and $\varepsilon$ are the mass and energy of the bombarding particle and $M_2$ is the mass of the lattice atom.

When all the collisions are elastic and the scattering of fast neutrons is isotropic, the average energy $\bar{\varepsilon}$ transferred in one collision is found from the expression

$$\bar{\varepsilon} = \frac{1}{2}\varepsilon_m.$$

Calculations based on these formulas are approximate because both measurements [15] and theory [16] show that neutrons of energies of the order of 1 MeV are scattered mainly in the direction of their motion. Fast reactor neutrons have on the average an energy of 1-2 MeV. Moreover, collisions may be inelastic. This means that the average energy transferred to the bombarded atoms may be less than the energy calculated above and a correction factor [10] of 0.5-0.9 may have to be applied.

As already mentioned, the first knocked-out atom can cause a collision cascade. The theory allows us to calculate the total number of atoms $N_d$ displaced per unit volume per unit time, which is found from the expression

$$N_d = n_1 \bar{v}. \tag{2}$$

Here, $\bar{\nu}$ is the total number of displaced atoms per each primary knocked-out atom averaged out over the energy of the primary knocked-out atoms; $n_1$ is the number of primary knocked-out atoms per unit volume per unit time:

$$n_1 = \Phi n_0 \sigma_d, \tag{3}$$

where $\Phi$ is the number of bombarding particles passing per unit area per unit time, usually called the flux and assumed to be known; $n_0$ is the number of atoms per unit volume of the irradiated crystal, also assumed to be known; and $\sigma_d$ is the cross section for a collision resulting in the displacement of one atom. For fast neutrons, this cross section can be assumed to be equal to the neutron-capture cross section, whose value usually lies in the range 1-10 barn (1 barn = $10^{-24}$ cm$^2$), and is found from tabulated data.

The value of $\bar{\nu}$ is calculated using the cascade process theory. There are several models of cascade processes: that of Snyder and Neufeld [13], a model proposed by Harrison, Seitz, and Koehler [12, 14], Kinchin and Pease's model [11], etc. The models are based on different assumptions, but they give the same result: when a moving atom or a particle collides with an atom at rest, the displacement of the latter atom uses up, on the average, half the energy of the moving atom or particle and each displacement requires an energy $\varepsilon_d$. The other half of the energy is used up in collisions in which the energy transfer of an atom at rest is less than this displacement threshold energy. To calculate $\bar{\nu}$, we can use, for example, the relationship deduced by Seitz and Koehler [12]:

$$\bar{\nu} = \left( 0.885 + 0.561 \ln \frac{X_m + 1}{4} \right) \frac{X_m + 1}{X_m}, \tag{4}$$

where $X_m + 1 = \varepsilon_m / \varepsilon_d$.

The value of $\varepsilon_d$ has been estimated approximately by Seitz, who assumed that in fast collisions the process is dynamic, i.e., irreversible [17]. The energy required to knock out an atom from a lattice site is then twice as large as in a reversible process, the latter energy being about twice the sublimation energy $\varepsilon_s$. Hence, it follows that $\varepsilon_d = 4\varepsilon_s$. Since for solids with strong bonds $\varepsilon_s \approx 5$-6 eV, Seitz has suggested $\varepsilon_d = 25$ eV as a suitable value for the calculations. Detailed calculations have been carried out for copper [18] and germanium [19]. The values obtained have been found to differ by a factor of 1.2-2 from the value proposed by Seitz. For silicon, the experimental value is $\varepsilon_d = 13$ eV [20].

Calculations carried out using Eqs. (1)-(4) showed that the total number of displaced silicon atoms $N_d$ was $1.4 \cdot 10^{18}$ cm$^{-3}$ for an integral neutron flux (dose) of $10^{18}$ neutrons/cm$^2$ and $N_d = 6.8 \cdot 10^{19}$ cm$^{-3}$ for a dose of $5 \cdot 10^{19}$ neutrons/cm$^2$. It was assumed in both cases that $\varepsilon_d = 13$ eV and $\sigma_d = 5$ barn. The integral fluxes were determined with an accuracy of 30%, and, therefore, the values of $N_d$ could not be calculated with a greater accuracy. Consequently, the error in the determination of $N_d$ for an integral neutron flux of $10^{18}$ neutrons/cm$^2$ was $\pm 0.3 \cdot 10^{18}$ atoms/cm$^3$, and for an integral neutron flux of $5 \cdot 10^{19}$ neutrons/cm$^2$ was $\pm 2 \cdot 10^{19}$ atoms/cm$^3$. Thus, the calculations showed that each incident neutron displaced approximately one silicon atom per cm$^3$.

## §3. Review of the Principal Investigations of Radiation Defects in Silicon Bombarded with Fast Neutrons

The investigations of radiation defects in silicon have mainly been concerned with the determination, using various methods, of the positions of local levels in the forbidden band of irradiated silicon and the characteristics of these levels (carrier—capture cross sections, rate of generation,† annealing, etc.).

In the majority of investigations [21-37], the incident particles have been fast electrons of energies in the range 0.5-6 MeV. The number of investigations of silicon irradiated with fast neutrons in a reactor is much

---

† By the rate of generation of local levels or centers, we understand the number of such levels or centers generated by radiation per incident particle in 1 cm of its path in a substance.

TABLE 1. Energy Levels of Radiation Defects in Neutron-Irradiated Silicon

| Method† | Dose, neutrons/cm$^2$ | Level positions, eV | | Author |
|---|---|---|---|---|
| T | $10^{14}$ | | $\varepsilon_V + 0.29$ $\varepsilon_V + 0.16$ $\varepsilon_V + 0.05$ | Klein [3, 41] |
| R | $10^{12}$-$10^{14}$ | Spectrum of levels from $\varepsilon_C - 0.16$ to middle of forbidden band | | Wertheim [42] |
| P | $10^{14}$-$10^{16}$ | $\varepsilon_C - 0.16$ | $\varepsilon_V + 0.45$ $\varepsilon_V + 0.38$ $\varepsilon_V + 0.30$ | Vavilov et al. [39, 40] |
| T | $10^{15}$-$10^{17}$ | Spectrum of levels from $\varepsilon_C - 0.30$ to middle of forbidden band | | Sonder [38] |
| P | ? | $\varepsilon_C - 0.16$; near middle of forbidden band | | Longo [43] |
| A | $10^{18}$-$10^{20}$ | $\varepsilon_C - 0.16$ $\varepsilon_C - 0.21$ | $\varepsilon_V + 0.30$ | Fan et al. [7] |

†Notation used: T — temperature dependence of the Hall effect; R — temperature dependence of the volume recombination velocity of nonequilibrium carriers; P — photoconductivity spectra; A — infrared absorption spectra.

less [7, 38-43]. The main results of these investigations are presented in Table 1. It is evident from this table that different authors found different sets of levels. Almost all of them found a level at $\varepsilon_V + 0.29 \pm 0.01$ eV. Klein found that the rate of generation of this level was 0.35 cm$^{-1}$. Klein also found [3, 41] two other levels: $\varepsilon_V + 0.16$ eV (generation rate of 0.35 cm$^{-1}$) and $\varepsilon_V + 0.5$ eV (generation rate less than 0.65 cm$^{-1}$). These two levels have not been observed in other investigations of neutron-irradiated silicon (Table 1).

Local levels at $\varepsilon_V + 0.38$ eV and $\varepsilon_V + 0.45$ eV have also been reported only once. They were detected by Plotnikov et al. [40] in p-type silicon with a low oxygen content (<10$^{16}$ atoms/cm$^3$). These levels were probably typical of irradiated silicon with a low oxygen content and that was why they were not observed by other workers who used different grades of samples.

A level near the middle of the forbidden band was deduced by Longo [43] from the step in the photoconductivity in the region of 2.25 $\mu$ (0.55 eV) in n- and p-type silicon samples whose Fermi level lay near the middle of the forbidden band after neutron irradiation. Unfortunately, the radiation dose used in Longo's experiments and the resistivity of his samples before irradiation were not available to the present author (the details are given in the Proceedings of Purdue University, USA). It would be interesting to know why the Fermi level in Longo's samples lay near the middle of the forbidden band of irradiated silicon samples (this indicated very high resistivity): whether the initial material had a very high resistivity or whether the radiation doses were very high. It is known that even very low resistivities of several thousandths of $\Omega \cdot$ cm can be increased by applying fast-neutron doses of the order of $10^{19}$-$10^{20}$ neutrons/cm$^2$, to values of hundreds of k$\Omega \cdot$ cm and even higher [44, 45].

Vavilov and Plotnikov, who also investigated the photoconductivity of irradiated silicon, found no levels in the middle of the forbidden band. They found that after irradiation with doses of $10^{13}$-$10^{15}$ neutrons/cm$^2$, the Fermi level of p-type silicon samples whose initial resistivity was 10-100 $\Omega \cdot$ cm was found to lie 0.1-0.25 eV above the top of the valence band at 100°K [39, 40, 46]. It is possible that the local level near the middle of the forbidden band is due to radiation defects which appear in silicon after the application of doses higher than $10^{15}$ neutrons/cm$^2$. Therefore, it has not been observed in investigations in which fluxes of $10^{12}$-$10^{15}$ neutrons/cm$^2$ have been employed.

Sonder [38] investigated the temperature dependence of the Hall coefficient and concluded that the forbidden band of silicon irradiated with neutrons had a spectrum of closely spaced levels extending from the middle of the forbidden band to $\varepsilon_c - 0.30$ eV. The presence of local levels near the middle of the forbidden band was reported by Wertheim [42], who investigated the temperature dependence of the volume recombination velocity of nonequilibrium carriers in silicon irradiated with uranium fission neutrons† ($10^{12}$-$10^{14}$ neutrons per $cm^2$). According to Wertheim, the spectrum of these closely spaced levels extends from the middle of the forbidden band to $\varepsilon_c - 0.16$ eV.

It follows from the investigations reported in [7, 39, 40] that neutron-irradiated silicon has a discrete level in the $\varepsilon_c - 0.16$ eV range. This local level at $\varepsilon_c - 0.16$ eV has been observed by many workers in electron-bombarded silicon. Experiments on the electron spin resonance in silicon irradiated with 0.5-1.5 MeV electrons have established that this level is associated with radiation defects known as A centers, each of which includes an oxygen atom [32-35]. There are as yet no electron-spin resonance data which would confirm the presence of such centers in neutron-irradiated silicon. Recently, two papers were published on the electron-spin resonance in silicon irradiated with fast neutrons [47, 48]. The authors of these papers report that the spectra obtained are very complex and do not resemble any of the known electron-spin resonance spectra of electron-irradiated silicon. In addition to the $\varepsilon_c - 0.16$ eV level, all the local levels found in neutron-irradiated silicon and listed in Table 1 have been observed, with the exception of the levels near the middle of the forbidden band, in electron-irradiated silicon (cf., for example, review [21]).

The existence of a local level at $\varepsilon_c - 0.21$ eV in the forbidden band of silicon irradiated with neutrons or electrons has been established only by Fan and Ramdas [7], who investigated the infrared absorption spectrum of irradiated samples. The paper of Fan and Ramdas is the most closely related to our investigations, and, therefore, we shall consider it in some detail.

Fan and his colleagues thoroughly investigated the infrared absorption spectrum and the photoconductivity of silicon irradiated with reactor neutrons [7, 44].

In the infrared spectrum of silicon irradiated with neutron doses of $10^{18}$-$10^{19}$ neutrons/$cm^2$, they found eight new absorption bands. At room temperature, the bands had absorption peaks at 1.8, 3.3, 3.9, 5.5, 6.0, 20.5, 27.0, and 30.1 μ.

In addition, they found that irradiation reduced the infrared absorption due to free carriers. This effect made it easier to investigate the absorption bands in samples which initially had a high density of free carriers.

In his investigations, Fan found that the irradiation shifted the absorption edge toward long wavelengths. The magnitude of this shift seemed to tend to saturation after prolonged irradiation, and the saturation value was about 0.1 eV. Fan explained this observation by a possible modification of the edges of energy bands due to radiation defects [49].

The radiation-induced absorption band at 1.8 μ was the first to be discovered [4-6]. When the temperature was lowered, this band became narrower, and its maximum shifted toward short wavelengths, but the total absorption remained practically constant. Its half-width at 80°K was equal to 0.1 eV. The band profile was almost Lorentzian. Fan was of the opinion that this absorption band was most probably due to the excitation and not the ionization of centers. The 1.8-μ band had been observed in high-resistivity n- and p-type samples. It had not been found in n-type samples whose Fermi level ($\varepsilon_f$) was 0.16 eV from the bottom of the conduction band after irradiation with $8.5 \cdot 10^{18}$ neutrons/$cm^2$, but it had been observed at room temperature in the spectrum of a sample for which $\varepsilon_f = \varepsilon_c - 0.22$ eV at 300°K. However, the spectrum of the same sample recorded at 90°K had no such band. Hence, it was concluded that the 1.8-μ band corresponded to an energy level lying approximately 0.21 eV below the conduction band and that this absorption band might not be observed if the level is occupied by electrons.

---

† In contrast to reactor neutrons, which have a wide range of energies from hundredths of 1 eV to several MeV, about 99% of fission neutrons usually have an energy of 12 MeV.

The 3.3-$\mu$ band was observed in irradiated n-type samples for which $\varepsilon_f = \varepsilon_c - 0.22$ eV at 300°K. However, this band was not observed after the irradiation of high-resistivity n- and p-type samples [7]. Hence, it was concluded that the 3.3-$\mu$ band was also associated with the $\varepsilon_c - 0.21$ eV level and that the absorption could be observed only when this level was occupied by electrons. At low temperatures, this band split into a series of peaks whose intensities gradually decreased in the short-wavelength direction. The strongest peaks were observed [7] at 3.6 $\mu$ (0.343 eV), 3.45 $\mu$ (0.359 eV), and 3.3 $\mu$ (0.374 eV). The separations between these levels (the distances between peaks expressed in electron-volts) were four times greater than the distances between the excited states of a donor impurity of group V in silicon, i.e., the absorption was that expected for the excitation of an electron bound to a doubly charged center.

The 5.5-$\mu$ band was observed by Fan and Ramdas in neutron-irradiated n-type silicon for which $\varepsilon_f = \varepsilon_c - 0.13$ eV at 200°K. This band was not observed in high-resistivity samples, and, hence, Fan and Ramdas concluded that this band was associated with centers which had an energy level at $\varepsilon_c - 0.16$ eV.

The absorption band at 3.9 $\mu$ was observed in low-resistivity p-type silicon. This band also appeared in the spectrum of high-resistivity p-type samples after some annealing when the Fermi level approached the valence band. Fan and Ramdas assumed that the absorbing centers responsible for this band had a level near $\varepsilon_v + 0.27$ eV.

The 3.9-$\mu$ band of p-type samples and the 5.5-$\mu$ band of n-type samples were also observed in the photoconductivity spectra. According to Fan and Ramdas, this indicated that they were associated with the photoionization of centers and not with transitions between local levels within the forbidden band.

A band was observed at 6 $\mu$ in neutron-irradiated low-resistivity p-type samples. Fan and Ramdas observed this band when the Fermi level was separated from the valence band by less than 0.13 eV. However, they found a sample with the Fermi level 0.08 eV from the valence band which did not exhibit this absorption band. It was concluded that the band was associated with certain structure defects which were present in some samples.

The 20.5-, 27-, and 30.1-$\mu$ bands observed by Fan and Ramdas in the long-wavelength part of the spectrum of neutron-irradiated silicon were not investigated in detail. It was established that the 20.5-$\mu$ band was observed in n- and p-type samples and was independent of the Fermi level position. Fan and Ramdas were of the opinion that this band could be associated with the lattice vibrations. There was no further information on the 27- and 30.1-$\mu$ bands.

Thus, from their investigations, Fan and Ramdas concluded [7] that neutron irradiation generated the following levels in the forbidden band: $\varepsilon_c - 0.16$ eV, $\varepsilon_c - 0.21$ eV, and $\varepsilon_v + 0.27$ eV.

The existence of the $\varepsilon_c - 0.16$ eV and $\varepsilon_v + 0.27$ eV levels, deduced from the absorption spectra, raised no controversy because similar levels were also observed by other workers (Table 1). There was some doubt about the $\varepsilon_c - 0.21$ eV level, which was ascribed by Fan and Ramdas to the absorption bands at 1.8 and 3.3 $\mu$.

The present author is of the opinion that it is not permissible to associate the 3.3-$\mu$ band with electron transitions from local levels at $\varepsilon_c - 0.21$ eV to the conduction band, or to affirm the existence of such a level solely from the fact that the 3.3-$\mu$ band is observed in irradiated samples for which $\varepsilon_f = \varepsilon_c - 0.22$ eV but not in high-resistivity samples. Unfortunately, Fan and Ramdas did not give the numerical values of the resistivity or the Fermi level position for high-resistivity samples. The presence of the absorption band at 3.3 $\mu$ in a sample with $\varepsilon_f = \varepsilon_c - 0.22$ eV may, in principle, be explained by electron transitions to some level above $\varepsilon_c - 0.22$ eV from a deeper level (transitions between discrete levels within one local center). For example, transitions to the level $\varepsilon_c - 0.07$ eV from the level $\varepsilon_c - 0.42$ eV may also be accompanied by absorption in the region of 3.3 $\mu$. The difference $[(\varepsilon_c - 0.42) - (\varepsilon_c - 0.08)]$ eV = 0.34 eV is exactly equal to the quantum of energy corresponding to $\varepsilon = 3.6$ $\mu$, which is the strongest component of the 3.3-$\mu$ band. This absorption band is observed, as in the case of Fan and Ramdas' investigations, in samples with $\varepsilon_f = \varepsilon_c - 0.22$ eV, i.e., when the level $\varepsilon_c - 0.42$ eV is occupied by electrons and the $\varepsilon_c - 0.08$ eV level is empty, but is not observed in high-resistivity samples when the Fermi level is lower than $\varepsilon_c - 0.42$ eV, i.e., when the $\varepsilon_c - 0.42$ eV level is not occupied by electrons. Moreover, the 3.3-μ band is effectively a group of very narrow bands whose half-widths are less than 0.01 eV. It is not very likely that such narrow bands are associated with electron transitions from a discrete energy level to an energy band.

This discussion of the 3.3-μ band observed by Fan and Ramdas shows that the conclusion about the presence of the $\varepsilon_c - 0.21$ eV level in the forbidden band of neutron-irradiated (or electron-irradiated) silicon is premature and requires further experimental verification.

This review shows that a complete spectrum of local levels in neutron-irradiated silicon has not yet been established. Certain levels have been observed by some workers and not others. This may be partly due to the fact that different workers investigated different samples with different impurity compositions and different concentrations of the main impurities, using different radiation doses and test temperatures. Moreover, the methods of investigation were different and hence the accuracy of the results, as well as the sensitivities, were also different. Thus, for example, measurements of the temperature dependence of free carriers and photoconductivity were most easily carried out on low-resistivity samples, i.e., when smaller radiation doses were used. In view of this, almost all the electrical measurements were carried out on samples irradiated with doses not exceeding $10^{14}$ neutrons/cm² and the temperatures employed were not lower than 70°K. However, it would be desirable to investigate the infrared absorption spectra of samples with higher resistivities, i.e., using larger doses, when the absorption by free carriers is small and more radiation defects are formed, so that the absorption bands associated with local levels of radiation defects could be detected more easily.

For this reason, Fan and Ramdas investigated samples irradiated with doses of the order of $10^{19}$ neutrons per cm² at temperatures close to liquid helium.

Investigations of the effects of large doses are of great interest, since such doses markedly alter the properties of semiconducting crystals. Knowing the nature of the changes induced by such doses is particularly important in connection with the prolonged use of semiconducting devices under conditions when they are subjected to ionizing radiation. There have been very few investigations of the effects of large doses on semiconductors. Only the investigations of Fan et al. referred to earlier have dealt with the determination of the positons of local levels in the forbidden band of strongly irradiated silicon. The present author's investigations of silicon irradiated with neutron doses of the order of $10^{17}$-$10^{20}$ neutrons/cm² may yield additional useful information.

## CHAPTER II

## Experimental Method

### §1.  Instruments Used to Determine Infrared Absorption Spectra

A study was made of the infrared absorption and transmission spectra of irradiated silicon single crystals. Most of the samples investigated were in the form of plane-parallel polished plates with an illuminated surface slightly less than 0.5 cm²; their thickness was 0.3-1.8 mm.

The spectra were investigated at room temperature and at liquid nitrogen and liquid helium temperatures.

The first attempts to determine the spectra using standard infrared spectrometers of the single-beam IKS-12 and double-beam IKS-14 type showed that these instruments were unsuitable for the determination of the spectra of small samples, particularly at low temperatures. This was especially true of the double-beam instrument, which could not be used to obtain a spectrum at liquid helium temperature. The following difficulties were encountered:

First, a spectrophotometer was used to determine the transmission spectrum, i.e., the ratio of the infrared flux I transmitted through the sample to the incident flux $I_0$ at a given wavelength, i.e., $t = I/I_0$. Using a single-beam instrument the signals $I_0$ and I were determined separately and then they were divided to obtain $t$.

In the case of the double-beam instrument, a direct record of the ratio t = $I/I_0$ (in percent for each wavelength) was obtained on the chart. The infrared fluxes in the two channels for direct recording of the ratio (expressed in percent) were equalized with photometric and compensating wedges made in the form of combs with wedge-shaped teeth, placed behind the entry slit of a monochromator in front of which a sample and an empty mount were located. The width of the wedge-shaped teeth varied from 0.3 to 3 mm. If the sample was small and its illuminated surface was a tenth of a $cm^2$, i.e., much smaller than the image of the source of light at the entry slit of the monochromator, the area of the wedge-shaped teeth of the comb covering the slit became comparable with the sample area. A slight shift of the sample away from the optical axis affected the magnitude of the flux obtained from the entry slit and this resulted in large errors and a poor reproducibility of the results.

Secondly, when small samples were placed immediately in front of a monochromator slit, only part of the incident flux was employed.

Thirdly, to carry out low-temperature experiments using liquid nitrogen or helium, it was necessary to place the investigated samples in special cryostats with windows. Usually, the samples were held in copper mounts. In our case, the mounts were in the form of polished plates 2 x 20 x 30 mm with an aperture for the sample of the order of 4 x 8 mm. If a cryostat with a sample was placed directly in front of an empty slit of a monochromator (and in a standard instrument there was no other place where the sample could be inserted), then the mounted sample was no longer at the focus of the infrared beam and almost the whole of the copper mount was within the aperture of the beam incident on the entry slit of the illuminator. Since the temperature of the mount was much less than the temperature of the radiation detector, a flux of radiant energy was emitted by the detector in the direction of the cryostat and this flux was reflected from the polished copper mount and did not reach the focusing system because the mount was outside the focus. The thermal radiation of the sample could be neglected, since its surface and reflection coefficient were considerably less than the reflecting surface and the reflection coefficient of the mount. The observed spectrum was thus the transmission spectrum of the sample under investigation combined with the emission spectrum of the detector, which very greatly complicated the analysis of the results. In practice, for a double-beam instrument this meant that when a spectrum of a sample was recorded at liquid helium temperature, the pen of the recorder tended to go beyond the zero line. This "negative radiation flux" effect has been considered in detail by Stepanov and Khvashchevskaya [50].

These difficulties were overcome by changing the illumination system in the infrared spectrometers referred to above in order to obtain a small focused image of the radiation source on a sample in a cryostat [51, 52]. An illumination attachment for a double-beam spectrometer is shown, together with the ray paths, in Fig. 1. The central (hottest) part of a Silit rod 1, which was the source of the infrared radiation, was focused and reduced in the ratio 3:1 by mirrors 2 (a plane mirror, 45 x 65 mm) and 3 (a spherical mirror, 75 x 95 mm, with a focal length F = 125 mm). The infrared beam was thus focused on the sample 4 and on the empty mount 5 with an aperture whose shape and area were equal to the illuminated area of the sample. Then, the image of the rod was magnified by mirrors 6 (a plane mirror, 45 x 65 mm) and 7 (a spherical mirror, 75 x 95 mm, F = 148 mm) and focused by the same mirrors onto the entry slits 8 of the illuminator. The focal lengths of the spherical mirrors and the positions of the mirrors were such that the dimensions of the intermediate image of the hot part of the Silit rod were 2.5 x 12 mm and the image of the rod at the entry slits of the monochromator completely covered these slits. Moreover, allowance was made for a cryostat to replace a sample. All the mirrors were mounted in special holders with adjustment screws and these were attached to a special plate which was fixed in place of the illuminator used before. This arrangement was used with the double-beam infrared spectrometer of the type described by Malyashev, Markov, and Shubin [53], which was an experimental model of the IKS-14; the arrangement was tested using this experimental model. However, the arrangement could also be used with the standard model of the IKS-14, because the attenuator wedge device, the distance between the entry slits of the two channels, and the dimensions of the slits themselves were the same as in the experimental model. Such an illuminator made it possible to use almost the whole infrared radiation flux in the investigation of relatively small samples and to determine their spectra at low temperatures.

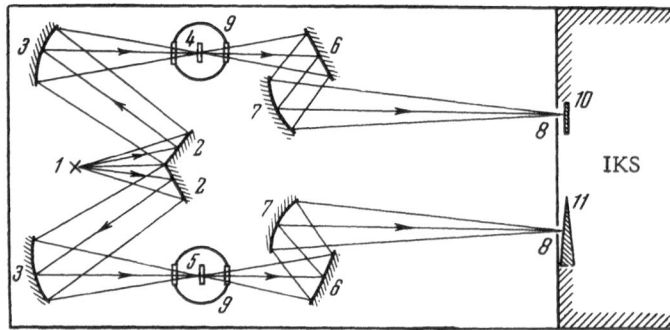

Fig. 1. Illumination attachment for a double-beam spectrometer. 1) Silit rod; 2, 6) plane mirrors; 3, 7) spherical mirrors; 4) mounted sample; 5) empty mount; 8) entry slit of monochromator; 9) cryostat; 10) compensator; 11) photometric wedge.

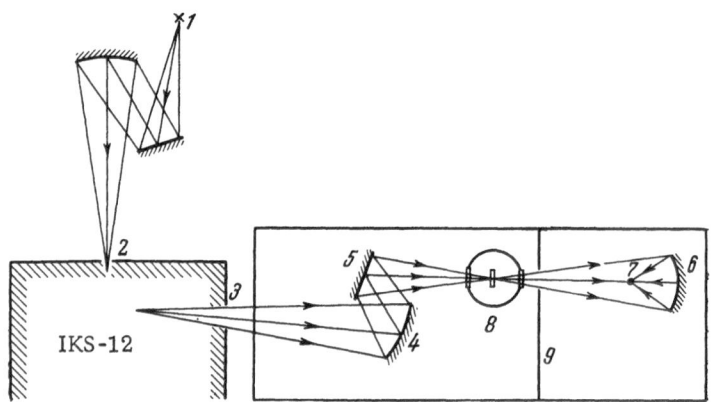

Fig. 2. Attachment for illuminating samples with monochromatic radiation. 1) Silit rod; 2) entry slit of monochromator; 3) exit window of monochromator; 4, 6) spherical mirrors; 5) plane mirror; 7) thermocouple; 8) cryostat with sample; 9) screen.

A similar attachment for recording spectra at low temperatures has been constructed for a single-beam IKS-12 spectrometer [54].

The transmission spectra were recorded with a double-beam infrared spectrometer. A single-beam IKS-12 spectrometer was used in those cases when it was important to know more exactly the transmission coefficient (for subsequent calculation of the absorption coefficient) and to determine more exactly the wavelength in the spectrum. The determination of the transmission coefficients and wavelengths by means of the double-beam instrument was less accurate than by means of the IKS-12.

A special attachment was also constructed for the IKS-12 spectrometer for the purpose of recording low-temperature spectra when a sample was illuminated with monochromatic light. The general arrangement used is shown in Fig. 2.

The radiation of a Silit rod 1 was focused with the standard illuminator system onto the entry slit 2 of the IKS-12. A monochromatic radiation from the source emerged from a monochromator through a special window 3. A radiation detector was placed outside the IKS-12 and was attached to a metal plate together with mirrors 4, 5, and 6. Using the spherical mirror 4 (F = 173 mm) and the plane mirror 5, the monochromatic radiation was magnified by a factor 0.6 and focused onto a sample in the cryostat 8; the radiation transmitted by the sample was focused with the spherical mirror 6 (F = 51 mm) onto the thermoelectric cell 7. To ensure stable operation of the cell by reducing the influence of thermal fluctuation fluxes (which always existed, since the temperature of the cryostat case was several degrees lower than the temperature of air in the laboratory), it was necessary to protect the cell from the cryostat by a screen 9 with an aperture for the transmission of light. The whole attachment was enclosed in a light-tight box with a circular aperture on top to admit the cryostat. Several spectra were recorded using this attachment. It was found that these spectra did not differ greatly from the spectra obtained when a sample was illuminated with nonmonochromatic light. Therefore, in the majority of cases, the monochromatic attachment was not employed, because the lower part of the cryostat with windows was inside the box and, therefore, it was very difficult to adjust the sample which was inside the cryostat.

Depending on the investigated region of wavelengths, the following prisms were used in the infrared instruments: a glass prism F-1 for the region 0.7-2.5 $\mu$ (the linear dispersion for the average wavelength $\lambda_{av}$ = 1.3 $\mu$ was 0.102 $\mu$/mm); an LiF prism in the region 1-6 $\mu$ (linear dispersion for $\lambda_{av}$ = 4.5 $\mu$ was 0.035 $\mu$/mm); an NaCl prism for the region 2-15 $\mu$ (linear dispersion for $\lambda_{av}$ = 11.8 $\mu$ was 0.186 $\mu$/mm).

The largest spectrometer slit width s for the 1- to 4-$\mu$ region was 0.1 mm in most cases, which corresponded to a spectral slit width $s_\lambda$ = 0.003 $\mu$ for the glass (F-1) and LiF prisms; in the region of about 9 $\mu$, the width s was on the average 1 mm, which corresponded to $s_\lambda$ = 0.06 $\mu$ for the NaCl prism.

## § 2. Cryostats

To carry out the optical investigations on solid samples at low temperatures, special metal cryostats were developed [54-56].†

There are two methods of cooling a sample in optical investigations: the sample is attached to a cooled copper rod (a cold duct) or is immersed in a chamber containing a cold gas in the form of hydrogen or helium (cf., for example, [56-62]).

When a sample is cooled by the second method, lower temperatures, practically identical with the temperatures of the surrounding gases, can be reached. However, the cryostat construction is then more complicated. If a sample is cooled by attaching it to a cold duct, the temperature of the sample differs from the temperature of the cooling liquid and of the copper rod by several degrees. Good thermal contact between the sample and the cold duct is important. In the present investigation, we used the cold-duct method of cooling. Moreover, the cryostat had to satisfy the following requirements in the low-temperature optical investigations:

1. The cryostat should be sufficiently strong to withstand transportation to a cryogenic laboratory in a different building some distance away, where the helium was poured in. Therefore, we used a metal cryostat.

2. The cryostat should be vacuum tight to avoid constant pumping during the process of filling with liquefied gas and during the optical investigations.

3. The cryostat casing should have two plane-parallel windows of suitable material for illuminating the sample placed within the cryostat.

4. The cryostat should be demountable to allow the placing of different samples in it.

5. The dimensions and volume of the cryostat should be sufficient to contain such an amount of liquid helium that the optical investigations could be continued for two to four hours.

Two variants of a metal cryostat satisfying these requirements were constructed.

---

† The author is deeply grateful to A. B. Fradkov, Doctor of Technical Sciences, for advice on low-temperature techniques.

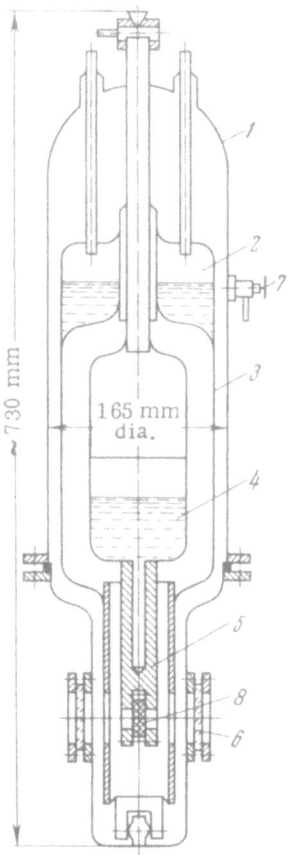

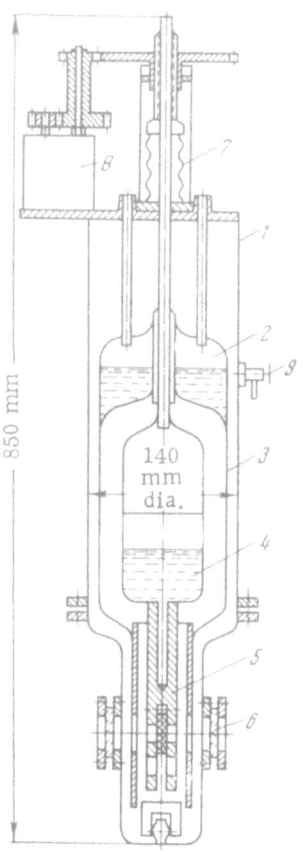

Fig. 3. First variant of the cryostat.
1) External casing; 2) nitrogen bath;
3) copper screen; 4) helium container; 5) cold duct; 6) window; 7)
vacuum valve; 8) sample.

Fig. 4. Second variant of the cryostat. 1) External casing; 2) nitrogen
bath; 3) screen; 4) helium bath; 5)
cold duct; 6) window; 7) sylphon
bellows; 8) motor; 9) vacuum valve.

Figure 3 shows the first variant schematically. A liquid container 4 with a copper rod (cold duct) 5 soldered to it was suspended by a thin-walled stainless-steel tube in a cylindrical casing 1. The lower part of the casing had two circular apertures of 30 mm diameter. Windows, which were glass F-1 or NaCl plates 5 mm thick, were attached to the casing by steel flanges with rubber seals. The cryostat was pumped through a vacuum valve with a sylphon bellows. To improve the thermal insulation when working with liquid helium, a nitrogen bath 2 was placed in the upper part of the cryostat and this bath cooled a copper screen 3 which surrounded the inner container and the cold duct. The nitrogen bath was suspended from the top cover of the casing by two thin-walled stainless-steel tubes through which the liquid nitrogen was poured in. The cryostat casing, the nitrogen bath, the screen, and the inner container were all made of polished copper tubes with walls 1-1.5 mm thick. The lower part of the casing and the screen were demountable to allow the sample to be changed. The various parts of the casing were joined by flanges and rubber seals. The lower part of the screen was joined to the upper part by screws or springs because this joint did not have to be hermetic. The lower part of the cold duct had a ground or threaded end to which a copper frame with two apertures was attached. The sample under investigation was placed in one of the apertures and the other (empty) aperture was used to determine the spectrum of the radiation source. Only relatively narrow beams (5 mm wide) and small samples could be used because the dimensions of the apertures in the copper frame were limited by the dimensions of the window.

The various joints in the cryostat were checked to ensure that they were hermetic, and the cryostat was washed with hot water and alcohol and then heated thoroughly before being pumped to outgas its walls. We used a cylindrical heater 210 mm in diameter and 50 mm long which surrounded the cryostat and heated it to 100°C. Boiling water was poured into the nitrogen and helium containers. The cryostat was pumped in this state for several tens of hours. Only after such conditioning did the cryostat become sufficiently vacuum-tight.

Before the helium was poured in, the cryostat was pumped to $10^{-3}$ mm Hg and then disconnected from the vacuum line. The nitrogen bath was filled with liquid nitrogen and the inner container with cold gaseous helium; the cryostat was left in this state for 2-3 hr. The volume of the inner container was designed to hold 1.3 liters of liquid; about 1 liter of liquid helium was used for cooling. The liquid helium was completely lost by evaporation from the cryostat in 8 hr; helium gas was produced at the rate of 2 liters/min. When the cryostat was placed in front of the entry slit of an infrared spectrometer and its windows were illuminated with focused radiation from a Silit rod (through which a current of 6.5 A was passed) the period during which the helium evaporated away was halved. Liquid hydrogen was retained in the cryostat for more than 72 hr.

We also constructed a second metal cryostat which differed from the first variant because the inner container with the cold duct could be moved vertically in an assembled cryostat filled with a cooling liquid [54]. This was achieved by means of sylphon bellows and a GASM-400 motor which moved the inner container a distance of 40 mm in 15 sec. In this variant, the volume filled with the cooling liquid was half that in the first variant. The cold duct had a flat end with two identical apertures, one above the other. These apertures were 4 x 8 mm. The sample under investigation was attached to a frame so that it covered one aperture while the other aperture was used to determine the spectrum of the Silit rod in a single-beam spectrometer and the 100% transmission line in a double-beam instrument. The other parts of the cryostat were the same in both variants. Figure 4 shows the general form of the second variant. The results obtained with the second variant had better reproducibility. This variant was particularly convenient in investigations using monochromatic illumination, since horizontal displacements were not required and this made it easier to construct a light-tight casing for the attachment. Moreover, the apertures for the samples could be made greater than in the first variant.

Both cryostats were also used to record the spectra of samples cooled with liquid nitrogen, which was then poured into the inner container.

The samples under investigation were attached to a frame or to the flat end of the cold duct using adhesives BF-6 or BF-2.

The temperature of a sample was determined with a calibrated 30-$\Omega$ carbon resistor. When a silicon sample 1 mm thick was attached to a frame and the frame was screwed onto the end of the cold duct of a cryostat filled with liquid helium, the temperature of the sample was found to be $14 \pm 0.3°K$ in the absence of infrared radiation and $18 \pm 0.5°K$ when the sample was illuminated with focused nonmonochromatic radiation from a Silit rod through which a current of 6.5 A was passing. When the same cryostat was filled with liquid nitrogen, the temperature of the sample was in all cases $78 \pm 0.3°K$.

When the cold duct had a flat end with apertures and the sample was attached directly to the cold duct and not to a frame, the temperature of the sample in the liquid helium case was about 5°K less. When we refer to "liquid helium temperature" we shall understand the temperature of a sample placed in a cryostat filled with liquid helium, i.e., 14-18°K.

## § 3. Determination of Absorption Spectra

Using an infrared spectrometer, we recorded the transmission spectrum, i.e., the dependence of the transmission coefficient $t = I/I_0$ on the wavelength $\lambda$. Greatest interest lay in the absorption coefficient $\alpha$ (cm$^{-1}$), which was determined on the assumption that the energy of an electromagnetic wave decreased in silicon by a factor of e in a distance $1/\alpha$ (cm).

For a solid plane-parallel plate of thickness d with a reflection coefficient R, a refractive index n, and an extinction coefficient k (which was the imaginary component of the total refractive index), the absorption coefficient $\alpha$ at a given wavelength $\lambda$ is related to the transmission coefficient t by the following expression, which allows for multiple reflection at the boundaries of the sample:

$$\frac{I}{I_0} = \frac{(1-R)^2 + 4R\sin^2\psi}{e^{-\alpha d} - R^2 e^{-\alpha d} - 2R\cos 2(\varphi + \psi)},$$

where

$$\varphi = \frac{4\pi n d}{\lambda}, \quad \psi = \arctan^{-1}\frac{2k}{n^2 + k^2 + 1}.$$

The term with $\cos 2(\varphi + \psi)$ represents the change in the intensity due to interference effects [49]. This term can be neglected for a thick sample and a broad emission band; $\psi$ is a small quantity, because outside the fundamental absorption band of semiconductors we have $k = (\alpha\lambda/4\pi) \ll 1$ and this quantity can usually be neglected [63]. For the same reason, in the case of normal incidence on a semiconductor, we may assume that

$$R = \frac{(n-1)^2}{(n+1)^2}. \tag{5}$$

In our case, since we were not employing very thin films, we could use the relationship

$$t = \frac{I}{I_0} = \frac{(1-R)^2 e^{-\alpha d}}{1 - R^2 e^{-2\alpha d}}. \tag{6}$$

The refractive indices of germanium and silicon are well known [63], so that R can be calculated from Eq. (5). For silicon [64] in the 1.5-11 $\mu$ range, we may assume that R is 0.300 ± 0.003; in the 1-1.2 $\mu$ range, R = 0.310 ± 0.004. We carried out control measurements of the reflection coefficient of freshly polished samples using a standard reflecting attachment for an IKS-12 spectrometer, and obtained R = 0.30 ± 0.01 for the 1-12 $\mu$ range. This was the value we used in our calculations. The same values of n and R were also valid at low temperatures. It has been demonstrated theoretically and experimentally [65, 66] that, for diamond-type lattices, such as that of silicon, the temperature dependence of the refractive index n is given by

$$\frac{dn}{dt} = a + bT, \tag{7}$$

where $a = 4.3 \cdot 10^{-6}$ deg$^{-1}$ and b = $7 \cdot 10^{-3}$ deg$^{-2}$. Since the temperature-induced changes in n lie within the limits of the accuracy of the measurement of its values used to calculate R, we could use the room-temperature values of n and R in the calculations of $\alpha$ from the spectra obtained at low temperatures. The dependence of $\alpha$ on $\lambda$ was the absorption spectrum.

In optical investigations of semiconductors, it is necessary to allow for the fact that these substances have a high refractive index and, therefore, are highly reflecting. Therefore, to obtain more accurate results, it was necessary to carefully select the thickness of the samples to be investigated. Solving Eq. (6) for $\alpha$, we obtain the expression

$$\alpha = \frac{1}{d}\ln\left\{\frac{(1-R)^2}{2T} + \sqrt{\left[\frac{(1-R)^2}{2T}\right]^2 + R^2}\right\}. \tag{8}$$

The relative error in $\alpha$ depends on d, R, T, $\Delta R$, and $\Delta T$, where $\Delta R$ and $\Delta T$ are the errors in the determination of R and T; this dependence is

$$\frac{\Delta\alpha}{\alpha} = \frac{1}{\alpha}\left[\left(\frac{\partial\alpha}{\partial R}\right)\Delta R + \left(\frac{\partial\alpha}{\partial T}\right)\Delta T\right]. \tag{9}$$

A detailed analysis of this expression yields the optimum conditions for the measurement of $\alpha d$.

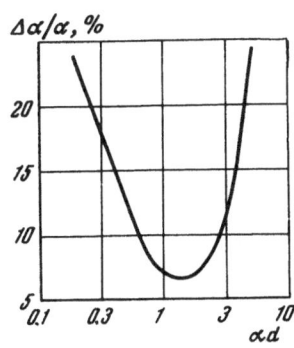

Fig. 5. Dependence of $\Delta\alpha/\alpha$ on $\alpha d$.

Figure 5 shows the dependence of $\Delta\alpha/\alpha$ on $\alpha d$, taken from [67]. This curve is calculated for $R = 0.3$ and $\Delta R = 0.015$, i.e., it is assumed that the error in the determination of R does not exceed 5% and the transmission coefficient is determined to within 2%.

It follows from Fig. 5 that the smallest error in the determination of $\alpha$ is obtained when the value of $\alpha d$ lies between 0.5 and 2.

In the present investigation, the absorption spectra were determined in most cases as a dependence of the transmission coefficient t on the wavelength (this form is used to present the spectra for a wide range of wavelengths) or in the form of a dependence of the absorption coefficient $\alpha$ on the wavelength.

Each spectrum was recorded 5-9 times. The reproducibility of the spectra at wavelengths of 1-6 and 8-14 $\mu$ was, on the average, within 5%; in the 6- to 8-$\mu$ region, which included many absorption bands of water and carbon dioxide, the reproducibility was on the average 8%.

## §4. Neutron Irradiation Technique

We mainly used silicon irradiated with 1- to 2-MeV neutrons. The irradiation was carried out in experimental reactors of the Academy of Sciences of the USSR at temperatures not higher than 80°C. During irradiation, the samples, in the form of polished plates, were placed in special cans made of pure aluminum. The irradiation lasted for periods of from several days to several months. In this way, doses from $10^{17}$ to $10^{20}$ neutrons/cm$^2$ were achieved. The values of the doses were determined to within 30%. After irradiation with doses greater than $10^{17}$ neutrons/cm$^2$, the silicon remained radioactive for a long time, and, therefore, had to be stored for periods up to several months before the measurements could be carried out.

We also investigated several samples bombarded with 1-MeV electrons from an electrostatic generator in the Semiconductor Laboratory of the Physics Institute, Academy of Sciences of the USSR. The temperature of the samples during irradiation with electrons did not exceed 40°C.

## §5. Electrical Measurements and Determination of the Fermi Level Position

In the determinations of the positions of the local levels in the forbidden band from the corresponding absorption bands, it was very important to know the Fermi level position. If an observed absorption band could be attributed to electron transitions from a definite local level in the forbidden band, we had to make sure that there were electrons at this local level. The probability that a level is occupied by an electron depends on its position relative to the Fermi level ($\varepsilon_f$). Levels located much lower than $\varepsilon_f$ are completely filled with electrons, while levels above the Fermi level are practically empty [68].

To estimate the Fermi level position at room temperature, we measured the resistivity $\rho$ of all the samples and the Hall effect of some of them.

Contacts were deposited before the measurements of the resistivity. If the resistivity was not too high, we deposited platinum electrodes by an electrolytic method; in the case of higher resistivities, we used gold electrodes deposited by evaporation. Generally, the resistivity was measured by the usual two-probe method, but we did use four probes in some cases.† To measure the Hall effect, we used cross-shaped samples whose length—width ratio was 4:1.

---

† The measurements of the resistivity and of the Hall effect were carried out using the apparatus in the Semiconductor Laboratory of the Physics Institute, Academy of Sciences of the USSR. The author takes this opportunity to express her deep gratitude to N. A. Penin and B. Zhurkin, who made the apparatus available.

The Fermi level was calculated from the relationship

$$\varepsilon_f = \frac{\varepsilon_0}{2} - kT_t \ln \frac{n}{n_i}, \tag{10}$$

where $\varepsilon_0$ is the effective forbidden band width; $\varepsilon_f$ is the position of the Fermi level measured from the bottom of the conduction band (for n-type samples) or from the ceiling of the valence band (for p-type samples); n is the density of free carriers, determined from the resistivity or from the Hall effect (the value of the carrier mobility in irradiated silicon was taken from [38]); $n_i$ is the density of free carriers in a pure crystal; k is Boltzmann's constant; and $T_t$ is the temperature of a sample.†

The average error in the calculation of the Fermi level position from Eq. (10) was, in our case, ±0.02 eV. This error was due mainly to the inaccuracy of the value of the forbidden band width of irradiated silicon.

The Fermi level position was determined at room temperature because at low temperatures, the resistivity of irradiated crystals increased to values of several $10^9 \, \Omega \cdot$ cm, so that special electrometric instruments were required to carry out electromagnetic measurements at low temperatures. In view of this, the attribution of the absorption bands associated with radiation defects to definite electron transitions was made using the room-temperature spectra, with the exception of a few cases which will be discussed specially. The low-temperature spectra were used mainly to find the structure of the bands.

## §6. X-Ray Structure Analysis of Irradiated Silicon

In order to determine the degree of damage of the lattice of irradiated crystals, we carried out an x-ray structure analysis of strongly irradiated single crystals. Laue and Debye diffraction patterns of silicon samples irradiated with $5.2 \cdot 10^{19}$ neutrons/cm² were obtained, at our request, at the Crystallography Institute of the Academy of Sciences of the USSR.‡

Analysis of the x-ray diffraction patterns showed that, to within one thousandth of an angstrom, the lattice constant of silicon ($a = 5.431$ Å) was not affected by the irradiation. Konobeevskii and Butra [69], who carried out an x-ray diffraction investigation of silicon single crystals irradiated with $5.5 \cdot 10^{19}$ fast neutrons per cm², reported that the lattice constant of silicon did not change by more than 0.1% after irradiation.

The Laue diffraction patterns of the irradiated silicon exhibited strong additional diffuse scattering. These diffuse maxima could be either due to thermal vibrations of atoms or, as pointed out by Huang [70], due to lattice defects (vacancies, interstitial atoms). However, the diffuse maxima of the irradiated silicon were observed to have the same intensity in the Laue diffraction patterns obtained at liquid nitrogen temperature [69]. Consequently, they could not be due to thermal vibrations but were due to lattice defects. Since the lattice constant of silicon was not greatly affected by irradiation, we concluded that the lattice defects were in the form of small disturbances representing single vacancies or interstitial atoms. Larger imperfections would have considerably altered the lattice constant. Radiation defects in the form of larger disturbances were not very stable and would be annealed during the irradiation itself (the irradiation temperature was 80°C). In view of this, it would have been interesting to carry out an x-ray diffraction study of silicon irradiated at low temperatures; however, such a study is fairly difficult.

---

†We have omitted the term $kT \ln(m_e/m_p)$, where $m_e$ and $m_p$ are the effective masses of electrons and holes, respectively; for silicon this term is less than 0.001 eV, i.e., it is much less than the error contributed by the other terms in the formula for $\varepsilon_f$.

‡ The author is very grateful to E. N. Belova of the Crystallography Institute, who obtained the Laue and Debye diffraction patterns of the irradiated silicon samples.

CHAPTER III

## Experimental Results and Discussion

We investigated the 1- to 14-$\mu$ infrared absorption spectra of irradiated n- and p-type silicon single crystals. The resistivity before irradiation ($\rho_0$) was within the limits 0.2-0.004 $\Omega \cdot$ cm. After irradiation ($\rho_{irr}$) the room-temperature resistivity ranged from several tens of $\Omega \cdot$ cm to several M$\Omega \cdot$ cm.

We mainly investigated samples irradiated with fast neutrons used in doses of $3 \cdot 10^{17}$, $6 \cdot 10^{17}$, and $9 \cdot 10^{17}$ neutrons/cm$^2$. We also prepared a batch of samples irradiated with a dose of $5.2 \cdot 10^{19}$ fast neutrons/cm$^2$.

More than 100 irradiated silicon samples differing in their main impurity content, type of conduction, and radiation dose were investigated. It was found that the form of the absorption spectra depended on the type of conduction, the temperature, and the Fermi level position. Figures 6 and 7 present typical transmission spectra obtained at room temperature and at low temperatures for n- and p-type silicon subjected to a dose of $6 \cdot 10^{17}$ neutrons/cm$^2$. The same figures include the spectra of unirradiated silicon crystals. Figure 7 shows, for comparison, the spectrum of high-resistivity unirradiated silicon in which the absorption by free carriers is small and which, therefore, is transparent to infrared radiation over the whole region investigated — right up to 14 $\mu$. It is evident from these spectra that, in the infrared region investigated, the free-carrier absorption is reduced considerably by irradiation and that irradiation gives rise to absorption bands not observed for unirradiated silicon. Moreover, the edge at which the absorption increases rapidly on the long-wavelength side of the fundamental absorption band is shifted by irradiation toward longer wavelengths.

The strongest absorption bands generated by irradiation lie at 1-8 $\mu$, and this region shows the greatest variety in the spectra.

In the 8- to 14-$\mu$ range, the spectra of the investigated samples differ little from one another: this region includes the bands associated with the lattice vibrations and with the presence of oxygen (the 9-$\mu$ band). For this reason, we shall consider separately the absorption bands in the 1- to 8-$\mu$ range and those in the 8- to 14-$\mu$ range.

### §1. Absorption Bands in the 1- to 8-$\mu$ Range

The infrared absorption spectra of irradiated silicon samples were found to have absorption bands at 7.2, 5.6, 4.0, 3.5, and 1.8 $\mu$ and a series of absorption bands in the 1.1- to 1.7-$\mu$ range.

a) Absorption Band at 7.2 $\mu$. The band at $7.2 \pm 0.2$ $\mu$ ($0.172 \pm 0.005$ eV) is only observed in the spectra of irradiated low-resistivity p-type silicon samples whose room-temperature Fermi level lies, after irradiation, not further than 0.16 eV from the ceiling of the valence band (Fig. 7). The 7.2-$\mu$ band is broad and fairly flat at room temperature, but at low temperatures it becomes stronger and has a well-defined maximum. The intensity of the 7.2-$\mu$ band is the same at nitrogen and helium temperatures. Weak absorption bands in the 6.5- to 8.2-$\mu$ region are superimposed on the 7.2-$\mu$ band; these weak bands are also observed in unirradiated silicon and are probably due to lattice vibrations. For this reason, accurate quantitative estimates of the band parameters (the position of the absorption maximum, the half-width, and the dependence of the intensity on the irradiation dose) cannot be obtained. A rough estimate of the width $\gamma$ gives 0.1 eV.

The 7.2-$\mu$ band is not observed at room temperature or at low temperatures in the spectra of samples whose Fermi level after irradiation lies higher than $\varepsilon_v + 0.20$ eV at room temperature (cf., for example, Fig. 8). Hence, we may conclude that the wide band at 7.2 $\mu$ is due to electron transitions from the valence band to a local level which is 0.17 eV above the top of the valence band, but no absorption is observed if this level is filled with electrons, i.e., when the Fermi level lies above this local level, as in the case presented in Fig. 8. It is natural to assume that electron transitions take place from the ceiling of the valence band $\varepsilon_v$ to this local level, and, therefore, the energy of quanta corresponding to the maximum of the absorption band in this case and in similar circumstances will be equated to the gap between the local level and the top of the valence band.

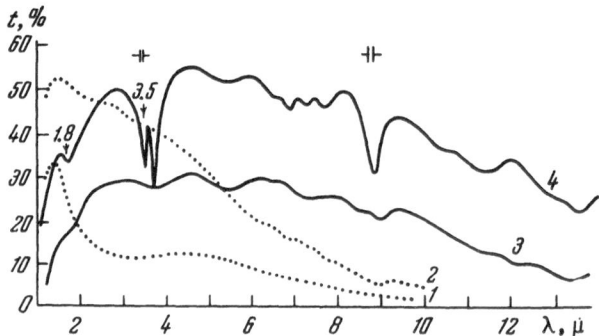

Fig. 6. Transmission spectra of n-type silicon at various temperatures: 298°K (1) and 14°K (2) before irradiation ($\rho_0 = 0.04 \ \Omega \cdot$ cm, d = 0.17 cm); 298°K (3) and 14°K (4) after irradiation ($\Phi = 6 \cdot 10^{17}$ neutrons/cm$^2$, $\varepsilon_f = \varepsilon_c - 0.24$ eV at 298°K).

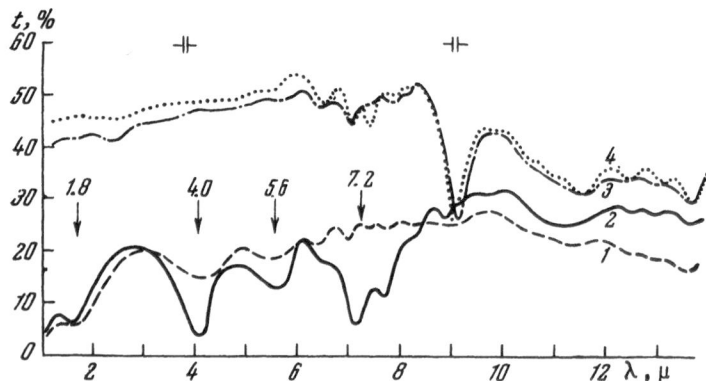

Fig. 7. Transmission spectra of p-type silicon at various temperatures: 298°K (1) and 14°K (2) after irradiation ($\rho = 0.007 \ \Omega \cdot$ cm, d = 0.1 cm, $\Phi = 6 \cdot 10^{17}$ neutrons/cm$^2$, $\varepsilon_f = \varepsilon_v + 0.16$ eV at 298°K); 298°K (3) and 14°K (4) before irradiation ($\rho_0 = 20 \ \Omega \cdot$ cm, d = 0.1 cm).

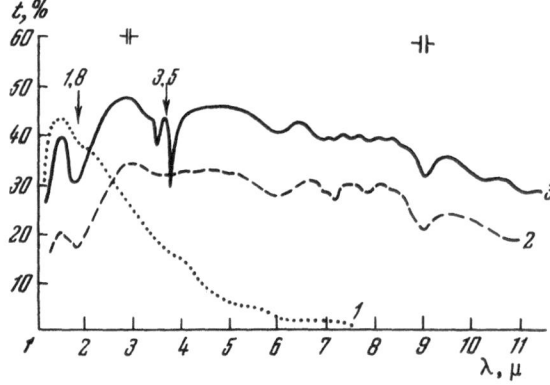

Fig. 8. Transmission spectra of p-type silicon at 298°K (1) before irradiation ($\rho_0 = 0.092 \ \Omega \cdot$ cm, d = 0.17 cm) and at 298°K (2) and 14°K (3) after irradiation ($\Phi = 6 \cdot 10^{17}$ neutrons/cm$^2$, $\varepsilon_f = \varepsilon_v + 0.58$ eV at 298°K).

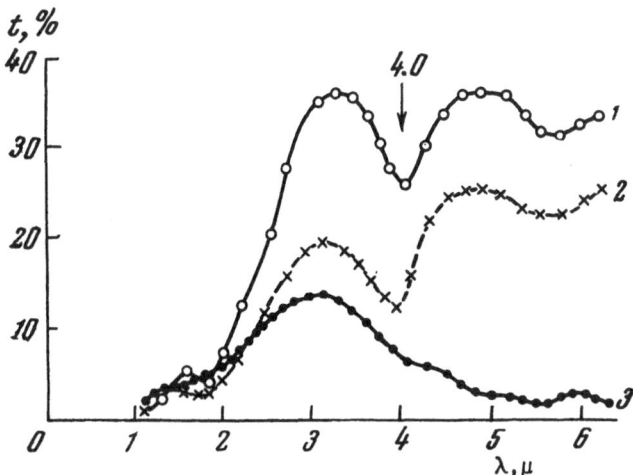

Fig. 9. Room-temperature transmission spectrum of p-type silicon irradiated
with $\Phi = 6 \cdot 10^{17}$ neutrons/cm$^2$: 1) before annealing; 2, 3) after annealing at
150°C (2 — after heating for 1.5 hr; 3 —after heating for 3 hr).

b) Band at 5.6 μ. The absorption band at 5.6 ± 0.1 μ (0.22 ± 0.1 eV), like the 7.2-μ band, is
observed only in irradiated p-type silicon samples whose Fermi level lies about 0.16 eV above the top of the
valence band (Fig. 7) but is not observed in the spectra of high-resistivity irradiated p-type samples (cf., for
example, Fig. 8). The 5.6-μ band is relatively weak and diffuse at room temperature; at low temperatures it
becomes stronger (Fig. 7). An approximate estimate of its width at helium temperature gives 0.1 eV, i.e., a
value typical of bands due to electron transitions from an energy band to a local level [71]. This observation,
as well as the fact that the 5.6-μ (0.22 eV) band is observed in the spectra of irradiated low-resistivity p-type
samples but not in the spectra of irradiated high-resistivity p-type samples or in n-type samples, indicates that
this band is due to electron transitions from the valence band to the $\varepsilon_v$ + 0.22 eV level.

The increase in the intensity of the 5.6-μ band at low temperatures, like the corresponding increase in
the 7.2-μ band, can be explained by the fact that when the temperature is lowered the Fermi level of samples
in which these two bands are observed (p-type samples) shifts closer to the top of the valence band and the
probability of the population of the corresponding local level with holes becomes greater. Consequently, the
probability of transitions of electrons from the valence band due to the absorption of corresponding quanta increases and, therefore, the absorption coefficient of the corresponding band becomes greater.

A weak band whose intensity and width is approximately equal to the corresponding room-temperature
parameters of the 5.6-μ band is observed in many of the investigated crystals in the 5.9- to 6-μ range. We
cannot ascribe this band to transitions associated with local levels of radiation defects because this band is independent of the Fermi level position in the investigated crystals and is also independent of the radiation dose.
Moreover, it is also observed in high-resistivity unirradiated crystals. It is very likely than the 5.9- to 6-μ
band is due to lattice vibrations because, when the temperature is lowered, it changes in the same way as the
bands in the 8- to 14-μ region, which are due to lattice vibrations.

c) Band at 4.0 μ. The absorption band at 4.0 ± 0.1 μ (0.31 ± 0.01 eV) is one of the strongest
bands in the spectrum of irradiated silicon. It is observed only in the spectra of p-type silicon. The 4.0-μ
band can be seen quite clearly at room temperature in those p-type silicon samples whose room-temperature
Fermi level lies below $\varepsilon_v$ + 0.30 eV after irradiation (cf., for example, Fig. 7); the intensity of this band increases considerably at low temperatures, but at nitrogen and helium temperatures the amplitude of the maximum is the same. The half-width of the band at low temperatures is about 0.1 eV. An increase in the radiation dose, which increases the resistivity and shifts the Fermi level toward the middle of the forbidden band, reduces the intensity of the 4.0-μ absorption. The 4.0-μ band is not observed in samples whose Fermi level lies

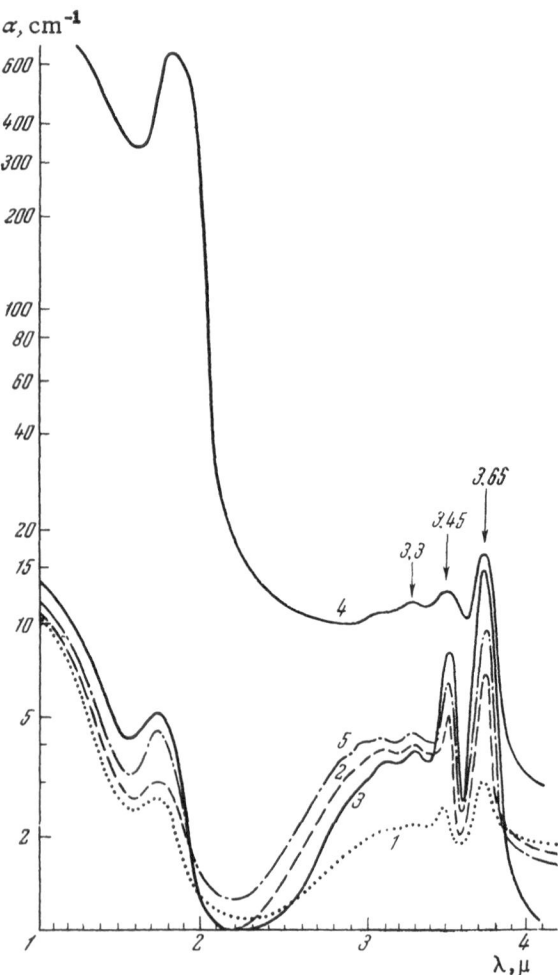

Fig. 10. Absorption spectrum of n-type silicon at 14°K after irradiation with various doses ($\rho_0 = 0.04$ $\Omega \cdot$ cm). (1) $\Phi = 3 \cdot 10^{17}$; (2) $6 \cdot 10^{17}$; (3) $9 \cdot 10^{17}$; (4) $5.2 \cdot 10^{19}$ neutrons/cm²; (5) electron flux of $10^{18}$ electrons/cm².

above $\varepsilon_V + 0.30$ eV after irradiation; for example, it is not observed in the spectrum of high-resistivity silicon (cf. Fig. 8). This allows us to conclude that the 4.0-$\mu$ absorption in irradiated silicon is associated with electron transitions from the valence band to a local level separated by 0.31 eV from the top of the valence band.

The 4.0-$\mu$ band is relatively little affected by heating (annealing). Figure 9 shows the absorption spectra in the 1- to 6-$\mu$ range of irradiated low-resistivity p-type silicon after heating in a vacuum furnace at 150°C for 1.5 and 3 hr. It is clear from Fig. 9 that heating for 1.5 hr has practically no effect on the 4.0-$\mu$ band but simply increases the overall absorption background which is due to free carriers. However, the 4.0-$\mu$ band is destroyed by 3 hr heating at the same temperature.

d) Band at 3.5 $\mu$. The 3.5-$\mu$ (0.35 eV) band is observed in strongly irradiated n- and p-type silicon crystals for certain positions of the Fermi level. At room temperature, this band is broad and flat. At low temperatures, it splits into several narrow bands whose intensities gradually decrease in the direction of short wavelengths. We can observe maxima at 3.65 $\mu$ (0.34 eV), 3.45 $\mu$ (0.36 eV), 3.3 $\mu$ (0.375 eV), 3.1 $\mu$ (0.40 eV), and 2.9 $\mu$ (0.43 eV). The narrowest and strongest maxima are those at 3.65 and 3.45 $\mu$ (cf., for example, Figs. 6, 8, and 10).

The 3.5-$\mu$ band is observed in the spectra of neutron-irradiated n-type silicon crystals whose Fermi level lies near the middle of the forbidden band. It is also observed in the spectra of irradiated n-type silicon crystals whose Fermi level after irradiation lies between $\varepsilon_C - 0.20 \pm 0.03$ eV and the middle of the forbidden band. The 3.5-$\mu$ band is not observed in irradiated n-type silicon whose Fermi level $\varepsilon_f$, after irradiation with a dose of $6 \cdot 10^{17}$ neutrons/cm², lies near $\varepsilon_C - 0.17$ eV, but is observed in the same sample irradiated with a larger dose ($9 \cdot 10^{17}$ neutrons/cm²), after which $\varepsilon_f = \varepsilon_C - 0.36$ eV. It is also observed in n-type irradiated silicon for which $\varepsilon_f = \varepsilon_C - 0.25$ eV (Fig. 6).

The 3.5-$\mu$ band is not observed in irradiated p-type silicon for which $\varepsilon_f = \varepsilon_V + 0.17$ eV after irradiation, or in the spectrum of irradiated p-type silicon for which $\varepsilon_f = \varepsilon_V + 0.40$ eV, but it is observed in the same sample after irradiation with a larger dose, when $\varepsilon_f = \varepsilon_V + 0.58$ eV (Fig. 8).

The 3.5-$\mu$ band cannot be associated with electron transitions to the conduction band from a level within the forbidden band because, judging by the value of the corresponding quantum (about 0.35 eV), this level should lie near $\varepsilon_C - 0.35$ eV and the 3.5-$\mu$ band should not be observed in irradiated samples whose Fermi level is near the middle of the forbidden band.

From the results obtained, we may conclude that the 3.5-$\mu$ absorption band is due to electron transitions from a local level near the middle of the forbidden band (between $\varepsilon_V + 0.40$ eV and $\varepsilon_V + 0.58$ eV) to a different local level, which may lie 0.35 eV above the former level. The conclusion that the 3.5-$\mu$ band is due to

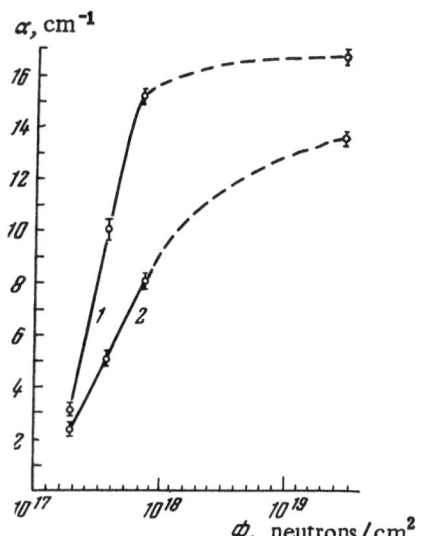

Fig. 11. Dependence of the absorption coefficient at the maxima of the (1) 3.65-μ and (2) 3.45-μ bands at 14°K on the radiation dose applied to n-type silicon with $\rho_0 = 0.04\ \Omega \cdot$ cm ($N_{imp} = 4 \cdot 10^{17}$ atoms/cm³).

transitions between two local levels is supported by its profile. As already mentioned, the 3.5-μ band is really a group of very narrow bands (whose components are narrower than 0.01 eV) which merge into one wide band at room temperature. The absorption bands associated with electron transitions from a local level within the forbidden band to the conduction band or from the valence band to a local level are usually about 0.1 eV wide (cf., for example, the absorption bands in silicon due to the main impurities [71]).

The fine structure of the 3.5-μ band may be associated either with the presence of several slightly differing centers with almost identical local levels or with the presence of several excited electron states associated with one type of center.

The intensity of the 3.5-μ band depends strongly on the radiation dose. It can be seen from the spectra displayed in Fig. 10 that the intensities of the maxima of the components of this band increase in proportion (within the limits of the experimental error) to the radiation dose $\Phi$ up to a certain limit $\Phi_{lim}$, above which saturation seems to set in. Figure 11 shows the dependence of the refractive index at the maxima of the 3.65- and 3.45-μ bands on the radiation dose applied to an n-type silicon sample which had $\rho_0 = 0.04\ \Omega \cdot$ cm and a phosphorus concentration $N_{imp} = 4 \cdot 10^{17}$ cm⁻³. It is evident from Fig. 11 that, in the case of this sample, the saturation begins at $\Phi_{lim} \approx 10^{18}$ neutrons/cm². Unfortunately, the value of $\Phi_{lim}$ could not be determined more accurately because samples irradiated with doses in the range $10^{18}$-$10^{19}$ neutrons/cm² were not available.

Experiments have shown that the profile of the 3.5-μ band of n-type samples, its fine structure, intensity, and the positions of the maxima are independent of whether silicon is doped with phosphorus or arsenic, i.e., these properties are independent of the nature of the donor impurity. But the intensity of the 3.5-μ band depends strongly on the concentration of the donor impurity, i.e., on the initial (preradiation) density of free electrons.

The 1- to 5-μ absorption spectra of silicon samples having different amounts of arsenic and irradiated with a dose of $5 \cdot 10^{19}$ neutrons/cm² are presented in Fig. 12. The same figure includes the spectrum of p-type silicon subjected to the same dose. The resistivity of all the investigated samples subjected to $5 \cdot 10^{19}$ neutrons per cm² was not less than several kΩ · cm at room temperature and, in some cases, reached several MΩ · cm, i.e., the Fermi level of these samples was close to the middle of the forbidden band. It follows from Fig. 12 that the higher the donor impurity concentration, the stronger the 3.5-μ absorption band.

The absorption coefficient in the region of 3.5 μ of samples with $N_{imp} = 10^{18}$ cm⁻³ exceeded 100 cm⁻¹ at room and helium temperatures. Consequently, we were unable to measure the band maxima for some samples because they were too thick (the thinnest of the samples irradiated with $5 \cdot 10^{19}$ neutrons/cm² was 0.06 cm thick). The relationship between the intensities of the maxima of the band components in those cases where they could be measured (cf. spectra 1, 2, and 5 at 14°K in Fig. 12) was not affected by an increase in $N_{imp}$.†

---

† To measure an absorption coefficient of the order of 100 cm⁻¹ with the smallest possible error, it is necessary to use samples whose thickness is less than 0.01 cm. The irradiation of such thin samples in a reactor or the reduction of the thickness of an irradiated sample is risky because silicon plates 0.01 thick are very brittle and can be fractured easily during their loading into a reactor, transporting after irradiation, attachment to the cold duct of a cryostat, etc.

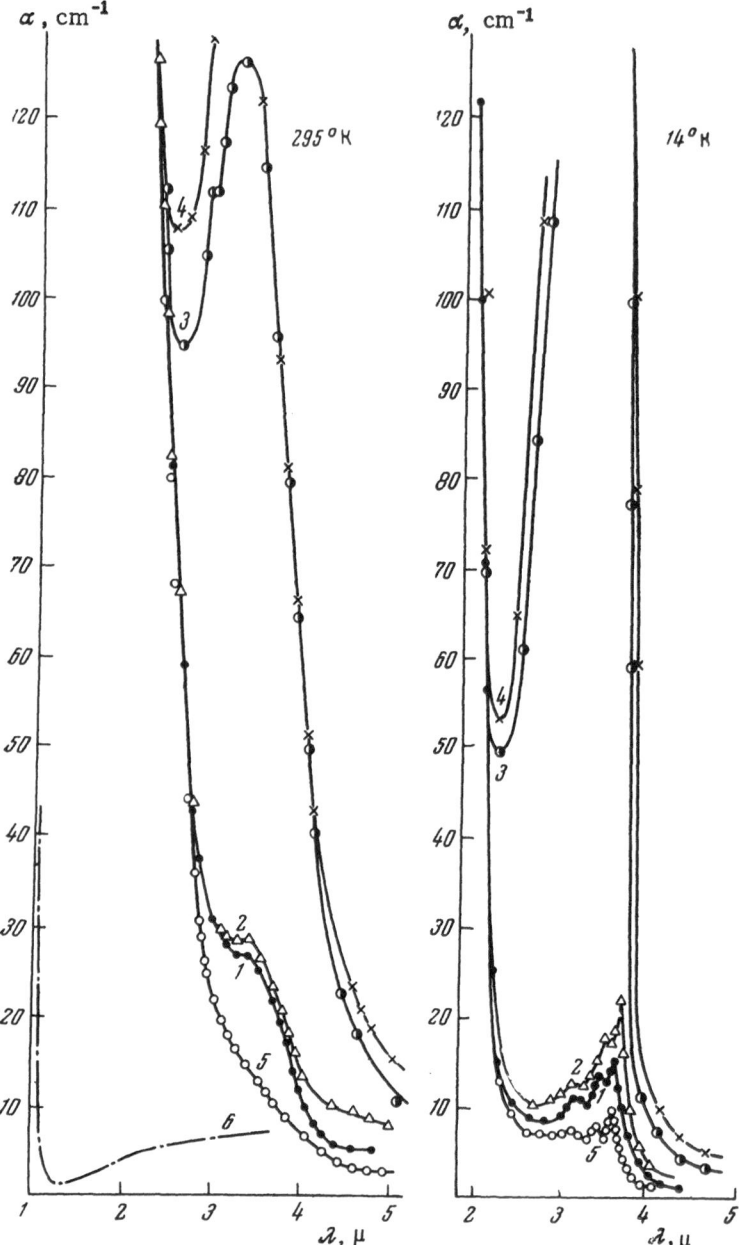

Fig. 12. 3.5-$\mu$ band at 14 and 295°K for irradiated silicon samples ($\Phi = 5 \cdot 10^{19}$ neutrons/cm$^2$) with various concentrations of the main impurity: arsenic — (1) $1 \cdot 10^{17}$; (2) $4 \cdot 10^{17}$; (3) $2 \cdot 10^{18}$; (4) $2 \cdot 10^{19}$ cm$^{-3}$; boron — (5) $6 \cdot 10^{18}$ cm$^{-3}$; curve 6 represents the spectrum of unirradiated silicon with $N_{imp} = 1 \cdot 10^{17}$ cm$^{-3}$ arsenic atoms.

The 3.5-$\mu$ band is much weaker for p-type silicon than for the n-type material and the spectrum in Fig. 12 (curve 5) is well defined at helium temperatures only.

Analysis of the results obtained shows that the absorbing center with which the 3.5-$\mu$ band is associated is an electron trap. It is evident from the spectra in Fig. 10 and from the plots in Fig. 11 that the absorption in a sample with $N_{imp} = 4 \cdot 10^{17}$ cm$^{-3}$ at the maximum of the 3.5-$\mu$ band increases proportionally to the radiation dose up to $\Phi_{lim} \approx 10^{18}$ neutrons/cm$^2$. For higher doses ($\Phi = 5 \cdot 10^{19}$ neutrons/cm$^2$) the absorption coefficient at the maximum of the 3.5-$\mu$ band of the same sample is still the same, i.e., a sample subjected to a stronger dose still contains the same number of absorbing centers although the number of structure defects should be larger for a bigger dose. When a sample with a higher concentration of the donor impurity $N_{imp} = 2 \cdot 10^{18}$ cm$^{-3}$, i.e., with a higher initial density of free electrons, is subjected to this large dose ($5 \cdot 10^{19}$ neutrons/cm$^2$), the absorption coefficient at the maximum of the 3.5-$\mu$ band increases by almost 100 (Fig. 12). This increase is even greater for a sample with $N_{imp} = 1.5 \cdot 10^{19}$ cm$^{-3}$. Thus, when a dose of about $10^{19}$ neutrons/cm$^2$ is applied, the number of defects associated with the absorption band at 3.5 $\mu$ is of the same order as the dose but the number of absorbing centers is governed by the initial (pre-irradiation) number of donor electrons. The number of simplest defects (vacancies and interstitial atoms) generated by a dose of the order of $10^{19}$ neutrons/cm$^2$, calculated from Eqs. (2)-(4), is also of the order of $10^{19}$ cm$^{-3}$. Consequently, we may assume that the 3.5-$\mu$ band is associated with imperfections consisting of one or two simple defects.

The 3.5-$\mu$ band disappears completely after being annealed for several minutes at 220°C. For example, 40-80 sec 220°C annealing of irradiated silicon ($N_{imp} = 2 \cdot 10^{18}$ cm$^{-3}$, $\Phi = 5 \cdot 10^{19}$ neutrons/cm$^2$, $\rho_{irr} = 8.5$ k$\Omega \cdot$ cm) reduces the intensity of this band considerably, while a 5-min heating at the same temperature almost completely destroys the 3.5-$\mu$ band, but the resistivity of the sample remains high: it decreases by a factor of 3 compared with the value before annealing (the resistivity after the 5-min annealing becomes 2.7 k$\Omega \cdot$ cm). The whole group of bands at 3.5 $\mu$ is destroyed by annealing at approximately the same rate. This indicates that this group is either associated with a single center or with several centers which differ very little from one another.

The 3.5-$\mu$ band is also observed in the spectrum of n-type silicon irradiated with a dose of $10^{18}$ cm$^{-2}$ electrons of 1-MeV energy. Figure 10 (curve 5) shows the spectrum of a silicon sample cut from the same material as the samples irradiated with neutrons, whose spectra are also given in the same figure. The order of magnitude of the absorption coefficient, the structure of the bands, and the distribution of intensities in the components of the band are the same for electron and neutron irradiations. Hence, we may conclude that the irradiation of silicon with fast electrons produces the same defects as neutron irradiation.

It is very likely that the radiation defects with which the 3.5-$\mu$ band is associated are the same centers which are observed in the ESR experiments ("II and III centers") reported recently by Jung and Newell [48]. "II and III centers" are annealed in the same way as the 3.5-$\mu$ band (after several minutes at about 200°C) and are observed only in silicon samples irradiated with large neutron doses ($10^{17}$-$10^{19}$ neutrons/cm$^2$), in which the Fermi level after irradiation lies near the middle of the forbidden band. Analysis of the spectra they obtained has led Jung and Newell to the conclusion [48] that the observed centers are defects consisting of several (most likely two) vacancies.

e) Band at 1.8 $\mu$. The band at 1.8 $\mu$ (0.69 eV) is observed in the spectra of irradiated n- and p-type silicon [72]. It is difficult to determine the dependence of this band on the Fermi level position, since it was observed in all the investigated p-type samples (low- and high-resistivity) and in almost all the n-type samples (Figs. 6-8).

The 1.8-$\mu$ band is not observed in n-type silicon for which $\varepsilon_f = \varepsilon_c - 0.17$ eV after irradiation, but it is observed in n-type silicon subjected to the same dose, for which $\varepsilon_f = \varepsilon_c - 0.24$ eV.

The 1.8-$\mu$ band cannot be ascribed to electron transitions from a local level to the conduction band. If this were so, such a local level would lie in the forbidden band near $\varepsilon_c - 0.69$ eV, i.e., approximately 0.40 eV from the top of the valence band, and the 1.8-$\mu$ band would not be observed when this level is not filled with electrons, i.e., it would not be observed in samples whose $\varepsilon_f$ lies in the forbidden band separated by less than

0.40 eV from the top of the valence band. However, we observed the 1.8-μ band at room temperature for a p-type sample for which $\varepsilon_f = \varepsilon_v + 0.20$ eV after irradiation (Fig. 7).

Neither can the 1.8-μ band be attributed to electron transitions from the valence band to a suitable local level in the upper half of the forbidden band about 0.40 eV from the bottom of the conduction band, because such an attribution is in conflict with the observation of the 1.8-μ band in a sample for which $\varepsilon_f = \varepsilon_v - 0.24$ eV after irradiation. The 1.8-μ band is associated with the energy transitions of electrons between discrete levels within a local center.

The 1.8-μ band is the strongest and best-defined band in the absorption spectrum of irradiated silicon, so that its width can be measured, its profile investigated, and the dependence of the absorption coefficient at the maximum of the band on the radiation dose can be found quantitatively.

It follows from Fig. 13, which shows the 1.8-μ band of n-type silicon, that the absorption coefficient at the maximum of the band increases when the temperature is reduced, and that the maximum shifts toward shorter wavelengths. At 78°K, the maximum lies at 1.7 μ, but a further drop in temperatures produces no more shift. The shift of the maximum of the band with temperature can be accounted for by the Franck—Condon principle.

At 300°K, the width of the 1.8-μ band is 0.16 ± 0.01 eV; at 77°K, it is 0.10 ± 0.01 eV.

For the convenience of comparison of the actual band profile with the theoretically calculated Gaussian and Lorentzian profiles, Fig. 14 shows the absorption spectrum in the 1- to 2-μ region against the energy of incident light quanta for p-type irradiated silicon. It follows from Figs. 13 and 14 that the exact determination of the profile is difficult because of the nonuniform absorption background on the short-wavelength side of the 1.8-μ band at the fundamental absorption edge. However, if we take into account the possible superposition of the background and of the long-wavelength tail of the band, we find that when the background is subtracted, the band observed at 1.8 μ is close to the Gaussian distribution.

When the radiation dose is increased, the absorption coefficient at the maximum of the 1.8-μ band increases (within the limits of the experimental error) linearly with the radiation dose. It follows from Fig. 14, which shows the 1.8-μ band for two radiation doses applied to the same sample ($3 \cdot 10^{17}$ and $9 \cdot 10^{17}$ neutrons per cm$^2$) that for a ratio of the doses of 1 : 3 the values of the absorption coefficient at the maximum of the band at nitrogen temperature are in the ratio 1 : 2.8. At room temperature, this ratio is 1 : 2.7. The results at nitrogen temperature are more accurate than those at room temperature because, at low temperatures, the 1.8-μ band is stronger and more clearly defined (cf. Figs. 13 and 14) and, consequently, at nitrogen temperature, the error in the calculation of the absorption coefficient at the maximum of the band by subtracting the background is less.

Since the intensity of the absorption in the 1.8-μ band is a linear function of the radiation dose, this absorption should be of the order of several hundreds of cm$^{-1}$ for a dose of $5.2 \cdot 10^{19}$ neutrons/cm$^2$. Such an absorption coefficient cannot be measured using samples about a millimeter thick, because the corresponding transmission coefficient would be less than a thousandth of one percent, which is outside the limits of the sensitivity of the apparatus employed. In fact, the transmission spectra of samples irradiated with a flux of $5 \cdot 10^{19}$ neutrons/cm$^2$ (Fig. 15) do not contain the 1.8-μ band even when the weaker 3.6-μ band is still visible; the 1.8-μ band seems to have merged with the fundamental absorption edge and the effective absorption edge seems to have shifted to 2-3 μ.

To check whether this shift of the edge is indeed due to the strong absorption in the 1.8-μ band, which cannot be measured using samples 0.06 cm thick (for which the spectra are shown in Fig. 15), we took the risk of fracture and ground a sample irradiated with a dose of $5 \cdot 10^{19}$ neutrons/cm$^2$ to a thickness of several microns. For this, we used a sample of the highest resistivity ($\rho_0 = 0.07 \ \Omega \cdot$ cm), since, according to our assumptions, the 1.8-μ absorption in such a sample should be less than in other samples of lower resistivity and, therefore, a sample of that required thickness could be prepared more easily. After grinding and polishing, we were left with a sample 60 μ thick and we investigated its absorption spectrum in the wavelength range 1-3 μ.

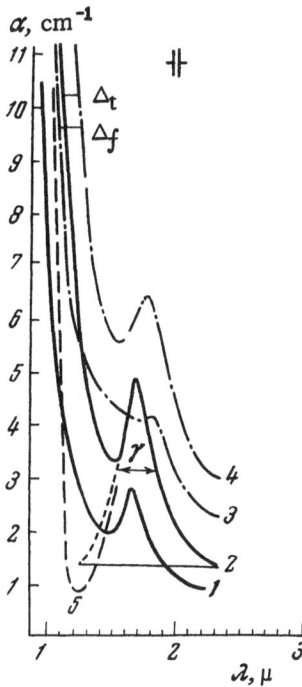

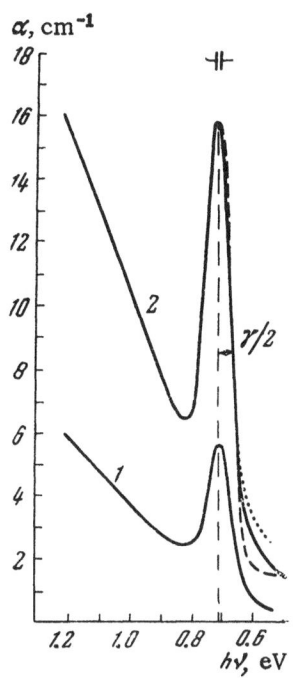

Fig. 13. 1.8-$\mu$ band at room and nitrogen temperatures for n-type silicon with $\rho_0 = 0.04\ \Omega \cdot$ cm, subjected to various neutron doses: 1) 78°K, $\Phi = 6 \cdot 10^{17}$ neutrons/cm²; 2) 78°K, $\Phi = 9 \cdot 10^{17}$ neutrons per cm²; 3) 293°K, $\Phi = 6 \cdot 10^{17}$ neutrons/cm²; $\varepsilon_f = \varepsilon_c - 0.22$ eV; 4) 293°K, $\Phi = 9 \cdot 10^{17}$ neutrons/cm², $\varepsilon_f = \varepsilon_c - 0.30$ eV; 5) 293°K, unirradiated original material; $\Delta_f$ and $\Delta_t$ represent the fundamental absorption edge shift as a function of the radiation dose and temperature, respectively.

Fig. 14. 1.8-$\mu$ band at 78°K for p-type silicon with $\rho_0 = 0.092$ $\Omega \cdot$ cm for various radiation doses: (1) $\Phi = 3 \cdot 10^{17}$ and (2) $9 \cdot 10^{17}$ neutrons/cm². The dashed curve represents a Gaussian distribution and the dotted curve represents a Lorentzian distribution

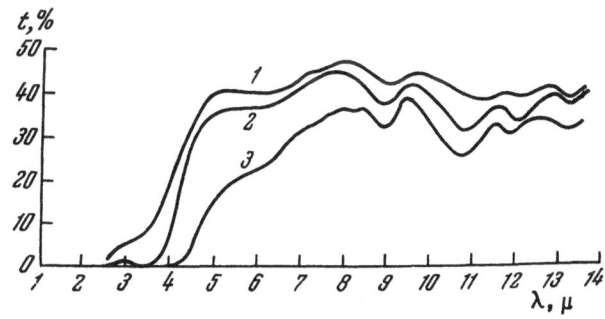

Fig. 15. Room-temperature transmission spectra of strongly irradiated silicon samples (d = 0.06 cm, $\Phi = 5.2 \cdot 10^{19}$ neutrons/cm²) with various concentrations of the main impurity (arsenic): $N_{imp}$ = (1) $4 \cdot 10^{17}$; (2) $2 \cdot 10^{18}$; and (3) $2 \cdot 10^{19}$ atoms/cm³.

The absorption spectra obtained at room temperature clearly showed the 1.8-μ band. The width of this band at room temperature was 0.16 eV, as in the spectra obtained for lower doses, and the band profile was close to a Gaussian distribution. The absorption coefficient at the maximum was 600 cm$^{-1}$ (the accuracy of this measurement was 10%), i.e., within the limits of the doses employed (up to $5 \cdot 10^{19}$ neutrons/cm$^2$) the absorption coefficient at the maximum of the 1.8-μ band increased approximately linearly with increasing dose. More accurate estimates were difficult to obtain because, as already mentioned, it was difficult to eliminate the influence of the background due to the edge of the fundamental absorption band.

The 1.8-μ band was less affected by annealing than the 3.5-μ band. For example, heating for 5 min at 220°C destroyed the 3.5-μ band but the 1.8-μ band was still observed. However, like the 4.0-μ band (cf. Fig. 9), the 1.8-μ band disappeared after 3-hr heating at 150°C.

In spite of the fact that the 1.8-μ band in the infrared absorption spectrum of irradiated silicon was observed some years ago in the very earliest investigations, there is as yet no agreed upon view about its origin. Fan and Ramdas are of the opinion that the 1.8-μ band is due to electron transitions from a level at $\varepsilon_v + 0.32$ eV to a higher level representing an excited state of an absorbing center and, therefore, should not be observed when there are no electrons at the $\varepsilon_v + 0.32$ eV level and when the absorption associated with electron transitions from the valence band to this level, i.e., the 4-μ band, is observed. In our experiments, the 4- and 1.8-μ bands were observed in the same sample (Fig. 7). These bands were exceptionally strong at low temperatures. The low-temperature Fermi level of irradiated samples could not be measured because of the very high resistivity of such samples, but at low temperatures we observed, simultaneously with the 4- and 1.8-μ bands, a strong band at 7.2 μ associated with electron transitions from the valence band to a level at $\varepsilon_v + 0.17$ eV. This meant that, in the case considered, the Fermi level could not lie higher than $\varepsilon_v + 0.17$ eV and the $\varepsilon_v + 0.32$ eV level could not be filled with electrons, i.e., the 1.8-μ band could not be associated with electron transitions from this level.

Plotnikov [73] attributed the 1.8-μ band to electron transitions from the valence band to the $\varepsilon_c - 0.4$ eV level. However, this explanation seems suspect because, first, the 1.8-μ band should not, in this case, be observed in a sample with $\varepsilon_f = \varepsilon_c - 0.24$ eV but, in fact, this band was observed. And, secondly, the $\varepsilon_c - 0.4$ eV level has not yet been observed in neutron-irradiated silicon (Table 1). The $\varepsilon_c - 0.4$ eV level has been found only in electron-irradiated silicon prepared by the floating vertical zone melting method ("zone-melted" silicon). The rate of generation of this level [3] is about 10 times less than the rate of generation of the $\varepsilon_c - 0.16$ eV level and about 6 times less than the rate of generation of the $\varepsilon_v + 0.30$ eV level. In contrast, the 1.8-μ band observed by the present author in silicon prepared by pulling from the melt in a quartz crucible and subjected to neutron or electron irradiation was one of the strongest bands.

Our results agree best with the conclusions of Kurskii [74], who ascribes the 1.8-μ band to defects in the form of complexes consisting of oxygen and vacancies (A centers). Investigations of electron-irradiated silicon [33] have shown that A centers contribute a local level at $\varepsilon_c - 0.16 \pm 0.01$ eV.

From the theoretical calculations and an analysis of the available experimental results, Kurskii concluded that the absorption band at 1.8 μ is associated with the excitation of the A centers and is always observed whenever the $\varepsilon_c - 0.17 \pm 0.01$ eV level is free of electrons. The absorption is of the same nature as the absorption by bound excitons. The value of the absorbed energy calculated by Kurskii, the Gaussian profile of the 1.8-μ band, the value of the width, and the temperature shift all agree with our results (within the limits of the experimental error).

The presence of bound excitons in semiconductors is confirmed by a steadily increasing number of experimental investigations [75, 76]. Analysis of the absorption bands associated with bound excitons shows that their profiles are Gaussian [75, 77, 78].

An investigation of the absorption band parameters is of great importance in the case of theories which account for the possible absorption processes in semiconductor crystals.

Since the 1.8-$\mu$ band is very strong and the absorption at the maximum of this band increases linearly as the radiation dose is increased, this band can be used in neutron dosimetry, especially as neutron dosimetry is not yet well developed and neutron fluxes are measured very inaccurately. According to our data, the coefficient of proportionality between the dose and the 1.8-$\mu$ band intensity is $(0.6 \pm 0.2) \cdot 10^{-15}$ cm/neutron for an n-type silicon sample with a $5 \cdot 10^{17}$ cm$^{-3}$ concentration of the main impurity atoms (phosphorus). In the case of p-type silicon with a $10^{17}$ cm$^{-3}$ concentration of the main impurity atoms (boron), the coefficient of proportionality is $(2.0 \pm 0.5) \cdot 10^{-15}$ cm/neutron.

## § 2. Infrared Absorption at the Long-Wavelength Edge of the Fundamental Absorption Band of Irradiated Silicon

a) Shift of the Edge in Neutron-Irradiated Silicon. The spectra obtained (cf. the spectra shown in Figs. 6 and 13) suggested that the fundamental absorption edge of silicon shifts toward long wavelengths after irradiation. It was difficult to estimate this shift quantitatively, since, first, we did not measure high values of the absorption coefficient and did not investigate the spectrum at $\lambda < 1 \mu$, i.e., we did not reach the region of direct transitions; and, secondly, a strong band at 1.8 $\mu$ near the absorption edge made it difficult to extrapolate the edge to the zero value of the absorption coefficient. However, we estimated, for a fixed value of the absorption coefficient, the shift of the edge $\Delta_f$ on the wavelengths and energy scales, using those regions of the absorption spectra near the edge where the absorption coefficient varied most rapidly.

Figure 13 shows how the edge shifts with the dose. When the dose is changed by $3 \cdot 10^{17}$ neutrons/cm$^2$, $\Delta_f = 0.18 \mu$, which should correspond to a change in the effective forbidden band width by 0.16 eV. The shift of the edge with the dose is, according to our results, nonuniform: it is less at small doses, has a maximum at approximately $10^{18}$ neutrons/cm$^2$, and apparently tends to saturation for doses greater than $10^{18}$ neutrons per cm$^2$. In the dose range $6 \cdot 10^{17}$-$9 \cdot 10^{17}$ neutrons/cm$^2$, the average shift is $(5 \pm 1) \cdot 10^{-19} \mu$ [or $(4 \pm 1) \cdot 10^{-19}$ eV] per unit dose. In order to estimate the correctness of this value of the shift (a large error could be due to, for example, the dispersion of the optical instrument being too small for the rapid change of the absorption near the edge or due to the slit being too wide for such measurements), we measured the temperature shift $\Delta_t$ for irradiated and unirradiated samples (cf., for example, Fig. 13). We found values of $(5.5 \pm 0.5) \cdot 10^{-4} \mu$/deg or $-(5.0 \pm 0.5) \cdot 10^{-4}$ eV/deg, which were in agreement, within the limits of the experimental error, with the published value of $-(4.0 \pm 0.5) \cdot 10^{-4}$ eV/deg.

The shift of the absorption edge in neutron-irradiated silicon has been observed by Fan et al. [5]. They point out that, up to the limit of the wavelength scale used in their measurements ($0.82 \mu$), the apparent shift of the absorption edge is proportional to the integral flux of the bombarding particles (about $2 \cdot 10^{-18}$ eV per neutron/cm$^2$) and tends to saturation at doses exceeding $5 \cdot 10^{16}$ neutrons/cm$^2$. The same conclusion has been reached by Plotnikov [73]. These results disagree quantitatively with those reported in the present paper. Judging by the published data of Fan et al., a study of the absorption near the edge of irradiated silicon was carried out, as in our experiments, on samples which were far too thick for such measurements, i.e., when the values of the absorption coefficient corresponding to the indirect transition range were small. Therefore, it was not clear whether the shift was observed in the direct transition region or was associated with a reduction in the effective width of the forbidden band.

One of the possible reasons for the shift of the absorption edge may be a change in the forbidden band width due to changes in the lattice constant caused by irradiation. According to Moss [63], a change in the lattice constant of silicon $\Delta a$ is related to a corresponding change in the effective width of the forbidden band ($\Delta \varepsilon_0$) by $\Delta a = 0.81 \Delta \varepsilon_0$ and hence the shift of the edge by 0.1 eV should correspond to $\Delta a = 0.08$ Å. Such a change in the lattice constant can be easily measured by x-ray structure analysis. However, analysis of the Debye patterns of strongly irradiated silicon samples ($\Phi = 5 \cdot 10^{19}$ neutrons/cm$^2$) shows no change of the lattice constant of silicon greater than a thousandth of an angstrom (cf. Chapter II). Consequently, the observed shift of the edge in the case of irradiated silicon cannot be explained by a change in the lattice constant.

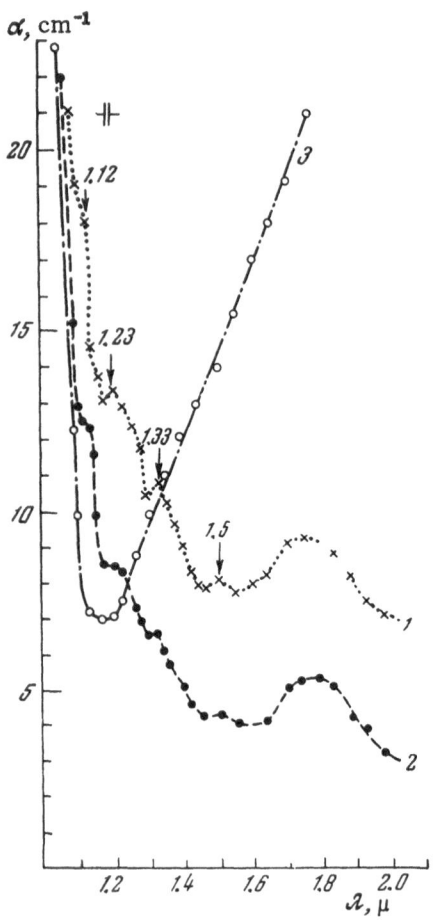

Fig. 16. Room-temperature absorption spectra in the region of 1-2 μ obtained for irradiated p-type silicon with various concentrations of the main impurity: 1) $N_{imp} = 2 \cdot 10^{18}$ boron atoms/cm$^3$, $\Phi = 10^{18}$ neutrons/cm$^2$, $\varepsilon_f = \varepsilon_v + 0.25$ eV; 2) $N_{imp} = 1 \cdot 10^{17}$ boron atoms/cm$^3$, $\Phi = 10^{18}$ neutrons/cm$^2$, $\varepsilon_f = \varepsilon_v + 0.40$ eV; 3) unirradiated silicon with $N_{imp} = 9 \cdot 10^{17}$ arsenic atoms/cm$^3$, 90% compensated with boron.

In our case, the edge shift in irradiated silicon could have been partly due to impurities, since the investigated samples had a fairly high concentration of the main impurity: $10^{17}$-$10^{19}$ cm$^{-3}$. At such concentrations, the effective width of the forbidden band may change due to a change in the energy-band curvature [79] and to the merging of the impurity bands with the edges of the conduction or valence bands [80]. However, a change in the energy-band curvature at high concentrations of the main impurity ($N_{imp}$) should shift the edge in the direction of short wavelengths and, for $N_{imp} = 10^{20}$ cm$^{-3}$, the shift should not exceed 0.05 eV in the case of n-type crystals and −0.03 eV in the case of p-type crystals [79], i.e., this type of shift cannot explain the observed shift in the direction of long wavelengths.

At impurity concentrations of $10^{18}$ cm$^{-3}$, impurity bands are formed out of discrete impurity energy levels in silicon [80], and, in the case of sufficiently high impurity concentrations, these bands may merge with the conduction and valence band edges, which reduces the forbidden band width.

In the case of unirradiated crystals with a high concentration of impurities, optical investigations in the infrared region near the fundamental absorption edge are difficult because of the strong absorption by free carriers.

In irradiated crystals, free carriers are localized at radiation defects, a crystal becomes more transparent, and the behavior of the absorption near the edge can be determined.

Electron transitions from the valence band to donor levels (or from acceptor levels to the conduction band) may give rise to a step in the absorption spectrum near the fundamental absorption edge. Since the ionization energies of the main impurities in silicon are of the order of 0.05 eV, the corresponding step may be observed near 1.1 μ. We observed such steps in some spectra of irradiated samples and we shall discuss them later. No detailed investigations of these steps were carried out, since such investigations would have required better resolution, higher dispersion, narrower slits, an especially strong source of light, etc., i.e., special experiments would have to have been carried out which were outside the scope of our study.

b)  A b s o r p t i o n   B a n d   N e a r   t h e   F u n d a m e n t a l   A b s o r p t i o n   E d g e.  When the fundamental absorption edge of irradiated silicon was investigated with a spectrometer fitted with an F-1 glass prism, which provided better resolution near the edge (about 1 μ) than LiF and NaCl prisms, a series of absorption bands was observed in the 1- to 1.7-μ region for irradiated n- and p-type silicon (Fig. 15). It was difficult to obtain numerical values of the characteristics of these bands because, on the one hand, they were very close to the fundamental absorption edge and were, therefore, difficult to separate from the edge; and, on the other hand, at high radiation doses these bands were overlapped by the short-wavelength wing of the strong band at 1.8 μ. In order to reduce the influence of the 1.8-μ band, we investigated the spectra in the region of 1-1.7 μ using several neutron-irradiated n- and p-type silicon samples whose 1.8-$\mu$ band was relatively weak. Figure 16 includes, for comparison, the spectrum of unirradiated silicon doped with $N_{imp} = 9 \cdot 10^{17}$ cm$^{-3}$ of arsenic 90% compensated with boron. The spectrum of this sample had no bands in the 1.1- to 1.7-$\mu$ region.

The room-temperature spectra of irradiated samples had weak bands in the form of steps at 1.12 $\mu$ (1.11 eV), 1.23 $\mu$ (1.01 eV), 1.33 $\mu$ (0.93 eV), 1.50 $\mu$ (0.83 eV), and 1.64 $\mu$ (0.75 eV). In principle, these bands could have been associated with electron transitions from the valence band to levels lying in the upper half of the forbidden band and separated by suitable gaps from the ceiling of the valence band. If we assume that the effective width of the forbidden band is $\varepsilon_0 = 1.18 \pm 0.02$ eV (according to our measurements, the fundamental absorption edge of the samples which exhibit bands in the region of 1-1.7 $\mu$ lies at $\lambda = 1.05 \pm 0.02$ $\mu$), these transitions should take place to the following levels: $\varepsilon_c - 0.07$ eV, $\varepsilon_c - 0.17$ eV, $\varepsilon_c - 0.25$ eV, $\varepsilon_c - 0.35$ eV, and $\varepsilon_c - 0.43$ eV, respectively. However, in such a case, the corresponding absorption can be observed only when these levels are free of electrons, i.e., when the Fermi level lies above them. We also observed these bands in n-type silicon in which the Fermi level after irradiation lay near $\varepsilon_c - 0.20$ eV, which meant that the 1.33-, 1.50-, and 1.65-$\mu$ bands could not be associated with transitions from the valence band to corresponding local levels. They must have been due to electron transitions to the conduction band from local levels at $\varepsilon_v + 0.25$ eV, $\varepsilon_v + 0.35$ eV, and $\varepsilon_v + 0.43$ eV, respectively. This interpretation is supported by the following results. First, the absorption at 1.33 and 1.50 $\mu$ is observed not only in n-type samples but also in high-resistivity p-type silicon samples whose Fermi level lies near $\varepsilon_v + 0.40$ eV (Fig. 16); the 1.65-$\mu$ band is also observed in irradiated p-type silicon whose Fermi level lies above $\varepsilon_v + 0.45$ eV. Secondly, levels at $\varepsilon_v + 0.35 \pm 0.03$ eV and $\varepsilon_v + 0.45 \pm 0.03$ eV have been reported in [42]. The $\varepsilon_v + 0.25 \pm 0.03$ eV level may be identified with the $\varepsilon_v + 0.22$ eV level deduced by us from the 5.6-$\mu$ band.

The bands at 1.12 and 1.23 $\mu$ were more difficult to interpret because we did not have irradiated samples with the Fermi level below $\varepsilon_v + 0.07$ eV, i.e., near the ceiling of the valence band, or above $\varepsilon_c - 0.07$ eV, i.e., very close to the bottom of the conduction band. The absorption at 1.12 $\mu$ could be associated with a level at $\varepsilon_v + 0.07$ eV or with a level at $\varepsilon_c - 0.07$ eV.

The values of the wavelengths of the band maxima which appear in the 1.1- to 1.7-$\mu$ region were determined with an accuracy of $\pm 0.01$ $\mu$, which corresponded to an error of 0.01 eV in the energy. If we take into account the error in the value of the forbidden band width (0.02 eV), we find that the positions of the corresponding levels with respect to the ceiling of the forbidden band are known with an accuracy of $\pm 0.03$ eV.

Generally speaking, shallow levels at the band edges in silicon irradiated with electrons, deuterons, and neutrons can be detected by nonoptical methods (cf. review in [3]). Silicon irradiated with neutrons (cf. Table 1) has only the $\varepsilon_v + 0.05$ eV level, whose generation rate is relatively rapid (about 0.65 cm$^{-1}$). It is possible that the weak band at 1.12 $\mu$ observed by us is associated with transitions from this level because, within the limits of the experimental error, the $\varepsilon_v + 0.05$ eV level coincides with the position of the level found from the 1.12-$\mu$ band. Shallow levels in irradiated silicon are ascribed to simple defects which are very unstable and are partly annealed at room temperature, but which can be generated rapidly [3]. Evidently, in our samples, irradiated with doses of the order of $10^{17}$-$10^{19}$ neutrons/cm$^2$, some of these defects are not destroyed and may give rise to the corresponding absorption.

It is very likely that the 1.12-$\mu$ step is associated with the formation of impurity bands, which we have mentioned earlier. To check the correctness of this hypothesis, it would be necessary to investigate the spectrum of heavily doped compensated silicon and to compare it with the data for irradiated silicon. In the compensated sample whose room-temperature spectrum is given in Fig. 16, the absorption by free carriers is still too strong at 1.12 $\mu$, and since the curve representing the absorption by free carriers and the fundamental absorption edge intersect in this region, we cannot establish whether the 1.12-$\mu$ band does or does not exist in the spectrum of compensated unirradiated silicon.

The absorption band at 1.23 $\mu$ may be associated with electron transitions from the valence band to an $\varepsilon_c - 0.17 \pm 0.01$ eV level or with electron transitions from an $\varepsilon_v + 0.17 \pm 0.01$ eV level to the conduction band. Similar levels, which can be generated relatively rapidly, have already been observed in irradiated silicon. In the case of irradiated n- and p-type crystals, when the Fermi level lies in the forbidden band between $\varepsilon_V + 0.17$ eV and $\varepsilon_C - 0.17$ eV, both transitions are, in principle, equally likely.

As already mentioned, the $\varepsilon_C - 0.17$ eV level observed in electron-irradiated silicon is ascribed to defects which contain oxygen (A centers). Therefore, if the 1.23-$\mu$ band of neutron-irradiated silicon is due to the presence of the $\varepsilon_C - 0.17 \pm 0.01$ eV level, associated with defects which also include oxygen, then this band should be absent or should be considerably weaker in the case of silicon prepared by floating vertical-zone melting ("zone-melted" silicon). To solve this problem and to determine the dependence of the 1.8-$\mu$ band on the presence of oxygen (to check whether this band is associated with radiation defects which include oxygen atoms), we investigated the spectrum of irradiated zone-melted silicon in the 1- to 5-$\mu$ range.

c) Results of an Investigation of Zone-Melted Silicon. We investigated the 1- to 5-$\mu$ region of the spectrum of zone-melted n-type silicon containing less than $10^{16}$ atoms/cm$^3$ of oxygen and doped with $10^{18}$ atoms/cm$^3$ of phosphorus, which was irradiated with a fast-neutron dose of approximately $10^{18}$ neutrons/cm$^2$. After irradiation, the Fermi level at room temperature was within the forbidden band, separated by about 0.22 eV from the bottom of the conduction band. For comparison, we recorded, under the same conditions, the spectrum of irradiated silicon prepared by pulling from the melt in a quartz crucible ("pulled" silicon) containing about $3 \cdot 10^{17}$ atoms/cm$^3$ of oxygen but with approximately the same phosphorus content as that of zone-melted silicon. In this case, the Fermi level in silicon subjected to the same radiation dose as that used for the zone-melted silicon lay near 0.23 eV from the bottom of the conduction band. The spectra were recorded at room and lower temperatures using LiF and glass F-1 prisms. At room temperature, the spectra recorded with an LiF prism were practically the same for both samples. At low temperatures, the pulled silicon exhibited bands at 1.8 and 1.23 $\mu$; in the zone-melted silicon these bands did not appear even at helium temperature. The transmission curves of these two samples, recorded at helium temperature, are shown in Fig. 17.

An investigation of the spectrum of the zone-melted silicon using F-1 prisms revealed absorption bands at the fundamental absorption edge which were also observed in the spectra of the pulled silicon samples: 1.12, 1.33, 1.5, and 1.64 $\mu$. The 1.33-, 1.5-, and 1.64-$\mu$ bands of the zone-melted silicon were considerably stronger than the corresponding bands of the pulled silicon. This indicated that the $\varepsilon_V + 0.25$ eV, $\varepsilon_V + 0.35$ eV, and $\varepsilon_V + 0.43$ eV levels with which these bands were associated were generated more rapidly in the zone-melted silicon. The 1.23-$\mu$ band was weaker in the zone-melted silicon than in the pulled material. Hence, we concluded, in agreement with our earlier assumptions, that the 1.23-$\mu$ band could have been due to electron transitions from the valence band to the $\varepsilon_C - 0.17$ eV level or transitions from the $\varepsilon_V + 0.17$ eV level to the conduction band. If the level $\varepsilon_C - 0.17$ eV is associated with defects each of which includes an atom of oxygen, then there should be fewer such defects, and consequently fewer $\varepsilon_C - 0.17$ eV levels in the zone-refined silicon, only one form of transition (from the $\varepsilon_V + 0.17$ eV to the conduction band) should predominate, and the 1.23-$\mu$ band should be weaker than it is in the pulled material or it may not be observed at all. Unfortunately, a wide range of irradiated zone-melted silicon samples with various dopant concentrations was not available, and, therefore, we were unable to follow the behavior of all the observed absorption bands in the infrared region as a function of the oxygen content.

The results obtained were far too few to draw general conclusions about the dependence of the infrared absorption spectrum of irradiated silicon on the oxygen content. However, the results did indicate that the 1.23- and 1.8-$\mu$ bands were due to radiation defects which were associated with oxygen impurities in silicon.

§ 3. Infrared Absorption of Irradiated Silicon in the 8- to 14-$\mu$ Range

a) Absorption Associated with Lattice Vibrations. In the 8- to 14-$\mu$ range, the infrared absorption spectra of irradiated and unirradiated silicon single crystals have absorption bands due to lattice atom vibrations [63, 81] (cf., for example, Figs. 6, 7, 15). Near 9 $\mu$, a band associated with the lattice vibrations overlaps a band due to the vibrations of oxygen atoms in silicon. The 9-$\mu$ band will be considered separately.

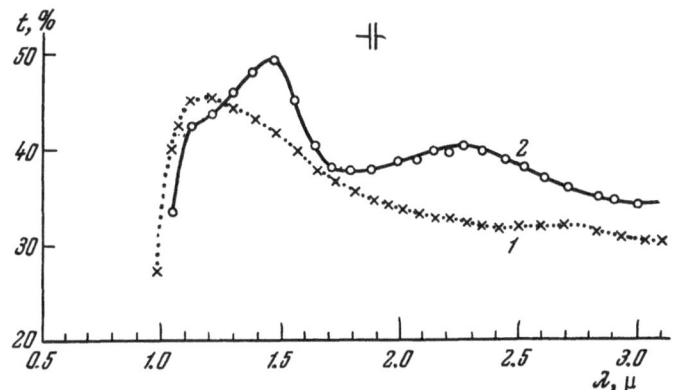

Fig. 17. Helium-temperature transmission spectra of irradiated n-type silicon containing various amounts of oxygen ($N_{imp} = 1 \cdot 10^{18}$ atoms/cm³, $\Phi = 1 \cdot 10^{18}$ neutrons/cm², d = 0.1 cm): 1) $N_{ox} < 10^{16}$ atoms/cm³; 2) $N_{ox} = 5 \cdot 10^{17}$ atoms/cm³.

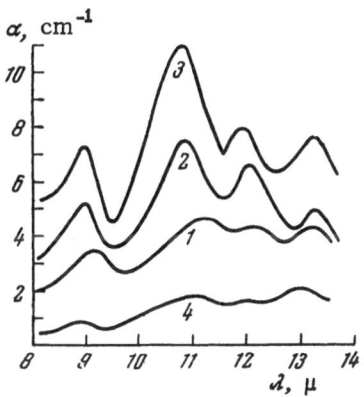

Fig. 18. Room-temperature absorption bands due to the lattice vibrations in unirradiated silicon and in irradiated n-type silicon ($\Phi = 5 \cdot 10^{19}$ neutrons/cm²) with various concentrations of the main impurity (arsenic): 1) $N_{imp} = 4 \cdot 10^{17}$; 2) $2 \cdot 10^{18}$; 3) $2 \cdot 10^{19}$ atoms/cm³; curve 4 represents unirradiated silicon, $N_{imp} = 4 \cdot 10^{17}$ atoms/cm³.

In covalent crystals such as silicon, the optical vibration mode should be inactive in the infrared region because such crystals do not have, in principle, an electric dipole moment.

However, the spectra of pure silicon crystals have absorption bands associated with the lattice vibrations which are due to an electric dipole moment of the second order [82].

Figure 18 shows the room-temperature absorption spectra in the 8- to 14-$\mu$ range for three n-type silicon samples differing from one another in the concentration of the main impurity (arsenic) but irradiated with the same neutron dose: $5 \cdot 10^{19}$ neutrons per cm². The same figure includes the absorption spectrum of an unirradiated pure silicon sample (a zone-melted sample with $\rho_0 = 200 \ \Omega \cdot cm$). It follows from Fig. 18 that the absorption in the 8- to 14-$\mu$ range is stronger in irradiated than in unirradiated silicon and that the value of the absorption coefficient increases as the main impurity concentration increases in the original material. This can be explained as follows: The lattice absorption in silicon crystals is due to a second-order dipole moment; it is very weak in pure crystals. Lattice defects and impurities can deform a crystal and the electron shells of silicon atoms, which may intensify the components of the second-order dipole moment and may give rise to first-order components [81]. This effect is of intrinsic interest and has recently been used in a detailed investigation and verification of the theory of vibrations of the covalent lattice carried out by Balkanski and Nasarewicz [83, 84], who observed, in silicon irradiated with a dose of $10^{19}$ neutrons/cm², absorption bands in the range 40-130 $\mu$ due to a first-order dipole moment.

Investigation of the vibrations of a lattice with defects is particularly easy in the case of irradiated silicon because irradiation considerably reduces the absorption by free carriers, and this makes it easier to observe the relevant absorption bands.

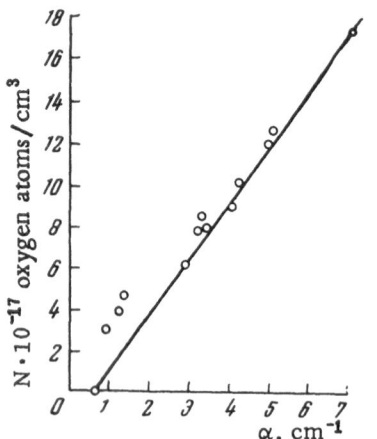

Fig. 19. Dependence of the absorption coefficient at the 9-μ band maximum on the concentration of oxygen atoms [87].

Fan and Collins [81] investigated the lattice vibration
spectrum in the range 7-35 μ using pure silicon and silicon
with a main impurity concentration of $3 \cdot 10^{18}$ atoms /cm$^3$
and the same concentration of defects (generated by nucleon
bombardment). According to their data, the intensity of the
absorption bands due to the lattice vibrations is independent
of the impurity concentration and irradiation. This conclusion of Fan and Collins disagrees with the results obtained in
the present investigation.

b) Absorption Associated with Oxygen
Atoms. As already mentioned, the absorption spectra of
silicon in the infrared region near 9 μ include a band associated with oxygen atoms. Kaiser et al. [85-87] have
shown that the 9-μ band is due to angular vibrations of the
valence bond between an interstitial oxygen atom and a
silicon atom at a lattice site. This band can be seen clearly in the spectra of the pulled silicon whose oxygen content
is $10^{17}$-$10^{18}$ atoms/cm$^3$, but is practically absent in silicon
crystals containing $<10^{16}$ atoms/cm$^3$ of oxygen (the zone-
melted silicon).

In the absence of free-carrier absorption near 9 μ, the lattice vibrations give rise to an absorption of 0.7-
0.8 cm$^{-1}$. Therefore, this absorption is treated as a background and the absorption associated with oxygen atoms
is measured from the level of this background. For the pulled silicon, the total absorption coefficient of the 9-
μ band, i.e., without allowance for the lattice background, lies between 1 and 6 cm$^{-1}$, depending on the concentration of oxygen atoms. Knowing the absorption coefficient of the 9-μ band for a given sample, we can
determine the concentration of oxygen in it using a graph of the dependence of the absorption coefficient at
the 9-μ band maximum on the number of oxygen atoms per cm$^3$. This graph has been plotted by Kaiser and
Keck [87] by comparing the absorption coefficient at the maximum of the 9-μ absorption band with the results
of an analysis of the oxygen gas evolved when the investigated sample was melted in vacuum; this graph is
presented in Fig. 19.

The value of the total absorption coefficient at the 9-μ maximum was within the range 1-3 cm$^{-1}$ for the
majority of our samples, and this value corresponded to oxygen concentrations of $1 \cdot 10^{17}$-$3 \cdot 10^{17}$ atoms/cm$^3$.

We found that the spectra of n- and p-type samples irradiated with a dose of the order of $10^{17}$ neutrons
per cm$^2$ had a band at 9.5 μ which was stronger than the band at 9 μ and which disappeared when the dose
was increased. In particular, when the dose was $6 \cdot 10^{17}$ neutrons/cm$^2$ or more, the 9.5-μ band was not observed, but it was seen in unirradiated silicon samples. The 9.5-μ band has been investigated quite thoroughly
by other workers. It is attributed to the vibrations of the bond between the oxygen and the silicon in the SiO$_4$
complexes, which are present in high concentrations in heat-treated silicon [88]. Figure 20 shows a region of
the absorption spectrum near 9 μ for two p-type silicon samples cut from the same single crystal which were
subjected to different doses: $3 \cdot 10^{17}$ and $6 \cdot 10^{17}$ neutrons/cm$^2$. The 9.5-μ band is observed for a sample irradiated with a dose of $3 \cdot 10^{17}$ neutrons/cm$^2$ but not in the spectrum of a sample irradiated with $6 \cdot 10^{17}$
neutrons/cm$^2$. The same figure shows the 9.5-μ band of an unirradiated sample.

When the radiation dose is increased and the 9.5-μ band disappears, the intensity of the 9-μ band begins to decrease. By way of example, Fig. 21 presents the 9-μ absorption band for three n-type silicon samples
cut from the same single crystal and irradiated with different fast-neutron doses. After irradiation with a dose of
$3 \cdot 10^{17}$ neutrons/cm$^2$, a certain number of free carriers still remained in the sample and this gave rise to an
additional absorption in the infrared region considered. Therefore, the total absorption at 9 μ for this dose was
higher than that after irradiation with doses of $6 \cdot 10^{17}$ and $9 \cdot 10^{17}$ neutrons/cm$^2$. The 9.5-μ band of the sample
subjected to the smallest dose (curve 1 in Fig. 21) had an absorption coefficient of 0.9 cm$^{-1}$ at the maximum

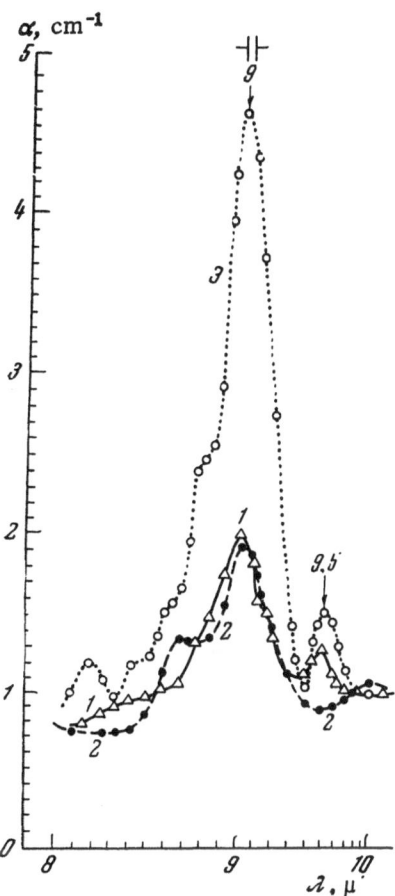

Fig. 20. Absorption in the region of the 9-μ band of p-type silicon irradiated with doses of (1) Φ = $3 \cdot 10^{17}$ and (2) $6 \cdot 10^{17}$ neutrons/cm², and the corresponding absorption in (3) high-resistivity unirradiated silicon.

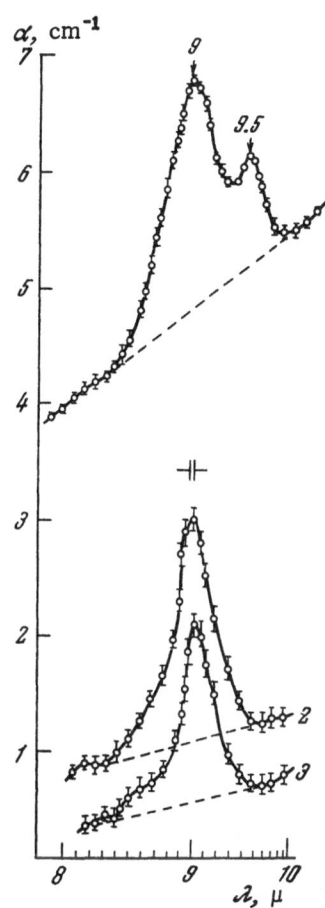

Fig. 21. Absorption in the region of 9 μ for n-type silicon irradiated with various doses: (1) Φ = $3 \cdot 10^{17}$, (2) $6 \cdot 10^{17}$, and (3) $9 \cdot 10^{17}$ neutrons/cm².

(measured from the background due to the absorption by free carriers), while the absorption coefficient at the maximum of the 9-μ band was, in this case, 1.95 ± 0.05 cm⁻¹. The application of the intermediate dose (curve 2) destroyed the 9.5-μ band and somewhat reduced the 9-μ band absorption, whose absorption coefficient at the maximum became 1.90 ± 0.05 cm⁻¹. After the largest dose (curve 3), the 9.5-μ band was again absent and the absorption coefficient at the maximum of the 9-μ band became 1.50 ± 0.05 cm⁻¹. We also obtained the spectrum of a sample which had been cut from the same single crystal as the samples whose absorption spectra near 9 μ are presented in Fig. 21; this sample was irradiated with a relatively large dose (5.2 · 10¹⁷ neutrons/cm²). The lattice absorption of 0.78 cm⁻¹ was observed, but not the 9-μ or 9.5-μ bands.

The results obtained are presented in the form of a graph of the dependence of the absorption coefficient at the 9- and 9.5-μ band maxima on the radiation dose. They can be used to determine the role of oxygen in the formation of radiation defects in silicon irradiated with neutrons [89].

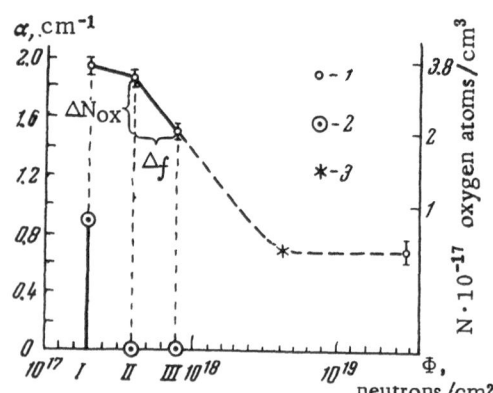

Fig. 22. Dependence of the absorption at the 9-
and 9.5-μ band maxima on the radiation dose:
1) experimental points for 9 μ; 2) experimental
points for 9.5 μ; 3) calculated points for 9 μ.

§4. Role of Oxygen in the Formation of
Radiation Defects in Neutron-Bombarded
Silicon. Determination of the Rate of
Generation of A Centers in Neutron-
Irradiated Silicon

If oxygen atoms take part in the formation of radiation defects, then an increase in the concentration of
such defects (we shall call them A centers by analogy
with the centers observed in electron-bombarded silicon)
when the radiation dose is increased should be accompanied by a decrease in the concentration of interstitial
oxygen atoms which give rise to the 9-μ band, because
some oxygen atoms are used to form the A centers. The
associated reduction in the absorption at 9 μ can be used
to determine the number of such centers per one incident
neutron, i.e., the rate of generation of these centers.

It follows from Fig. 22 that the weakening of the
9-μ band is nonuniform. The weakening is very slight
as long as the 9.5-μ band associated with the oxygen in the $SiO_4$ complexes still exists. This may be explained as follows: The binding of the silicon to the oxygen atoms in $SiO_4$ complexes is weaker than the binding of interstitial oxygen to silicon (the former bonds are easier to break and the A centers are formed primarily from oxygen atoms yielded by such complexes). Interstitial oxygen atoms begin to take part in the formation of the A centers (and the corresponding absorption at 9 μ begins to decrease) only when all the oxygen in
the complexes is exhausted. The rate of generation of the A centers can be determined from the region of the
curve which represents this process. Knowing the dose increment $\Delta\Phi$ which results in a decrease in the ab-
sorption coefficient by $\Delta\alpha$, we can find $\Delta\alpha$ from Fig. 19 and thus calculate $\Delta N_{ox}$. In our case, $\Delta N_{ox} = 1.2 \cdot$
$10^{17}$ atoms/cm³ for $\Delta\Phi = 3 \cdot 10^{17}$ neutrons/cm², and hence the rate of generation of the A centers in silicon irradiated with neutrons is $0.4 \pm 0.1$ cm⁻¹. Bearing in mind that the dose is determined with an error up to 30%,
the rate of generation is obtained with the same accuracy. Hence, we can find the radiation dose which can
produce the maximum number of defects in a sample with a given concentration of oxygen atoms. In the
samples investigated by us, the initial oxygen content was $5 \cdot 10^{17}$ atoms/cm³. Consequently, saturation could
be expected at $\Phi \approx 1 \cdot 10^{18}$ neutrons/cm² $(5 \cdot 10^{17}/0.4 = 1.2 \cdot 10^{18})$ when all the interstitial oxygen had become exhausted; at higher doses, only the lattice absorption in the region of 9 μ should be observed, as indeed
it was for a dose of $5 \cdot 10^{19}$ neutrons/cm².

## Conclusions

1. The investigations reported in the present paper have shown that the irradiation of silicon single crystals with reactor neutrons produces infrared absorption bands which are associated with the local levels of radiation defects.

Table 2 lists the wavelengths of the maxima of the observed bands and the corresponding quantum energies hν, as well as the transitions (represented by arrows) responsible for these bands. For comparison, Table 2
also includes the results of Fan and Ramdas [7]. The system of energy levels and electron transitions corresponding to the observed absorption bands is presented schematically in Fig. 23.

Table 2 shows that the 7.2-μ band and the bands in the region of 1.1-1.7 μ have been observed for the
first time in the present investigation. No data on these bands are given in the papers of Fan et al.

TABLE 2. Absorption Bands Due to Radiation Defects in Neutron-Irradiated Silicon

| Results of present study | | | Fan's results [7] | |
|---|---|---|---|---|
| $\lambda$, $\mu$ | $h\nu$, eV | Electron transition, eV | $\lambda$, $\mu$ | Electron transition, eV |
| 7.2 | 0.17 | $\varepsilon_v \to \varepsilon_v + 0{,}17$ | | |
| 5.6 | 0.22 | $\varepsilon_v \to \varepsilon_v + 0{,}22$ | 5.5 | $\varepsilon_c - 0{.}16 \to \varepsilon_c$ |
| 4.0 | 0.31 | $\varepsilon_v \to \varepsilon_v + 0{,}30$ | 3.9 | $\varepsilon_v \to \varepsilon_v + 0{,}27$ |
| 3.5 {3.65 | 0.34 | Transitions from level near | 3.5 | $\varepsilon_c - 0{.}21 \to \varepsilon_c$ |
| 3.45 | 0.36 | middle of forbidden band to | | |
| 3.3 | 0.38 | levels between $\varepsilon_C - 0.16$ | | |
| 3.1 | 0.40 | and $\varepsilon_C - 0.25$ | | |
| 2.9 | 0.43 | | | |
| 1.8 | 0.69 | Excitation of A centers | 1.8 | $\varepsilon_v + 0{.}27 \to \varepsilon_c - 0{,}21$ |
| 1.64 | 0.75 | $\varepsilon_v + 0{,}43 \to \varepsilon_c$ | | |
| 1.50 | 0.83 | $\varepsilon_v + 0{,}35 \to \varepsilon_c$ | | |
| 1.33 | 0.93 | $\varepsilon_v + 0{,}22 \to \varepsilon_c$ | | |
| 1.23 | 1.01 | $\varepsilon_v \to \varepsilon_c - 0{,}17$ | | |
| 1.12 | 1.11 | $\varepsilon_v + 0{,}07 \to \varepsilon_c$ | | |

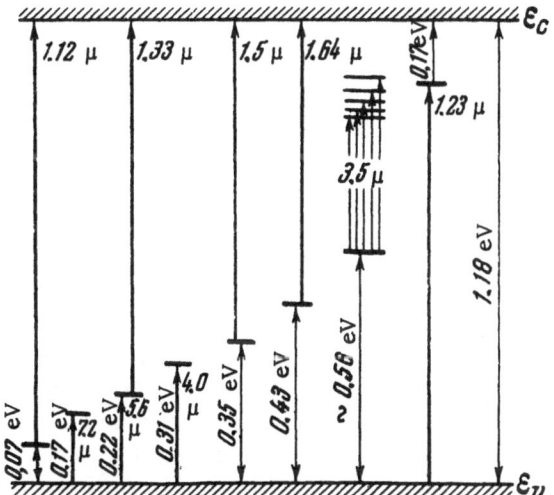

Fig. 23. Scheme of energy levels and electron transitions corresponding to the absorption bands observed in the spectrum of neutron-irradiated silicon.

2. The absorption bands at 7.2, 5.6, and 4 $\mu$ observed for irradiated low-resistivity n-type silicon samples are due to electron transitions from the valence band to levels at $\varepsilon_v + 0.17$ eV, $\varepsilon_v + 0.22$ eV, and $\varepsilon_v + 0.30$ eV, respectively.

A local level near $\varepsilon_v + 0.17 \pm 0.01$ eV had been observed in neutron-irradiated silicon by only one worker (Table 1) and, therefore, there had been doubts about it. Our results, which were obtained independently using a completely different method, confirm the presence of a level near $\varepsilon_v + 0.17$ eV in the forbidden band of irradiated silicon.

The level at $\varepsilon_v + 0.22 \pm 0.01$ eV, deduced by us from an analysis of the absorption band at 5.6 $\mu$ of irradiated low-resistivity p-type silicon and from the 1.33-$\mu$ band of irradiated high-resistivity n- and p-type samples, has not been reported before for neutron-irradiated silicon.

However, a level near $\varepsilon_v + 0.22$ eV has been observed in electron-irradiated silicon by Vavilov et al. [28, 29]. The absorption spectrum of p-type silicon irradiated with fast electrons has not yet been investigated. It would have been interesting to carry out such an investigation and to compare the results with those reported in [28, 29].

A local level at $\varepsilon_v + 0.30 \pm 0.01$ eV has been observed in neutron-irradiated silicon by many workers and there are no doubts about its existence. Our investigation shows that it is associated with stable defects which are destroyed by annealing for several hours at 150°C.

3. Analysis of the absorption band at 3.5 $\mu$, which represents a group of narrow bands with maxima clearly visible at low temperatures, has shown that this band is associated with electron transitions from a local

level near the middle of the forbidden band lying slightly below $\varepsilon_V + 0.58 \pm 0.03$ eV.

Local levels near the middle of the forbidden band in silicon irradiated with large neutron doses have also been observed in several other investigations (Table 1). The results of our investigation have confirmed the existence of these levels.

The experiments on the annealing of irradiated samples exhibiting the 3.5-$\mu$ band show that the group of bands at 3.5 $\mu$ is associated with the same radiation defect, i.e., it is due to electron transitions within the same local center, which may possibly be excited.

Our investigations show that Fan's interpretation of the 3.5-$\mu$ band, which he assigns to electron transitions from a local level at $\varepsilon_C - 0.21$ eV to the conduction band, is incorrect. Therefore, it seems doubtful whether the $\varepsilon_C - 0.21$ eV level deduced by Fan from an analysis of the 3.5-$\mu$ band does exist.

4. The results of our investigation of the 1.8-$\mu$ band and our comparison of the experimental data with Kurskii's theoretical work [74] indicate that this band is associated with the excitation of the A centers, which are radiation defects each including an atom of oxygen. The absorption at 1.8 $\mu$ has the same nature as the absorption by bound excitons.

Analysis of the 1.8-$\mu$ band shows that an investigation of the absorption band parameters (width, profile, etc.) is important in the development of theories which account for the possible absorption processes in semiconducting crystals.

The linear dependence of the absorption coefficient at the maximum of the 1.8-$\mu$ band on the radiation dose makes it possible to use silicon in the dosimetry of neutron fluxes in reactors.

5. Our investigations confirm the existence in neutron-irradiated silicon of local levels near $\varepsilon_V + 0.43$ eV and $\varepsilon_V + 0.35$ eV, which have been reported earlier in [40], as well as a shallow level at $\varepsilon_V + 0.05$ eV (according to our data, $\varepsilon_V + 0.07 \pm 0.03$ eV), reported in [3].

Thus, an analysis of our data and of the published results (Table 1) shows that the existence of the following local levels in neutron-irradiated silicon can be regarded as definitely established: $\varepsilon_V + 0.07$ eV, $\varepsilon_V + 0.17$ eV, $\varepsilon_V + 0.22$ eV, $\varepsilon_V + 0.30$ eV, $\varepsilon_V + 0.35$ eV, $\varepsilon_V + 0.43$ eV, $\varepsilon_C - 0.17$ eV; in addition, there are levels near the middle of the forbidden band.

6. Our investigation shows that the shift of the fundamental absorption edge in the direction of long wavelengths observed in the infrared spectra of irradiated silicon cannot be due solely to a change in the effective forbidden band width. The shift of the rapid rise of the absorption at 1 $\mu$, which looks like the shift of the fundamental absorption edge, may be due to the presence of absorption bands of irradiated silicon at 1.1-1.7 $\mu$; these bands are reported for the first time in the present paper. This can also account for the quantitative difference between the values of the edge shift obtained in the present study and those reported in [5, 7, 73], since electron transitions giving rise to the absorption band in the region of 1.1-1.7 $\mu$ have not been observed.

To finally solve the problem of the edge shift in irradiated silicon, it would be necessary to investigate the spectrum in the region of high absorption and shorter wavelengths, i.e., in the region where direct transitions begin.

7. Investigations of the absorption band at 9 $\mu$ in irradiated silicon have shown that oxygen atoms can take part in the formation of radiation defects. The rate of formation of centers which include oxygen atoms, determined from the reduction in the intensity of the 9-$\mu$ absorption band maximum as the radiation dose is increased, is 0.4 cm$^{-1}$.

8. Our study demonstrates that the absorption associated with the lattice vibrations becomes much stronger in irradiated silicon. The greater the radiation dose and the higher the dopant concentration, the higher is the intensity of this absorption. This means that radiation defects, like impurities, can deform the electron shells of silicon atoms, and this may result in the appearance, in covalent silicon crystals, of a first-order dipole moment and intensification of the second-order components.

The investigation of this problem is itself of intrinsic interest.

9. Investigation of the infrared spectra of irradiated silicon shows that such irradiation can be used as a method of increasing the transparency so that optical methods can be applied to study heavily doped silicon samples (lattice vibrations, formation of impurity bands, etc.) in such samples.

10. The investigation carried out shows that irradiated silicon slabs can, because of their strong absorption in the region of 1.8 and 3.5 $\mu$, be used as filters which remove the whole visible part of the spectrum and transmit infrared radiation with a transmission edge in the region of 2-4 $\mu$. Filters with such characteristics are very useful in infrared technology in order to suppress the interference due to solar radiation.

In conclusion, the author expresses her gratitude to N. N. Sobolev for directing this investigation and for his constant interest, and to V. S. Vavilov for suggesting the subject and for his interest.

The author thanks N. A. Penin, L. V. Keldysh, and Yu. A. Kurskii for valuable discussions of the results obtained.

## Literature Cited

1. K. Lark-Horovitz, Phys. Rev., 73:1256 (1948).
2. W. E. Jonson and K. Lark-Horovitz, Phys. Rev., 76:442 (1949).
3. C. A. Klein, J. Appl. Phys., 30:1222 (1959).
4. K. Lark-Horovitz, M. Becker, R. Davis, and H. Y. Fan, Bull. Am. Phys. Soc., 25:29 (1959).
5. M. Becker, H. Y. Fan, and K. Lark-Horovitz, Phys. Rev., 85:730 (1952).
6. V. S. Vavilov, A. F. Plotnikov, and G. V. Zakhvatkin, Fiz. tverd. tela, 1:976 (1958).
7. H. Y. Fan and A. K. Ramdas, J. Appl. Phys., 30:1127 (1959).
8. J. W. Glen, Advance Phys., 4:318 (1955).
9. The Effects of Radiation on Materials (ed. by I. I. Harwood). Reinhold, New York (1959).
10. G. J. Dienes and G. H. Vineyard, Radiation Effects in Solids. Interscience Publ., New York (1957).
11. G. H. Kinchin and R. S. Pease, Rep. Progr. Phys., 18:1 (1955).
12. F. Seitz and J. S. Koehler, Solid State Phys., 2:307 (1956).
13. W. S. Snyder and J. Neufeld, Phys. Rev., 97:1636 (1955); 99:1326 (1955); 103:862 (1956).
14. W. A. Harrison and F. Seitz, Phys. Rev., A98:1530 (1955).
15. M. Walt and H. H. Barschall, Phys. Rev., 93:1062 (1954).
16. H. Feshbach and V. F. Weisskopf, Phys. Rev., 76:1550 (1949).
17. F. Seitz, Disc. Faraday Soc., 5:271 (1949).
18. H. B. Huntington, Phys. Rev., 93:1414 (1954).
19. B. E. Watt, Phys. Rev., 87:1037 (1953).
20. J. J. Loferski and P. Rappaport, Phys. Rev., 98:1861 (1955); 100:1261A (1955).
21. G. N. Galkin, Author's Abstract of Dissertation. Physicotechnical Institute, Academy of Sciences of the USSR, Leningrad (1962).
22. G. Wertheim, Phys. Rev., 105:1730 (1957); 110:1272 (1958).
23. G. Wertheim, J. Appl. Phys., 30:1195 (1959).
24. G. Wertheim and D. N. E. Buchnan, J. Appl. Phys., 30:1232 (1959).
25. D. Hill, Phys. Rev., 114:1414 (1959).
26. G. N. Galkin, N. S. Rytova, and V. S. Vavilov, Fiz. tverd. tela, 2:9 (1960).
27. G. N. Galkin, Fiz. tverd. tela, 3:630 (1961).
28. V. M. Malovetskaya, G. N. Galkin, and V. S. Vavilov, Fiz. tverd. tela, 4:1372 (1962).
29. V. S. Vavilov, G. N. Galkin, V. M. Malovetskaya, and A. F. Plotnikov, Fiz. tverd. tela, 4:1969 (1962).
30. H. Saito and M. Hirate, Japan. J. Appl. Phys., 2:678 (1963).
31. N. A. Vitkovskii, G. P. Lukirskii, G. V. Mashovets, and V. I. Myakota, Fiz. tverd. tela, 4:1140 (1962).
32. G. Bemski, B. Szymanski, and K. Wright, J. Phys. Chem. Solids, 24:1 (1963).

33. G. Watkins, J. Corbett, and R. Walker, J. Appl. Phys., 30:1198 (1959).
34. J. H. Corbett, G. D. Watkins, R. M. Chrenko, and R. S. McDonald, Phys. Rev., 121:1013 (1961).
35. G. Bemski, J. Appl. Phys., 30:1195 (1959).
36. J. A. Baicker, Phys. Rev., 129:1174 (1963).
37. E. Sonder and J. L. Templeton, J. Appl. Phys., 31:1279 (1960).
38. E. Sonder, J. Appl. Phys., 30:1186 (1959).
39. V. S. Vavilov and A. F. Plotnikov, Fiz. tverd. tela, 3:2455 (1961).
40. A. F. Plotnikov, V. S. Vavilov, and L. S. Smirnov, Fiz. tverd. tela, 3:3253 (1961).
41. C. A. Klein and W. D. Straub, Bull. Am. Phys. Soc., Ser. II, 3:375 (1958).
42. G. Wertheim, Phys. Rev., 111:1500 (1958).
43. T. A. Longo, Thesis. Purdue Univ. (1957).
44. W. G. Spitzer and H. Y. Fan, Phys. Rev., 109:1011 (1958).
45. É. N. Lotkova, V. S. Vavilov, and N. N. Sobolev, Physical Problems in Spectroscopy, Vol. 2. (1963), p. 224.
46. V. S. Vavilov, E. N. Lotkova, and A. F. Plotnikov, J. Phys. Chem. Solids, 22:31 (1961).
47. M. Nisenoff and H. Y. Fan, Phys. Rev., 128:1605 (1962).
48. W. Jung and G. S. Newell, Phys. Rev., 132:648 (1963).
49. H. Y. Fan, Rep. Progr. Phys., 19:107 (1956).
50. V. I. Stepanov and Ya. S. Khvashchevskaya, Opt. i spektroskopiya, 5:394 (1958).
51. É. N. Lotkova and A. M. Shustov, Pribory i tekhn. éksp., 6:133 (1962).
52. É. N. Lotkova and A. M. Shustov, Cryog., 2:41 (1964).
53. V. I. Malyshev, M. N. Markov, and A. A. Shubin, Doklady Akad. Nauk SSSR, 86:273 (1962).
54. É. N. Lotkova, V. S. Vavilov, and N. N. Sobolev, Opt. i spektroskopiya, 13:216 (1962).
55. É. N. Lotkova and A. B. Fradkov, Pribory i tekhn. éksp., 1:188 (1961).
56. É. N. Lotkova and A. B. Fradkov, Cryog., 6:238 (1961).
57. H. O. McManon, R. M. Haines, and G. H. King, J. Opt. Soc. Am., 39:786 (1949).
58. V. Roberts, J. Sci. Instr., 31:251 (1954); 32:294 (1955).
59. V. P. Babenko, V. L. Broude, and V. S. Medvedev, Pribory i tekhn. éksp., 3:99 (1956).
60. W. Duerig and J. Mador, Rev. Sci. Instr., 23:421 (1952).
61. L. J. Schoen, L. E. Kuentzel, and H. P. Broida, Rev. Sci. Instr., 29:633 (1958).
62. F. E. Geiger, Rev. Sci. Instr., 26:383 (1955).
63. T. S. Moss, Optical Properties of Semiconductors, 1959, [Russian translation, IL (1961)].
64. C. D. Salzberg and I. J. Villa, J. Opt. Soc. Am., 47:244 (1957).
65. E. Antoncik, Czech. J. Phys., 6:209 (1956).
66. H. Lukes, Czech. J. Phys., 10:317 (1960).
67. F. Oswald, Optics, 16:527 (1959).
68. A. F. Ioffe, Physics of Semiconductors. Izd. Akad. Nauk SSSR, Moscow-Leningrad (1957).
69. S. T. Konobeevskii and F. P. Butra, Atomnaya energiya, 5:572 (1958).
70. K. Huang, Proc. Roy. Soc. (London), A190:102 (1947).
71. G. Picus, E. Burstein, and B. Henvis, J. Phys. Chem. Solids, 1:75 (1956).
72. É. N. Lotkova, Fiz. tverd. tela, 6:1905 (1964).
73. A. F. Plotnikov, Dissertation. P. N. Lebedev Physics Institute, Moscow (1962).
74. Yu. A. Kurskii, Fiz. tverd. tela, 6:2263 (1964).
75. J. R. Haynes, Phys. Rev. Letters, 4:361 (1960).
76. D. G. Thomas and J. J. Hopfield, Phys. Rev., 128:2135 (1962).
77. A. S. Davydov, Uspekhi fiz. nauk, 82:393 (1964).
78. S. I. Pekar, Investigations in the Electron Theory of Crystals. Gostekhizdat (1961).
79. G. B. Dubrovskii and V. K. Subashiev, Fiz. tverd. tela, 4:3018 (1962).
80. E. Conwell, Phys. Rev., 103:51 (1956).
81. R. J. Collins and H. Y. Fan, Phys. Rev., 93:674 (1954).
82. M. Lax and E. Burstein, Phys. Rev., 97:39 (1955).

83.  M. Balkanski and W. Nasarewicz, J. Phys. Chem. Solids, 23:573 (1962).

84.  M. Balkanski, W. Nasarewicz, and E. Da Silva, Reprint.

85.  H. J. Hrostowski and R. H. Kaiser, Bull. Am. Phys. Soc., 1:295 (1956); Phys. Rev., 107:966 (1957).

86.  W. Kaiser, P. H. Keck, and C. F. Longe, Phys. Rev., 101:1264 (1956).

87.  W. Kaiser and P. H. Keck, J. Appl. Phys., 28:882 (1957).

88.  T. Arai, J. Phys. Soc. Japan., 17:246 (1962).

89.  É. N. Lotkova, Fiz. tverd. tela, 6:1559 (1964).

# ELECTRICAL AND OPTICAL PROPERTIES
# OF ELECTROLUMINESCENT CAPACITORS BASED ON
## ZnS:Cu

## Yu. P. Chukova

## Introduction

Electroluminescence is that form of luminescence in which a medium acquires energy directly from an electric field. For a long time, the term electroluminescence was used only for the luminescence of gaseous systems (gas-discharge luminescence); the electroluminescence of solids was only discovered much later. The occurrence of electroluminescence in liquids has not yet been proved satisfactorily. In the present paper, we shall consider only the electroluminescence of solids.

The luminescence of solids in an electric field was first observed by Losev [1]. The luminescence appeared in the region of a rectifying contact on a silicon carbide crystal. In 1936, Destriau [2] discovered the luminescence of zinc sulfide suspended in a liquid dielectric and placed between the electrodes of a capacitor to which an alternating electric field was being applied. In 1952, Haynes and Briggs [3] observed luminescence in forward-biased p−n junctions in germanium and silicon, and in 1955, Newman et al. [4, 5] published papers on the luminescence in reverse-biased p−n junctions in silicon. It is now known that many crystalline substances are capable of electroluminescence. In addition to the compounds already mentioned, there are ZnO, CdS, AlN, $Al_2O_3$, diamond, GaAs, BN, GaP, and others [6-13].

The electroluminescence of solids is frequently divided into the Losev effect and the Destriau effect, the former being the luminescence of a crystal in direct contact with electrodes, and the latter being luminescence without direct contact. Such a distinction is purely formal and does not reflect the true processes of electroluminescence in various substances. Depending on the energy structure of a crystal and on the structure of the local field, the electroluminescence may be due to a number of very different mechanisms.

Investigations of the electroluminescence excited by the Losev method are very promising in solid-state physics because they provide an opportunity of studying simultaneously all the electrical and optical properties of a substance; moreover, the interpretation of the phenomena observed is less complicated by side effects than in other methods. Investigations of the electroluminescence excited by the Destriau method are also being rapidly developed. This is because of the promising possibilities of technical applications of such electroluminescent capacitors [14, 15] (bright, uniformly emitting sources of light; electroluminescent television screens; light amplifiers; various types of indicators; etc.). Most investigations of this type concern zinc sulfide.

CHAPTER I

# Principal Mechanism of Electroluminescence

In spite of the fact that the electroluminescence of dielectric ZnS-base crystal phosphors has been known for 30 years, and in spite of the enormous number of investigations carried out so far, many important problems have not yet been solved. This is because of the complexity of the process of electroluminescence itself and the complexity of the heterogeneous layers used in the capacitors. The electroluminescence of relatively large single crystals is simpler to interpret, because the influence of additional components in a capacitor layer is eliminated. However, even in the case of single crystals, investigations of the electroluminescence meet with considerable difficulties because there are many possible combinations of excitation, energy transfer, and radiation processes. One must bear in mind that the luminescence is the radiation emitted in excess of the thermal radiation of a body. The magnitude of the deviation from the thermal radiation conditions is governed by the processes of energy transfer in a luminescent body from the point of absorption to the radiation-emitting centers. Therefore, in investigations of the electroluminescence, we need to consider not only those transitions which result in the emission of light, but the process as a whole, beginning from the moment of absorption of the energy.

In the electroluminescence of crystals, the electric field energy is transformed into radiation. The whole process can be divided into three basically different stages [16]:

1. Excitation. An electric field transfers electrons to an excited state whose energy differs from the ground state by an amount approximately equal to the forbidden band width. The excited state of a crystal represents not only the increase in the number of free carriers compared with the equilibrium density (generation of nonequilibrium carriers) or the increase in the kinetic energy of carriers in the conduction and valence band, but also includes the excited states of the impurities. However, the mechanisms of excitation of the impurities by an electric field do not differ basically from the mechanisms of generation of nonequilibrium carriers, although the excitation of impurities may not be accompanied by an increase in the conductivity. Consequently, we shall only consider the processes of nonequilibrium carrier generation.

2. Migration of the Excitation Energy. The transfer of energy along a crystal to a point at which deexcitation occurs may play an important role in electroluminescence. This is because, in many cases, the energy is transferred by free carriers, which are affected by the electric field. In the most general case, we must allow not only for the drift of free carriers, but also for the capture of these carriers, for the motion of excitons, and for the resonance energy transfer.

3. Luminescence. A radiative transition to an unexcited state usually takes place with the participation of crystal defects. In addition to such transitions, we can have interband and intraband transitions.

This schematic division of the process of electroluminescence into stages is generally agreed upon. However, when we consider the details of electroluminescence in actual substances, we find that different authors reach different conclusions. This is not surprising, because the excitation, the migration of energy, and the radiation may take place in different ways, and the relative influence of a particular process may be very different under different conditions

In the electroluminescence of dielectric zinc sulfide crystals, one of the most controversial problems is that of the method of excitation of electrons in a crystal up to optical energies. In view of this, we shall consider in detail the possible methods of generating nonequilibrium carriers and the associated problems.

## §1. Methods of Generating Nonequilibrium Carriers

In the absence of external radiation, nonequilibrium carriers can be generated in the following ways by the application of an electric field [17, 18]: 1) internal field emission of electrons; 2) impact ionization; 3) injection of minority carriers; 4) accumulation of carriers. We shall discuss these mechanisms in detail and obtain the corresponding expressions for the current density, which can be used to determine the dependence of these processes on the parameters of the exciting field and on the properties of a substance.

## A. Internal Field Emission of Electrons (Tunnel Ionization)

In the internal field emission of electrons in semiconductors and dielectrics, the following types of tunnel transition are possible [19]: a) from the valence band to the conduction band of a crystal (the Zener effect); b) from local energy levels to the conduction band (autoionization); c) from the cathode to the conduction band of a crystal (tunnel leakage of electrons), and from the valence band to the anode (tunnel leakage of holes); and d) between local energy levels of impurity centers.

1) Z e n e r   E f f e c t.   The possibility of the transfer of a valence electron to the conduction band in an electric field was first pointed out by Zener [20]. Later, the results of Zener, which were valid only for dielectrics with a narrow forbidden band, were extended by Franz [21] to substances with a much wider forbidden band. In the most general case, the number of valence electrons which undergo tunnel transitions to the conduction band per unit volume per unit time and without change in the crystal momentum is given by the formula [22]:

$$n_1 = N_1 \frac{eEd}{2\pi\hbar} \exp\left(-\frac{\pi}{2e\hbar E}\sqrt{2m}\,\Delta^{3/2}\right), \tag{I.1}$$

where E is the electric field intensity, $\Delta$ is the forbidden band width, m is the effective mass of electrons, d is the lattice period of a crystal, and $N_1$ is the number of valence electrons per unit volume of a crystal. However, usually the highest state of the valence band and the lowest state of the conduction band have different values of the crystal momentum, and, therefore, significant leakage begins in fields considerably higher than those that would be expected from Eq. (I.1).

On the other hand, the presence in the leakage process of an interaction with can change the crystal momentum of an electron permits transitions from the highest state of the valence band to the lowest state of the conduction band. Such interactions include the interactions of valence electrons with lattice phonons, slow conduction electrons, and with impurities. Since, in the interaction process, phonons can transfer not only a momentum but also an energy $\hbar\omega$ to an electron, the interaction is equivalent to a reduction in the barrier by $\hbar\omega$.

Keldysh [23] obtained the following formula for the number of pairs generated in a one-phonon tunnel transition per unit volume per unit time:

$$n_0 = F\left[N_k + (1 + N_k)\exp\left(-\frac{4\sqrt{2m^*\Delta}}{e\hbar E}\hbar\omega\right)\right]\exp\left[-\frac{4\sqrt{2m^*}}{3e\hbar E}(\Delta - \hbar\omega)^{3/2}\right], \tag{I.2}$$

where F is a weak function of temperature and of the field, $m^*$ is the reduced effective mass of electrons and holes, and $N_k$ is the Planck distribution:

$$N_k = \left[\exp\left(\frac{\hbar\omega}{kT}\right) - 1\right]^{-1}. \tag{I.3}$$

The formula (I.2) represents a product of the probability of an exchange of energy $\hbar\omega$ between an electron and a lattice and the probability of a leakage through a barrier which has been reduced by an amount equal to this energy; it allows for the phonon absorption and emission processes. The principal difference between the formulas (I.2) and (I.1) is the explicit temperature dependence. In (I.1) a weak temperature dependence appears due to the temperature dependence of the parameters of a crystal: $\Delta$, $m^*$, and d. The relative change in these quantities for $\Delta T = 1°$ is of the order of the linear expansion coefficient. For example, for zinc sulfide, $d\Delta/dt$ is equal to $-4.6 \cdot 10^{-4}$ eV/deg at 77°K and $-8.5 \cdot 10^{-4}$ eV/deg at 800°K [24]. This implicit temperature dependence is also retained in the formula (I.2), but the Planck distribution applies to the temperature dependence of the probability of generation of electron−hole pairs in the case of indirect tunnel transitions. At low temperatures $T < T_0$, where

$$T_0 = \frac{e\hbar E}{4k \sqrt{2m^* \Delta}},$$  (I.4)

we find that $n_0$ is practically constant.

At high temperatures $T > T_D$, where $T_D$ is the Debye temperature, related to the maximum phonon energy $\hbar \omega_m$ by

$$T_D = \frac{\hbar \omega_m}{k},$$  (I.5)

we obtain a weak linear dependence of $n_0$ on T. In the range $T_0 < T < T_D$, we observe an exponential rise of $n_0$. Since the lattice has phonons with various values of $\hbar \omega$, it follows that the quantity $n_0$ in the temperature range $T_0 < T < T_D$ may represent a superposition of expressions of the (I.2) type, i.e., a sum of exponentials corresponding to different values of $\hbar \omega$.

The dependence of the probability of a multiphonon tunnel transition on the voltage and temperature [23] is given by the following formula:

$$n_M = \frac{e\hbar E}{2 \sqrt{m^* kT}} \exp\left[ -\frac{\Delta}{kT} + \frac{1}{24m^*} \frac{(e\hbar E)^2}{(kT)^3} \right].$$  (I.6)

This formula is valid when the following condition is satisfied:

$$\Delta - \left(\frac{e\hbar E}{kT}\right)^2 \bigg/ 8m^* \geqslant 0.$$  (I.7)

In still higher fields, we must use the formula (I.2).

A formula similar to (I.6) is also obtained when a valence electron interacts with conduction electrons. In this case, T has the meaning of an effective conduction-electron temperature.

The tunnel effect when there is an interaction with charged impurities is considered in [25].

The number of electrons which tunnel from the valence band to the conduction band is given by the formula

$$n_{imp} = F_1 E^{5/2} N_{imp} \exp\left( -\frac{4 \sqrt{2m^*}}{3e\hbar E} \Delta^{3/2} \right),$$  (I.8)

where $F_1$ is a phonon defined by the parameters of a crystal and universal physical constants and $N_{imp}$ is the concentration of the charged impurity with which electrons interact. This effect may be important in degenerate semiconductors.

2)  Ionization of Local Levels.  A calculation of the probability of a tunnel transition of a localized electron from an impurity level in the forbidden band to the conduction band has been carried out by Franz [26]. He has obtained the following expression:

$$W = \exp\left( -\frac{4 \sqrt{2m^*}}{3e\hbar E} \Delta_i^{3/2} \right),$$  (I.9)

where $\Delta_i$ is the depth of the local level.

Comparing the emission from the valence band with the emission from an impurity level, we note that the latter is smaller for a given field $\Delta_i$ than the emission from the valence band in the same field if the forbidden band width is equal to $\Delta_i$.

3)  Emission from Electrodes.  The tunneling of electrons from the cathode to the conduction band of a crystal is, in many respects, similar to the external field emission from a metal considered by Fowler and Nordheim [27]. However, in the calculation of the emission in a dielectric, we must allow for some additional factors. The problem has been solved by Franz [28]. The emission current density calculated in this

way is

$$j = \frac{e^3 E^2}{8\pi h \Delta_c} \xi \exp\left(-\frac{4\sqrt{2m^*}}{3\hbar eE}\Delta_c^{3/2}\zeta\right).$$ (I.10)

Here, $\Delta_c$ is the difference between the vacuum work functions of a metal and a given crystal and $\xi$ and $\zeta$ are correction factors of the order of unity. Franz has also described a method for solving a similar problem for a tunnel transition from the valence band to the anode.

Thus, the main conclusion which can be drawn from all types of internal field emission is: the density of the emission current has the form

$$j = AE^n \exp\left(-\frac{\delta}{E}\Delta^{3/2}\right).$$ (I.11)

The quantities A and $\delta$ depend on the actual nature of the tunnel transition, and n lies usually between 1 and 3.

## B. Impact Ionization

Carriers whose energy is higher than the forbidden band width may generate new nonequilibrium carriers by impact ionization. The kinetic energy of carriers can be increased by the application of an external electric field. The idea of impact ionization in solids was suggested by Ioffe [29] in connection with the problem of breakdown in dielectrics. Subsequently, this effect has been investigated in detail by a number of workers. Seitz [30] has obtained the following formula for the probability of impact ionization in ionic crystals:

$$W = \delta_0 e^{-\frac{E_c}{E}},$$ (I.12)

where $\delta_0$ is a quantity of the order of the effective cross section of the interaction during ionization and $E_c$ is a characteristic field. Seitz has calculated the probability of impact ionization as a probability of that fluctuation $p_f$ of the free path of electrons for which an electron traverses a path $l_E$ without collision along the field direction and acquires an ionization energy E:

$$p_f \sim \exp\left(-\frac{l_E}{\bar{l}}\right),$$

where $\bar{l}$ is the mean free path. However, Seitz has not allowed for the fact that in ionic crystals the mean free path increases with the electron energy because of a reduction in the effectiveness of the interaction with the lattice phonons [31]. A more correct formula has been deduced by Chuenkov [32] by solving the transport equation

$$W' = \delta_0'\left(\frac{E}{E_c}\right)\exp\left(-\frac{E_c}{E}\right)^2.$$ (I.13)

The temperature dependence of the characteristic field is found from the expression

$$E_c = E_c^0\left[\frac{\exp\left(\frac{\hbar\omega_0}{kT}\right)+1}{\exp\left(\frac{\hbar\omega_0}{kT}\right)-1}\right]^{1/2},$$ (I.14)

where $E_c^0$ is the characteristic field at T = 0 and $\hbar\omega_0$ is the optical phonon energy.

Chuenkov has also shown that the formula (I.12) is valid only in a weak field,

$$E \ll \frac{2kT}{el_0},$$ (I.15)

where $l_0$ is the path traveled by a thermal electron. The formula for the probability of impact ionization in covalent crystals, analogous to (I.13), has been obtained in [33] from the transport equation. It should be mentioned that, in addition to impact ionization in the valence band, we can also have impact ionization of impurities, which requires much weaker fields.

## C. Injection of Minority Carriers

We have considered transitions of carriers through a potential barrier by the tunnel leakage. In addition to this process, we can also have suprabarrier carrier transitions [34]. It should be stressed that, in this case, the term "injection" means the transfer of minority (with respect to the bulk of a crystal) carriers over and not through a potential barrier. The injection may increase not only the density of minority carriers, but also, in the same degree, the density of majority carriers, so that the total charge may remain equal to zero and no phenomena associated with the formation of a space charge should be expected. However, if there is no source of excess minority carriers, the injection will be limited by the formation of a space charge.

When the recombination of carriers is radiative, injection produces radiation. This is the radiation observed by Losev in 1923. The injection of minority carriers may take place within a p—n junction or at the boundary between a metal electrode and a semiconductor provided the external voltage is applied in the forward direction. Then, the value of the current flowing through a junction [35] is

$$I = I_s \left( e^{\frac{eV}{kT}} - 1 \right).  \tag{I.16}$$

Here, I is the current through the junction, V is the energy applied to the junction, and $I_s$ is the saturation current in the reverse-bias case.

The electroluminescence observed when minority carriers are injected differs considerably from the two preceding cases in that it requires relatively weak fields, and its dependence on the external voltage is described by a formula analogous to (I.16).

The injection electroluminescence has been observed in semiconductor rectifiers made of germanium [3], silicon [3, 36], silicon carbide [37], cadmium telluride [38], gallium antimonide, gallium arsenide, and indium phosphide [11], and other substances.

From the nature of the dependence of the brightness on the voltage we may conclude that the electroluminescence of dielectric zinc sulfide phosphors is not due to the injection of minority carriers. There is disagreement on the question of which of the other two methods of generation of nonequilibrium carriers applies in the case of zinc sulfide electroluminescent capacitors but, in both cases, the explanation of the excitation of an electroluminescent phosphor is based on the hypothesis of electric field concentration.

## § 2.  Concentration of an Electric Field

The idea of the concentration of an electric field, first proposed by Piper and Williams [39], postulates the presence in an electroluminescent capacitor of points where the electric field intensity is considerably higher than the average intensity. This hypothesis is justified by the following observations.

1. The average field intensity in an electroluminescent capacitor is not only less than that required for the tunnel excitation of zinc sulfide but is also less than the intensity at which electrons can be accelerated to optical energies.

2. A microscopic investigation of electroluminescent powders has shown that crystallites luminesce only at separate points [40-42]. This characteristic of the luminescence is not a property of powders alone, but is also observed for single crystals in which the emission of light takes place at single points or in narrow strips [43-45].

3. Frankl [46] measured the distribution of the potential along the surface of a single crystal and thus directly demonstrated the existence of a cathode region with a higher field intensity.

4. Strong magnetic fields, of up to $1.7 \cdot 10^5$ Oe intensity, directed at right angles to the electric field do not quench the brightness of electroluminescence [47, 48]. Calculations show that this is possible only when local fields are considerably higher than the average field.

The hypothesis of the concentration of a field is now fully established. We shall consider in detail the causes of field inhomogeneity, of which there are several.

## A. Barrier Layers

Blocking and antiblocking layers can appear at the boundary between a metal and a semiconductor for certain relationships between the work functions and for certain types of conduction in the semiconductor [49, 50]. The appearance of such layers is due to the formation of a space charge which equalizes the Fermi level in the substances which are in contact. A blocking layer is characterized by a density of free carriers lower than that in the interior of a crystal, in contrast to an antiblocking layer, in which the density of free carriers is higher. Since the density of these carriers governs the value of the electrical resistivity, it follows that anti-blocking layers have a resistivity lower than the rest of the crystal, whereas blocking layers have a higher resistivity. The presence of a narrow low-resistivity layer is simply equivalent to a small reduction in the length of a sample. Conversely, the presence of a layer with higher resistivity may result in a considerable part of the voltage drop being concentrated in that layer. If the thickness of this layer is small (and this is always true in the case of metal—semiconductor contacts), the field intensity in this layer will be considerably higher than the average field intensity in the sample. Thus, blocking layers may concentrate the electric field.

The actual characteristics of a blocking layer depend on the energy structure of the semiconductor. We shall consider in detail some forms of these barrier layers.

A Mott — Schottky barrier [51] is formed at a metal—semiconductor boundary by ionization of a uniformly distributed impurity whose energy level lies at a certain depth in the forbidden band. Solving Poisson's equation [51] for a uniform charge density in a barrier,

$$\rho(x) = eN = \text{const,} \tag{I.17}$$

where $\rho$ is the density of charge in the barrier, N is the concentration of the ionized impurity, and e is the electron charge, we can obtain the principal characteristics of the barrier.

The potential curve is parabolic:

$$V(x) = \frac{2\pi\rho}{\varepsilon}(L-x)^2. \tag{I.18}$$

Here, $\varepsilon$ is the permittivity and L is the width of the barrier, defined as the distance between the boundary with the metal to a region where the potential can be regarded as constant.

The maximum field intensity in the barrier is proportional to the square root of the voltage across the barrier, and decreases linearly in the direction of the interior of a crystal:

$$E = \sqrt{\frac{8\pi\rho}{\varepsilon}V}. \tag{I.19}$$

The thickness of the space-charge layer increases as the square root of the voltage applied to the barrier:

$$L = \left(\frac{V\varepsilon}{2\pi\rho}\right)^{1/2}. \tag{I.20}$$

These Mott—Schottky barriers, known as exhaustion barriers, were first proposed by Piper and Williams as regions where the field is concentrated [39].

Rose-type barriers [53] appear when there is ionization of impurity levels lying at various depths in the forbidden band; these levels are, therefore, partly ionized. Such barriers, known as depletion barriers, have the following properties:

The potential decreases exponentially in the direction of the interior of the semiconductor:

$$V(x) = V_0 \exp\left[-\left(\frac{4\pi e N_2}{\varepsilon}\right)^{1/2} x\right],$$ (I.21)

where $N_2$ is the concentration of the impurity per 1 $cm^3$ per unit energy interval.

The density of the space charge decreases exponentially in the direction away from the boundary with a metal:

$$\rho(x) = e N_2 V(x).$$ (I.22)

The maximum field intensity in a barrier is proportional to the potential drop across it:

$$E = \left(\frac{4\pi e N_2}{\varepsilon}\right)^{1/2} V(x)$$ (I.23)

and it decreases exponentially in the direction of the interior.

The characteristic thickness of the space-charge layer is independent of the potential at the boundary with a metal:

$$L_1 = \left(\frac{\varepsilon}{4\pi e N_2}\right)^{1/2}.$$ (I.24)

p − n Junctions.† If an abrupt p−n junction is a boundary between two regions of a semiconductor with high free-carrier densities, its characteristics are similar to those found for Mott−Schottky exhaustion barriers. The most important difference between Mott−Schottky barriers and p−n junctions is that the negative space-charge region may be comparable in size with the positive space-charge region. If the impurity concentrations in p- and n-type regions are $N_a$ and $N_d$, respectively, the width of the p−n junction is [57]:

$$L = \left[\frac{\varepsilon(N_a + N_d)}{2\pi e N_a N_d} V\right]^{1/2}.$$ (I.25)

The other formulas can be obtained quite easily. In real p−n junctions, there is no abrupt boundary between the p- and n-type regions and, therefore, an impurity concentration gradient is possible, so that the relationships between the barrier characteristics may be more complex.

Semiconductor heterojunctions. Antiblocking and blocking layers capable of concentrating an electric field can also appear at the boundary between two semiconductors with different forbidden bands. The type of conduction of the material with the narrower forbidden band is not important. The nature of the transition layer (blocking or antiblocking) is governed only by the relationship between the work functions and by the type of conduction of the material with the wider forbidden band. This is illustrated clearly in Fig. 1, which shows the energy band schemes of a heterojunction between $Cu_2S$ and ZnS for two types of conduction of both materials and different relationships of their work functions. Here, $\varphi_1$ denotes the work function of $Cu_2S$ and $\varphi_2$ that of ZnS.

It should be mentioned that the electroluminescence of dielectric zinc sulfide is ascribed in [58, 59] to $Cu_2S$−ZnS heterojunctions which can exist on the surfaces of particles or in the interior at dislocations or at other imperfections.

## B. Distribution of an Electric Field in an Inhomogeneous Dielectric Medium

Since electroluminescent capacitors are complex heterogeneous systems, the field in them is strongly inhomogeneous. We can calculate the field distribution in a capacitor only if we know the distribution of the

---

† It should be mentioned that the suggestion advanced in [54-56] of the possibility of field concentration in n−n' and p−p' junctions is incorrect because n−n' and p−p' junctions are antiblocking layers and, consequently, do not concentrate the field.

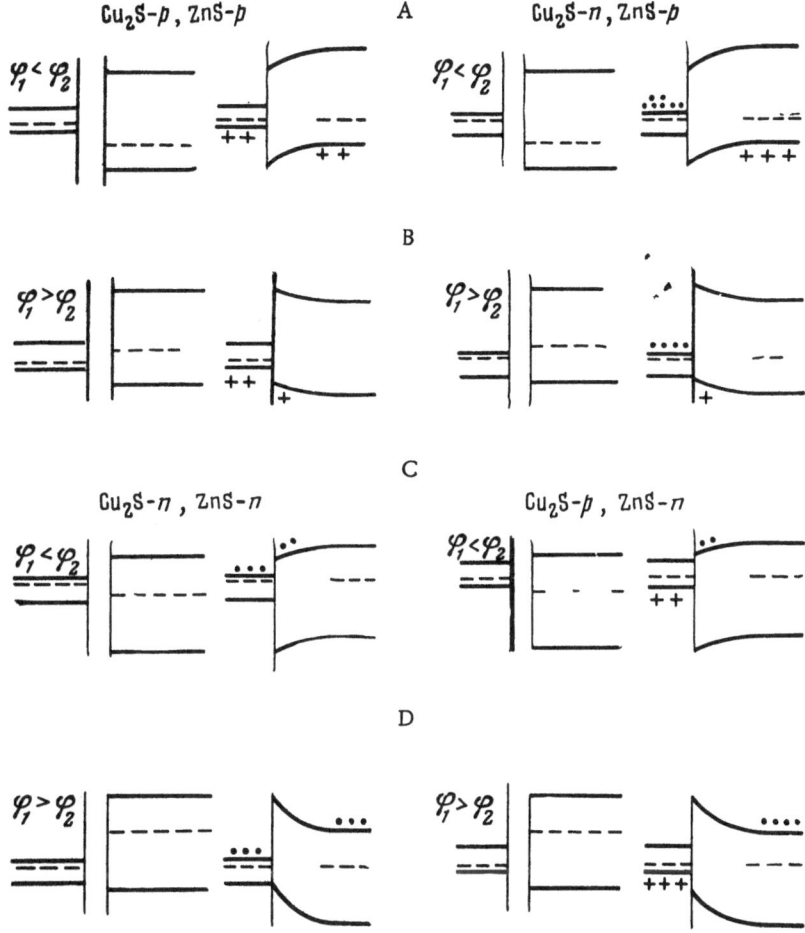

Fig. 1. Energy band schemes of a $Cu_2S - ZnS$ heterojunction for two types of conduction in the two substances and different relationships between the work functions. A, D) Blocking layer; B, C) antiblocking layer.

inhomogeneities. However, basically correct results can be obtained [60] by assuming the simplest model of a two-layer dielectric. In a two-layer capacitor, the dc field intensities are inversely proportional to the conductivities of the layers:

$$\frac{E_1}{E_2} = \frac{\sigma_2}{\sigma_1}.$$

(I.26)

In the ac case, the amplitudes of the field intensities are inversely proportional to the moduli of the admittances:

$$\frac{E_1}{E_2} = \frac{\sqrt{\sigma_2^2 + \left(\frac{\varepsilon_2\omega}{K}\right)^2}}{\sqrt{\sigma_1^2 + \left(\frac{\varepsilon_1\omega}{K}\right)^2}},$$

(I.27)

where K is a constant. For all good dielectrics, the capacitative component of the admittance is considerably larger than the ohmic component at frequencies close to the power frequency as well as at higher frequencies, i.e.,

$$\frac{E_1}{E_2} = \frac{\varepsilon_2}{\varepsilon_1}.$$

(I.28)

The concentration of a field due to the differences in the permittivities of different parts of a sample is observed most clearly in the electroluminescence of ferroelectric materials [61], for which the permittivity in the interior can be made very high because of the ferroelectric polarization, while at the surface the permittivity can have the optical value, i.e., it can be of the order of several units.

The formulas (I.26)-(I.28) are only approximately applicable to electroluminescent capacitors filled with powders. The field in electroluminescent phosphor particles has been calculated [62] on the assumption that these particles are spherical, uniformly distributed in a dielectric, and sufficiently far apart:

$$\frac{E}{\frac{U}{d}} = \frac{3\,\frac{\varepsilon_2}{\varepsilon_1}}{\left(1 + 2\,\frac{\varepsilon_2}{\varepsilon_1}\right) - c\left(1 - \frac{\varepsilon_2}{\varepsilon_1}\right)}\,. \qquad (I.29)$$

Here, $E$ is the field intensity inside a phosphor particle; $U$ is the voltage applied to the capacitor; $d$ is the thickness of the capacitor; $\varepsilon_2$ and $\varepsilon_1$ are the permittivities of the binder and of the electroluminescent phosphor, respectively; and $c$ is the concentration of the phosphor. Maeda [56] calculated the possible inhomogeneity of the field for conducting particles in the shape of elongated and oblate spheroids.

Because of the irregularities of the crystallite shape, the electric field intensity at projections of such crystallites is high. The field is particularly high at very sharp projections. All these forms of field concentration may apply to an electroluminescent capacitor filled with a phosphor powder. The hypothesis of the field concentration is almost universally used to explain the Destriau effect. However, there is no agreement about the cause of the field concentration. The view that the field concentration is associated with blocking layers is held by the majority. However, there are investigators [63, 64] who are of the opinion that the field is concentrated mainly at sharp projections of the conducting phase.

Thus, summarizing Chapter I, we can say that there are several opinions about the nature of the excitation of electroluminescent phosphors. Also, some problems concerning the subsequent stages in the electroluminescence process are not yet solved. However, if we consider the electroluminescence of dielectric zinc sulfide crystals excited with an alternating electric field, the whole process can be represented as follows: After the excitation of an electroluminescent phosphor, when the activator centers near the cathode are ionized, free carriers leave the excitation region and either reach capture centers or are extracted from the crystal. In the next half-period, they return to the positive space-charge region and recombine so as to emit light.

CHAPTER II

Experimental Method

### §1.  Preparation of Capacitors

Electroluminescent zinc sulfide phosphors can be synthesized in the form of single crystals [65], powders [66-68], or thin sublimated films [69-71]. The present paper deals with electroluminescent phosphor powders.

An electroluminescent capacitor contains a heterogeneous suspension of a phosphor in a liquid or solid dielectric. Such capacitors have a wide range of practical applications. Electroluminescent phosphors used in this way have an inhomogeneous granulometric composition. Lehmann [72] describes the particle size distribution by the following formula:

$$y = ax^2 \exp\left(-\frac{2x}{x_0}\right),$$

where $y$ is the number of particles of diameter $x$; $x_0$ is the diameter of particles of which there is the greatest number; and $a$ is a constant.

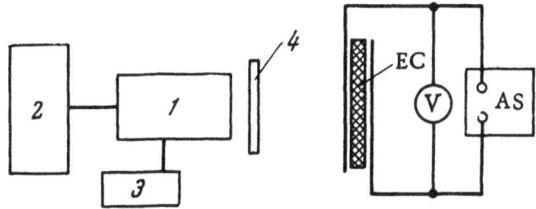

Fig. 2. Block diagram of the photometric apparatus.

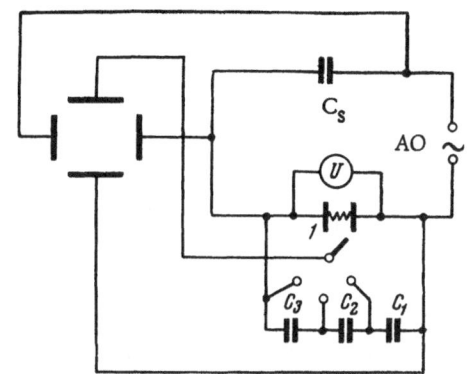

Fig. 3. Basic circuit for the measurement of power. AO—audio oscillator.

The luminescence spectra of zinc sulfide phosphors can assume various forms and is governed by the activator. For example, the introduction of copper [73] usually produces green-blue luminescence and the introduction of manganese [74, 75] produces green luminescence. Almost-white light can be obtained by mixing phosphors with different activators.

The majority of samples investigated in the present study were in the form of solid electroluminescent capacitors on glass substrates. A film of $SnO_2$ deposited on glass was used as the transparent electrode. An electroluminescent layer was deposited on this electrode by atomization. A second (aluminum) electrode was deposited by evaporation in vacuum. The electroluminescent layer was a heterogeneous system consisting of a ZnS phosphor suspended in a dielectric. The phosphor was usually unfractionated ZnS:Cu emitting green luminescence. The granulometric composition of this phosphor included 20% crystallites whose size was up to 5 μ and 72% crystallites whose size was up to 15 μ. The maximum of the luminescence spectrum was observed at 510 mμ. When the frequency of the exciting electric field was increased, the spectrum shifted slightly toward shorter wavelengths. The spectrum was not affected by the intensity of the external field.

In a few of the experiments, we used zinc sulfide emitting blue luminescence. Electroluminescent capacitors were prepared using various dielectric binders: an epoxide resin ÉP-096, a mixture of melamineformaldehyde and glyptal resins of the ML-92 type, polystyrene with nitrocellulose, polyethylene terephthalate, etc. The electrical properties of these dielectrics differed considerably. For example, for polystyrene, $\varepsilon = 2.5$ and $\tan \delta = 0.0002$, whereas, for the ML-92 resin, $\varepsilon = 9$ and $\tan \delta = 0.045$.

The method of preparation of the suspension and of the samples differed with the nature of the dielectric used. Polystyrene and nitrocellulose were dissolved in butyl acetate and amyl acetate, so that the ratio of these components was 1:1:5:5. In this case, the phosphor and the dielectric were used in equal quantities by weight. In some investigations, we used capacitors with different ratios of ZnS and of the dielectric binder. A heterogeneous layer was deposited using a special atomizer in several batches. After each atomization, the capacitor was dried at room temperature. In the final stage, the polystyrene was dried out at 80-90°C.

When the resins ML-92 and ÉP-096 were used, the weight ratio of the zinc sulfide and the dielectric was 1.5:1, and the necessary amount of the liquid dielectric was calculated from the dry residue. To improve the coating of the phosphor with the resin, we mixed the charge in a ball mill for one hour. Acetone was used as a "filler" in the atomization. The samples were dried at 70-80°C for 5 hr.

To ensure reliable contacts, some capacitors were provided with silver electrodes which were prepared by firing a silicate—silver paste at 400°C. A special procedure was followed to reach this temperature.

The method of preparation of the capacitors using polyethylene terephthalate was quite different. A polyethylene terephthalate film 10-μ thick was deposited on the conducting surface of polished glass. The film was then melted in an oven at 225°C; it wetted the glass well so that good adhesion to the glass was obtained after cooling. An atomizer was then used to deposit a suspension of ZnS in water on the cold surface of the film. After the evaporation of the water, another polyethylene terephthalate film was deposited, this time on the ZnS layer, and the sample was placed in an oven and again heated at 225°C. Next, the sample was cooled slowly and an aluminum electrode was deposited on it by evaporation in vacuum.

## §2. Measurement Methods

The electroluminescent capacitors were excited with a sinusoidal ac voltage. For this purpose, we used oscillators of the ZG-10, ZG-12, and ZG-4A types. The voltage applied to a capacitor was measured with an electrostatic voltmeter.

The brightness was recorded photometrically using the apparatus shown schematically in Fig. 2. The following designations are used in that figure: EC, electroluminescent capacitor; AS, audiofrequency source of sinusoidal voltage; V, voltmeter; 1, photomultiplier; 2, high-voltage source supplying the photomultiplier; 3, recording instrument; 4, optical filters. We used a VEI photomultiplier with an antimony−cesium photocathode, whose spectral sensitivity was almost constant in the range of wavelengths of the luminescence spectra of the capacitors. A set of optical filters 4 was used only in those cases when it was necessary to make the spectral sensitivity of the apparatus coincide with the spectral sensitivity of the human eye. The apparatus was calibrated in absolute light units.

The luminescence spectra were recorded with a UM-2 monochromator. An FÉU-18 photomultiplier was used as the radiation detector. The recording device in the photomultiplier circuit was a ballistic galvanometer of the M 21/1 type with a universal shunt.

a) Measurement of Average Power. During electroluminescence, the distribution of electrons in the energy levels may differ from the distribution in the absence of a field. It follows that the conductivity of crystals may depend strongly on the value of the external exciting field, i.e., electroluminescent capacitors are nonlinear elements [40, 76]. This must be allowed for in the selection of the measurement method. It has been shown [77, 78] that the active current through an electroluminescent capacitor differs widely from the sinusoidal form although the exciting voltage is sinusoidal. Therefore, the Q-meter method [79] and the method of phase shifts between current and voltage [80], which are frequently used to measure the average power, are unsuitable in this case. We used the method of Lissajou figures to measure the power absorbed in sinusoidal and nonsinusoidal processes over a wide range of frequencies [81]. The basic measurement circuit is shown in Fig. 3. The following designations are used in that figure: 1, electroluminescent capacitor, $C_s$, standard capacitor; U, voltmeter; $C_1$, $C_2$, $C_3$, capacitance divider.

To measure the power with an oscillograph, it was necessary to apply to its plates deflecting voltages proportional to the instantaneous values of the current and voltage in an electroluminescent capacitor. Let us assume that the voltage from an electroluminescent capacitor $U_0\varphi(t)$ is applied to the vertical plates of an oscillograph and that the sensitivity of these plates is g; the deflection along the y axis is then

$$y = \frac{U_0\varphi(t)}{g} . \tag{II.1}$$

The deflection of the beam in the x direction is proportional to the voltage across the standard capacitor $C_s$. If the current in the circuit varies with time in accordance with the law $I_0\psi(t)$, the charge in the capacitor is

$$q = \int I_0\psi(t)\,dt, \tag{II.2}$$

and the potential difference is

$$U_{C_s} = \frac{\int I_0\psi(t)\,dt}{C_s} . \tag{II.3}$$

Let us assume that the sensitivity of the oscillograph in the horizontal direction is j. The horizontal deflection is then

$$x = \frac{\int I_0\, \psi(t)\, dt}{j C_s}. \qquad (\text{II}.4)$$

If $\varphi(t)$ and $\psi(t)$ are harmonic functions of different frequencies, the combined oscillation represents the motion along an elliptic trajectory. In the case of arbitrary continuous functions, this motion will take place along some closed curve. We can show that the area of the resultant figure is proportional to the average power supplied by the current:

$$dS = y\,dx = \frac{W}{jg C_s}\, dt, \qquad (\text{II}.5)$$

$$W_{\mathrm{av}} = \frac{1}{T}\int_0^T W\,dt = jg\nu C_s S, \qquad (\text{II}.6)$$

where $S$ is the area of this figure and $\nu$ is the frequency of the external voltage.

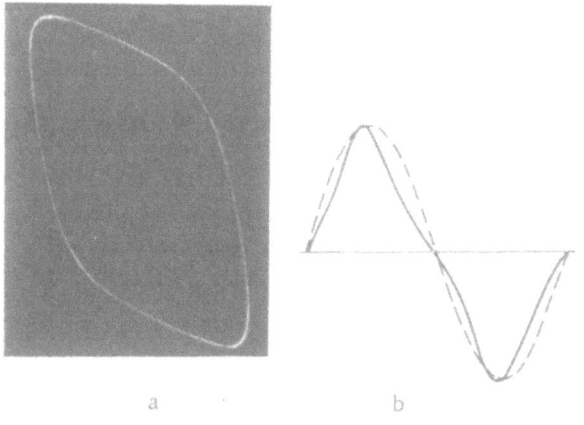

Fig. 4. Lissajous figure for an electroluminescent capacitor (a) and the current (continuous curve) and voltage (dashed curve) for the same capacitor (b).

Since the standard $C_s$ is in the electroluminescent capacitor circuit, to avoid considerable increase in the power consumption, this capacitor should have negligibly small losses. Therefore, we used air and mica capacitors. To extend the working range of the apparatus in respect to the voltage, we connected a capacitance divider in parallel with the electroluminescent capacitor. This made it possible to carry out measurements in the voltage range 30-300 V. The apparatus was carefully checked for the presence of stray phase shifts. The absolute calibration of the apparatus was carried out using radioengineering resistors whose resistances were first measured using a universal UM-2 bridge.

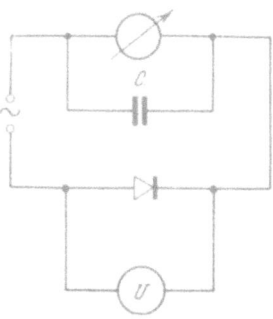

Fig. 5. Basic circuit for measuring the rectified current. The electroluminescent capacitor is represented as a detector.

Figure 4a shows a Lissajous figure obtained in the measurements of the power, while Fig. 4b shows the current through the sample derived from this figure. Both figures clearly demonstrate the nonsinusoidal nature of the current in an electroluminescent capacitor when a sinusoidal external voltage is applied to it. The departure from the sinusoidal waveform was greater at low frequencies and high voltages. This departure was not a defect of the apparatus, since the replacement of the electroluminescent capacitor in the circuit by an equivalent radioengineering capacitor and resistor suppressed this distortion.

The deviation from the sinusoidal waveform appeared at different frequencies and voltages for different capacitors, but the waveform was usually

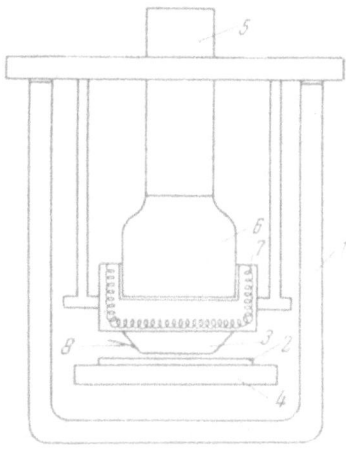

Fig. 6. Thermostat.

symmetrical (Fig. 4b) although the Lissajou figures were sometimes asymmetric. This was probably due to rectification, which will be considered in Chapter V. In some capacitors, we observed a sinusoidal current over the whole working range of frequencies and voltages. However, the majority of capacitors exhibited a considerable departure from the sinusoidal waveform.

b) Measurement of the dc Component. Figure 5 shows the basic circuit used to detect rectification of the current by an electroluminescent capacitor. An alternating sinusoidal voltage was applied to the capacitor and the instrument included in the circuit was of the magnetoelectric type, i.e., it recorded only the dc component of the current. A capacitor C was connected in parallel with the instrument and the capacitance was selected so that its ac impedance was, over the whole range of frequencies, less than the internal resistance of the instrument.

We also measured the dc current through the sample when a constant voltage was applied to it. The circuit for such measurements is not described because of its extreme simplicity. A UIP-1 device was used as the constant voltage source. The dc current was recorded with type M-198 and M-82 instruments.

c) Temperature Dependences. The temperature dependences were determined in a thermostat of the type described in [55] and shown in Fig. 6. Only the walls were silvered in the molybdenum glass Dewar flask 1. The luminescence emitted by a capacitor 2 was observed through the polished flat glass bottom of the flask. The aluminum electrode of the capacitor was in contact with a polished brass disk 3, against which it was pressed by a frame 4. To cool the capacitor, liquid nitrogen was poured into a Textolite tube 5, which ended with a copper can 6. A heater 7 was switched on when the temperature of the sample needed raising. The temperature of the sample was measured with a copper-constantan thermocouple 8.

The cold junction of the thermocouple was kept at a constant temperature in a Dewar flask filled with oil. The thermocouple emf was measured with a P-1 potentiometer.

When a sample was excited with a sinusoidal voltage, this apparatus made it possible to measure simultaneously the current rectified by the capacitor (using the circuit shown in Fig. 5), and the brightness of the luminescence (using the circuit shown in Fig. 2).

CHAPTER III

# Frequency Characteristics of an Electroluminescent Capacitor

## § 1. Brief Review of the Literature

The frequency dependence of the brightness has been investigated, experimentally and theoretically, more thoroughly than other frequency characteristic.

The average brightness of an electroluminescent capacitor decreases as the exciting voltage frequency is decreased if the value of the external field applied to the capacitor remains constant. In the audio-frequency range, the brightness is almost a linear function of the frequency but the dependence is weak [40, 82]. When the frequency is increased, the brightness passes through a maximum [83] or exhibits saturation [42, 84, 85].

In the case of capacitors filled with electroluminescent ZnS suspended in a dielectric, the saturation of the brightness is usually observed at frequencies of several kilocycles or several tens of kilocycles. The brightness maximum usually lies in the same range of frequencies. When the external voltage is increased, the brightness maximum shifts toward higher frequencies and so does the region where the brightness is saturated, i.e., when the external field intensity is increased, the range of frequencies in which the brightness depends linearly on frequency becomes wider.

These results have been obtained using alternating fields whose waveform is sinusoidal. In the case of pulse excitation or nonsinusoidal waveforms of the exciting voltage, the basic behavior is still the same, but the frequencies at which the brightness maximum or saturation is observed may be different [83, 86].

Since the majority of investigations of the frequency dependence of the brightness have been concerned with the audio-frequency range, the investigations of Rebane and Tal'viste [87] are of special interest, because they carried out measurements in the frequency range $10^{-2}$ cps $\leq \nu \leq 10^2$ cps. These measurements showed that down to about 1 cps the brightness decreased almost linearly as $\nu$ decreased, i.e., the behavior of the average brightness if the same as at audio frequencies. In the $10^{-2}$ cps $\leq \nu \leq 1$ cps range, the decrease in the brightness as the frequency is reduced was considerably more rapid, but it gradually slowed down on approach to $10^{-2}$ cps. Rebane and Tal'viste carried out these measurements on heterogeneous capacitors.

Oranovskii et al. [65] have investigated the frequency dependence of the brightness of ZnS: Cu single crystals, and their results at low frequencies were different. When the frequency was increased from 200 cps to 2.5 kc, the brightness increased almost linearly with the frequency, but the rise was slower at low frequencies. At frequencies of less than 100 cps, the brightness was found to be constant.

Thus, the most typical frequency dependence of the brightness is as follows: the brightness is constant at low frequencies, it increases almost linearly in the audio-frequency range, and becomes saturated or has a maximum at higher frequencies.

Different results have also been reported. For example, Diemer [44] has found that the brightness of ZnS: Cu single crystals is proportional to $\sqrt{\nu}$.

The frequency dependence of the brightness of phosphors doped with Ni and Co impurities [88, 89] may be superlinear in the frequency range between 20 and $10^5$ cps. In some cases, this dependence is almost quadratic. However, in the usual unquenched phosphors, a superlinear dependence of brightness on frequency is never observed at frequencies higher than 20 cps.

It is interesting to note that if a phosphor has two luminescence bands (for example, a blue and a green band), the frequency dependences of these bands are different. The saturation of the green band occurs when the intensity of the blue band is still increasing linearly [42]. The ratio of the intensities of these bands $B_b/B_g$ increases as the frequency rises, tending to a constant value.

The brightness of the radiation emitted by an electroluminescent capacitor is usually divided into a steady component and an alternating component ("brightness waves") [90]. When the frequency is varied, these components behave differently: the steady component increases as the frequency is increased, but the alternating component decreases [84].

The theoretical interpretation of the frequency dependence of electroluminescence is difficult: there is as yet no theory of the frequency dependence of the brightness valid in a wide range of frequencies. There are only a few calculations which approximate well the experimental results over a narrow range of frequencies or which predict some of the features. We shall now consider these calculations.

Curie [91] has calculated the dependence of the average brightness on the frequency of an external field. Using a simple bimolecular recombination scheme, he has obtained the formula

$$\frac{1}{B} = \frac{1}{kn_0^2\alpha}\left(1 + \frac{n_0\alpha}{2\nu}\right).\tag{III.1}$$

Here, k = const, $\alpha$ represents the rate of change in the electron density, and $n_0$ is the initial number of electrons in the conduction band. Comparison with the results of Weymouth [92] and Matossi [83] shows that the Curie formula is valid in the frequency range from 50 cps to 1 kc; at higher frequencies, the deviations of the calculated values from the experimental are very large.

The formula derived by Destriau [93, 94] is also valid in this range of frequencies. Taking into account the polarization at low frequencies, Destriau found a relationship between the applied field E' and the internal field E in the form

$$E = \frac{E'}{\sqrt{1 + \left(\frac{4\pi}{\varepsilon\rho\omega}\right)^2}},\tag{III.2}$$

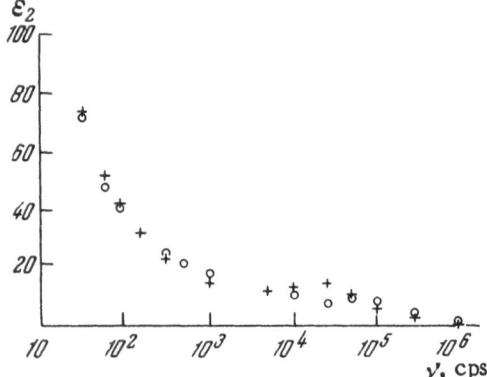

Fig. 7. Frequency dependence of the imaginary component of the complex permittivity $\varepsilon_2$.

where $\varepsilon$ is the permittivity and $\rho$ the resistivity. Assuming that the field is proportional to the voltage, Destriau has found that

$$B = B_0 \nu e^{-\frac{b}{\nu E'}} \qquad \text{(III.3)}$$

for $\nu$ close to zero when the unity in (III.2) can be neglected. However, we must point out major flaws in this theory:

1. "Rectification" of the frequency dependence of the brightness, plotted in the coordinates $\log(B/\nu)$, $1/\nu$, in the frequency range up to 1 kc, forces us to assume that the resistivity of crystal phosphors is very low, which contradicts the experimental results [95].

2. At low frequencies, no frequency dependence of the index of the exponential function is observed [72] (cf. § 2, Chapter IV).

The linear rise of the brightness with the frequency is usually explained on the basis of very simple considerations [39]. If we assume that the integral radiation emitted in a period is independent of the period of the applied field over a sufficiently wide range of frequencies but is governed solely by the intensity of this field and its ability to liberate electrons from donor levels, the average brightness should increase linearly with the frequency. In fact, the brightness usually increases sublinearly with the frequency.

An approximate calculation of the recombination interaction between blue and green luminescence centers in a strong electric field has been carried out by Fok [96]. He has obtained a relationship between the green and blue luminescence bands which is in good agreement with experiment.

This review of the experimental investigations and of the calculations shows that some features of the frequency dependence of the brightness can be explained satisfactorily but that no general explanation is as yet available.

The data on the frequency dependence of the absorbed power and of the optical yield are much more scarce.

Thus, according to the measurements of Zalm [40], tan δ of electroluminescent capacitors decreases linearly when the frequency is increased from 100 cps to 10 kc, while Ince and Outley [79] report that the imaginary component of the complex permittivity decreases slowly in the frequency range from 20 to $2.5 \cdot 10^6$ cps (Fig. 7). The circles in Fig. 7 represent the experimental results and the crosses theoretical calculations, which we shall consider below.

According to Zalm, Diemer, and Klasens [82], the optical yield $\eta$ of electroluminescent capacitors decreases at frequencies higher than 700-1000 cps, but at lower frequencies this yield is almost constant. The frequency dependence of the optical yield $\eta$ of the phosphors themselves [97] may assume various forms when the frequency is increased from 20 cps to 10 kc. For example, a ZnS:Cu phosphor emitting blue luminescence has a constant value of $\eta$ over this whole range of frequencies, while a ZnS:Cu, Cl phosphor with green luminescence has a well-defined maximum near 1 kc. These are typical of the experimental observations. Some of them (for example, the frequency dependence of $\eta$ for different phosphors) have not yet been accounted for. The decrease in the optical yield of electroluminescent capacitors at high frequencies has been ascribed by Zalm to the absorption in the series resistance of the electrodes, but he has not produced any convincing proof of this conclusion.

Thus, summarizing the results just reviewed, we can say that there is as yet no general explanation of the process and no model which can give a comprehensive description of the frequency characteristics of an electroluminescent capacitor in a wide range of frequencies.

## §2.  Selection of Equivalent Circuit

We have mentioned the complications in the interpretation of the results obtained for capacitors with heterogeneous dielectrics (phosphor + binder), which is due to the fact that there is as yet no clear picture of the influence of the separate components.  Such an influence may be fairly important in some cases [40, 62].

In the present chapter, we shall attempt to identify the properties of the phosphor itself and separate them from the characteristics which are due to the influence of other components of a capacitor.

The problem is most conveniently solved by considering the frequency characteristics of a|capacitor, because of the very strong dependence of the characteristics of the phosphor on the voltage, which masks the influence of other components of the capacitor.  Since electroluminescence is excited with an alternating electric field, the most convenient method is that of equivalent circuits.

This method has been used previously by a number of investigators [82, 86, 87, 98, 99] to describe various processes in an electroluminescent capacitor.  However, the method of equivalent circuits has been used only in a qualitative manner in the description of absorption processes.  The exception to this rule is the investigation by Ince and Outley [79].  The method of equivalent circuits has not yet been used at all to interpret the frequency dependence of the brightness.

Any capacitor subjected to an alternating voltage behaves as a circuit consisting of resistances and capacitances, and, therefore, it can be represented by an equivalent circuit [60].  In the case of a uniform or layered dielectric, these circuits are fairly simple.  In the case of a heterogeneous dielectric, the situation becomes complicated.  An electroluminescent capacitor is not only a heterogeneous but a strongly nonlinear system [76].  Equivalent circuits (even complex ones) represent the properties of such a system only approximately because such a system has continuously distributed capacitances and resistances [100].  Therefore, it is not surprising that different authors describe the processes occurring in an electroluminescent capacitor using equivalent circuits of different degrees of complexity with different numbers of effective parameters which may have different meanings in different circuits.  All these circuits are shown in Fig. 8.

In subsequent sections of this chapter, we shall be interested in only two characteristics of the electroluminescent capacitor: the absorbed energy and the energy lost by radiation.

The aim in this section is to determine which equivalent circuit with constant parameters and with the smallest possible number of effective parameters best describes the frequency dependences of these energies.

As the characteristic of the absorption in a capacitor, we shall use the value of the energy Q absorbed in one period, which is found from the expression

$$Q = \frac{W}{\omega} , \qquad \text{(III.4)}$$

where W is the active power absorbed by the capacitor and $\omega$ is the frequency.  If the current through the capacitor is sinusoidal (like the external voltage) then Q is proportional to the value of tan $\delta$ of the capacitor.

Instead of the energy radiated by the capacitor, we shall use the brightness B to which this energy is proportional.  In the majority of cases, the brightness increases as the frequency is increased, and begins to fall only at very high frequencies (cf. §1, Chapter II).

The experimental values of Q and B are compared with the calculated values of Q and $W_{R_0}$.  To obtain Q theoretically, it is necessary to calculate the active power W from the formula

$$W = \frac{1}{2} |U|^2 \operatorname{Re} \frac{1}{Z} , \qquad \text{(III.5)}$$

where Z is the total impedance of the capacitor.

The value of $W_{R_0}$, which we shall call the useful absorbed power, represents the power dissipated in a resistance $R_0$ which, in the majority of equivalent circuits, represents the active component of the impedance

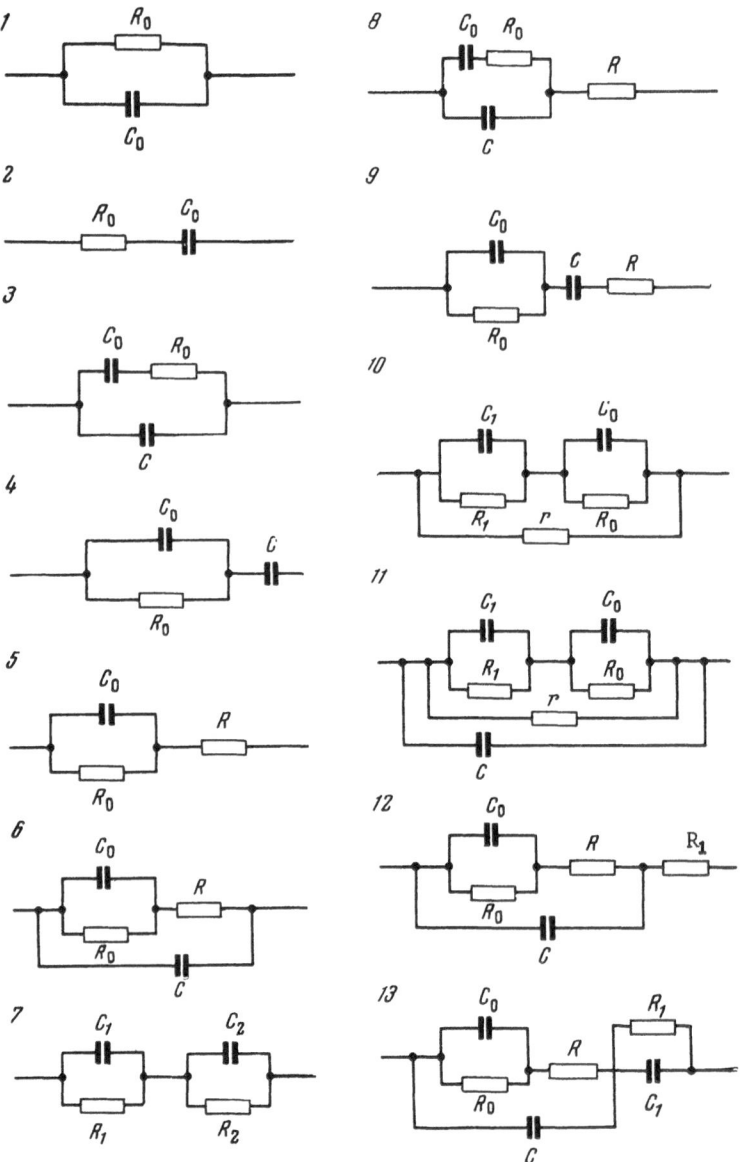

Fig. 8. Possible equivalent circuits for an electroluminescent capacitor.

of an electroluminescent phosphor. This is the only component of the power which can be transformed into light, because the absorption of the power in all other elements of the equivalent circuit of the capacitor should be regarded as parasitic from the point of view of electroluminescence. We note, however, that $W_{R_0}$ can be represented by the value of the brightness B only when the energy yield of an electroluminescent phosphor is constant.

We shall consider 13 possible equivalent circuits, beginning from those with two parameters and finishing with circuits with six constant parameters (cf. Fig. 8). We shall not give all the formulas because of their complexity, only the final result. It is found that several circuits with different numbers of parameters can explain qualitatively the frequency dependence of the energy absorbed in one period. However, circuit No. 5 is better than the others because it gives satisfactory qualitative agreement with experiment using the smallest number (three) of constant parameters.

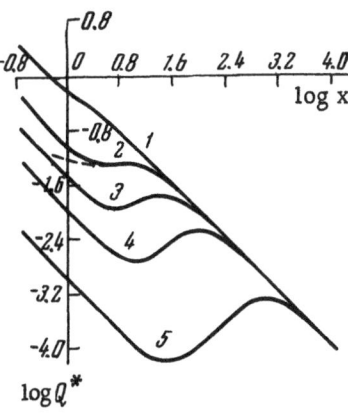

Fig. 9. Dependence of the dimensionless energy absorbed in the period on the dimensionless frequency plotted on a double logarithmic scale for a circuit with a constant $R_0$. Values of $\varkappa$: 1) 1; 2) 10; 3) 30; 4) 100; 5) 1000.

We shall consider this circuit in detail. Here, $R_0$ and $C_0$ are, respectively, the resistance and capacitance of the heterogeneous filling of the capacitor (phosphor + dielectric), and R is the resistance of the capacitor electrodes. For convenience, we shall introduce a dimensionless energy absorbed in one period

$$Q^* = \frac{2Q}{|U|^2 C_0 \varkappa} \qquad \text{(III.6)}$$

and a dimensionless frequency

$$x = \omega C_0 R_0. \qquad \text{(III.7)}$$

Then, the dependence of $Q^*$ on the frequency has the following form:

$$Q^* = \frac{1 + \varkappa + x^2}{[(1 + \varkappa)^2 + x^2] x}, \qquad \text{(III.8)}$$

where

$$\varkappa = \frac{R_0}{R}. \qquad \text{(III.9)}$$

A family of such dependences is shown on a double logarithmic scale in Fig. 9 (continuous curves). The nature of these curves can be explained as follows: At low frequencies, when $1/\omega C_0 \gg R_0$, an electroluminescent cell behaves as a pure resistance, and, if $R_0 \gg R$ (this is practically always satisfied), then

$$W = \frac{|U|^2}{2R_0} = Q\omega. \qquad \text{(III.10)}$$

For pure resistance, W is independent of the frequency, and, therefore, the frequency dependence of Q on the double logarithmic scale is a straight line with a slope the tangent of which is equal to unity. At high frequencies ($1/\omega C_0 \ll R$), the impedance of a capacitor is also ohmic, i.e., W = const, and again we have a linear region. At frequencies to the right of the maximum, practically all the power is absorbed by the electrodes, but the influence of the electrodes begins at frequencies to the right of the minimum because of the presence of a capacitance $C_0$ parallel to $R_0$, which starts to shunt $R_0$ and reduces the impedance of the cell.

The experimental curve is shown dashed in Fig. 9. It follows from this figure that the general nature of the experimental and theoretical curves is similar: linear parts are observed at low and high frequencies; however, the slopes of the experimental and theoretical curves are different at low frequencies. Thus, using an equivalent circuit, we can account qualitatively for the frequency dependence of the power absorbed in the capacitor. We must stress the nature of the discrepancies: at low frequencies, the experimental curve is flatter than the theoretical curve. The nature of the discrepancy between the experimental curve and curves calculated using other more complicated equivalent circuits is the same.

We shall introduce a dimensionless useful power $W_{R_0}^*$ by

$$W_{R_0}^* = \frac{2W_{R_0} R_0}{|U|^2 \varkappa^2}, \qquad \text{(III.11)}$$

where $W_{R_0}$ is the power absorbed by a resistance $R_0$, i.e., the useful power. Then, the dependence of $W_{R_0}^*$ on the dimensionless frequency is described by the formula

$$W_{R_0}^* = \frac{1 + x^2}{(1 + \varkappa + x^2)^2 + x^2 x^2}. \qquad \text{(III.12)}$$

$W_{R_0}^{\bullet}$ is constant at low frequencies but decreases when the frequency is high. Consequently, the frequency dependence of $W_{R_0}^{\bullet}$ does not agree even qualitatively with the frequency dependence of the brightness, because the brightness of an electroluminescent capacitor increases as the frequency is increased.

Comparison of the experimental and theoretical results (Fig. 9) shows that the resistance $R_0$ depends on the frequency. If $R_0 = $ const, then the tangent of the angle of slope should be −1 in double logarithmic coordinates; if the tangent of the angle of slope is $p - 1$, then $R_0$ depends on the frequency in accordance with the law

$$R_0 = \frac{R_0^{\bullet}}{v^p}, \tag{III.13}$$

where $R_0^{\bullet} = $ const and p = const.

### § 3. Allowance for the Hyperbolic Frequency Dependence of the Resistance $R_0$ [101]

Using the same equivalent circuit as before, but assuming that the pure resistance depends on the frequency in accordance with the law (III.13), we shall calculate the frequency dependence of the useful power and energy absorbed in one period (which we shall denote by the subscript "1").

In terms of dimensionless quantities, the frequency dependence of the energy absorbed in one period has the following form:

$$Q_1^{\bullet} = \frac{1 + ax^{\frac{p}{p-1}} + x^2}{\left[\left(1 + ax^{\frac{p}{p-1}}\right)^2 + x^2\right] x^{\frac{1}{1-p}}}, \tag{III.14}$$

where

$$Q_1^{\bullet} = \frac{4\pi RQ}{|U|^2 \gamma^{\frac{1}{1-p}}}, \tag{III.15}$$

$$\gamma = 2\pi C_0 R_0^{\bullet}, \tag{III.16}$$

$$a = \varkappa_1 \gamma^{\frac{p}{1-p}}, \tag{III.17}$$

$$\varkappa_1 = \frac{R_0^{\bullet}}{R}, \tag{III.18}$$

and

$$x = \gamma v^{1-p}. \tag{III.19}$$

The quantity $Q_1^{\bullet}$ depends on two parameters: p and $a$. The parameter $a$ has the same effect on $Q_1^{\bullet}(x)$ as $\varkappa$ in the case of $R_0 = $ const (Fig. 9). When $a$ increases, the separations between the minima and maxima along the $\log x$ and $\log Q_1^{\bullet}$ axes increase. The value of $a$ does not affect the slope of the linear parts of the curves. However, a change in the parameter p alters the slope of the linear part at low frequencies.

In Fig. 10, the abscissa represents $\log v$ with an accuracy to within a constant $\log \gamma$ [cf. Eq. (III.19)] and the ordinate in Fig. 10 represents $\log Q_1^{\bullet}$. The family of curves in Fig. 10 depends parametrically on p; $a = 3 \cdot 10^4$ for all the curves. It follows from Fig. 10 that, at low frequencies,

$$\frac{d(\log Q_1^{\bullet})}{d(\log v)} = p - 1. \tag{III.20}$$

In terms of dimensionless quantities, the useful power depends on the frequency as follows:

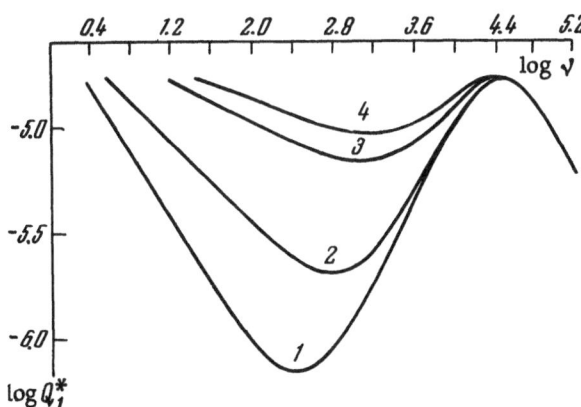

Fig. 10. Dependence of $Q_1^*(\nu)$ for several values of the parameter p: 1) $\frac{1}{5}$; 2) $\frac{1}{2}$; 3) $\frac{3}{4}$; 4) $\frac{4}{5}$.

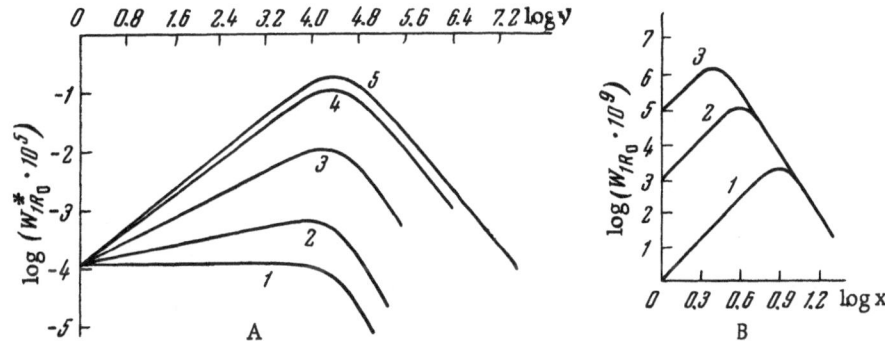

Fig. 11. Frequency dependence of $W_{1R_0}^*$. A) For $a = 3 \cdot 10^4$ and various values of p: 1) 0; 2) $\frac{1}{5}$; 3) $\frac{1}{2}$; 4) $\frac{3}{4}$; 5) $\frac{4}{5}$. B) For $p = \frac{4}{5}$ and various values of $a$: 1) $3 \cdot 10^{-4}$; 2) $10^{-3}$; 3) $10^{-2}$.

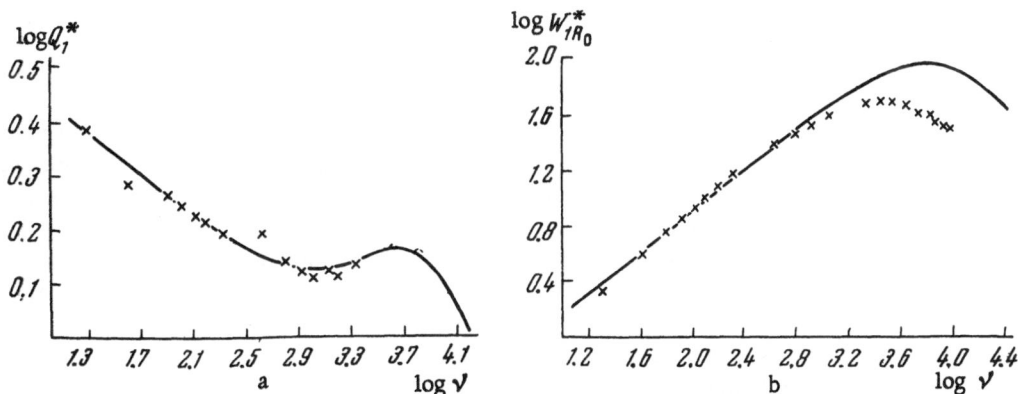

Fig. 12. Comparison of the experimental and theoretical data for capacitor No. 1. a) Frequency dependence of the energy on a double logarithmic scale; b) frequency dependence of the brightness on a double logarithmic scale.

$$W^{*}_{1R_0} = \frac{x^{\frac{p}{p-1}}}{\left(1 + ax^{\frac{p}{p-1}}\right)^2 + x^2},$$

(III.21)

where

$$W^{*}_{1R_0} = \frac{2W_{1R_0}R}{|U|^2 a}.$$

(III.22)

The dependence of $W^{*}_{1R_0}$ on the parameters p and $a$ is not very clear and, therefore, it is shown graphically in Fig. 11. The parameter p strongly affects the shape of the curves. For p = 0, i.e., for $R_0$ independent of the frequency, we find that the useful power is constant in a certain range of frequencies and then falls. However, for all p > 0, the value of $W^{*}_{1R_0}$ rises as the frequency is increased. At some frequency (which increases as p gets larger), the rise in $W^{*}_{1R_0}$ is replaced by a rapid fall. It must be stressed particularly that, at low frequencies,

$$\frac{d(\log W^{*}_{1R_0})}{d(\log \nu)} = p.$$

(III.23)

The change in the nature of $a$ does not affect the form of the curve $W^{*}_{1R_0}(x)$, but simply shifts it along the axes (Fig. 11B).

We shall now compare the experimental results with the theoretical calculations.

The crosses in Fig. 12a represent the experimental results for the capacitor No. 1 subjected to a voltage of 100 V. The tangent of the slope of the experimental curve plotted on a double logarithmic scale can be used to determine the exponent (index) of the hyperbola p [cf. Eq. (III.20)] and to calculate a family of curves which depend on one parameter $a$. The continuous curve in Fig. 12a is one of the curves of this family, which best fits the experimental dependence. If for given values of p and $a$ we now calculate the dependence of $W^{*}_{1R_0}$ on the frequency, we obtain the continuous curve shown in Fig. 12b. The experimental points obtained in the measurements of the brightness at low frequencies fit this curve well, but at high frequencies they lie somewhat below it. We note that the curves can be made to coincide by shifting along the ordinate axis only, because the relationship between x and $\nu$ is given by Eq. (III.19): here, $\gamma$ is a parameter whose value is found by superimposing on one another the curves representing the energy absorbed in one period.

Thus, using the $Q_1(\nu)$ curves to determine the parameters p, $\gamma$, and $a$, which describe the experimental dependence of the average power of the frequency, we can obtain a frequency dependence of the brightness which fits the experimental data over a wide range of frequencies. Since the measurements of the brightness are independent of the measurements of the power, the agreement between the theoretical and experimental results for the brightness can be regarded as a proof of the correctness of the equivalent circuit used in these calculations.

Comparing the degree of agreement between the experimental results and calculations, we note that, in the case of the actively absorbed power in the capacitor, the agreement is observed over the whole range of frequencies. As far as the brightness is concerned, at high frequencies, where the brightness passes through a maximum, the experimental values of $W^{*}_{1R_0}$ (which is the dimensionless value of the usefully absorbed power with which brightness is being compared) are found to be lower than those expected from calculations (Fig. 12). Thus, $W^{*}_{1R_0}$ approximates well the frequency dependence of the brightness at low frequencies only. The reason for the discrepancy at high frequencies is as follows: The brightness B can be represented by the power $W^{*}_{1R_0}$ absorbed by the heterogeneous layer only if the energy yield of the luminescence of the phosphor is constant. In fact, the energy yield depends very strongly on the voltage applied to a given particle of a phosphor. This question will be discussed in detail in Chapter IV.

§4.  Parameters of a Capacitor and the Generality
of the Proposed Circuit

a. Parameters of a Capacitor. In the equivalent circuit, we have assumed that R is the resistance
of the electrodes and $R_0$ and $C_0$ are the resistance and capacitance of the heterogeneous layer. Generally speaking, these parameters can be interpreted in several ways. For example, $R_0$ and $C_0$ can be used to describe the
phosphor itself, and R to describe the resistance of the dielectric. Because of this ambiguity, we shall consider
the problem in more detail.

To prove that R is the electrode resistance, we prepared a capacitor in which contacts with the electrodes
were located close to one another. If the resistance of the electrodes cannot be neglected, then the potential
difference applied to the heterogeneous layer will be different at different points, i.e., the brightness of the
luminescence at the surface of a capacitor will not be uniform and, with an increase in the frequency, the
luminous region will shrink to the contact area, as is indeed observed experimentally. Another way of proving
the point experimentally is to proceed as follows. If we prepare a batch of capacitors which differ from one
another only in the resistance of the transparent electrode R, then the extrema of the Q($\nu$) curve for capacitors
with large values of R should lie at lower frequencies. Such a dependence has been observed experimentally
for capacitors P-92, P-96, P-97, P-98. This dependence is shown in Fig. 15 and we shall consider it later (R increases as the capacitor number gets larger).

To prove that $C_0$ is the capacitance of the heterogeneous layer, we proceed as follows: The value of the
parameter $C_0$ which occurs in the calculations of the absorbed energy can be found by applying Eq. (III.16) to the
investigated samples. For sample P-92, it is found that $C_0 = 5 \cdot 10^{-9}$ F. On the other hand, the capacitance of
a parallel-plate capacitor of known dimensions can be easily estimated. Such an estimate for the same sample
gives the value $4.5 \cdot 10^{-9}$ F. The agreement between these two values is good. Similar results have been obtained for other samples. Usually, the value of the capacitance obtained from the parallel-plate capacitor
formula is slightly less. This is probably because of the inaccurate measurement of the thickness of the layer,
which was determined with a micrometer, and is larger than the true value of the thickness because of the surface roughness.

The hyperbolic frequency dependence of the resistance $R_0$ has been introduced on the basis of the experimental data, but it is in excellent agreement with the main ideas about the mechanism of electroluminescence.

The principal processes which result in electroluminescence take place in surface barrier layers which
exist at the boundaries separating components of the heterogeneous system. The application of an alternating
voltage alters the volume distribution of the charge in these components, and this may result in a frequency dependence of the pure resistance of the barrier.

For different capacitors, the values of the exponent (index) of the hyperbola p can be very different. The
upper limit of this parameter can be estimated theoretically. The condition of an extremum for the usefully
absorbed power $W_{1R_0}^*$ is as follows:

$$a^2 x_m^{\frac{2p}{p-1}} + \left(1 - \frac{2}{p}\right) x_m^2 = 1. \tag{III.24}$$

This equation can be rewritten by expressing the parameter $a$ in terms of the elements of the equivalent circuit
$R_0$, R, and $C_0$:

$$C_0 = \sqrt{\frac{\left(\frac{R_0^*}{R}\right)^2 - v_m^{2p}}{\frac{2}{p} - 1}} \cdot \frac{1}{2\pi v_m R_0^*}. \tag{III.25}$$

Hence it is clear that the parameter p should satisfy the condition p < 2; experimentally, we never found p > 1.
Usually, for good capacitors, p was close to 0.9-0.8. Numerous published papers on the frequency dependence

of the brightness, which is slightly sublinear, indicate that these values of p are highly typical [18, 84].

Experiments on quenched electroluminescent phosphors [88, 89] indicate a superlinear increase in the brightness with frequency. However, even in this case, the value of p is always less than 2, which is in full agreement with our estimate.

Data on the frequency characteristic of the barrier regions are presented in [102]. Friauf [103, 104] has carried out calculations for AgBr crystals with metal electrodes assuming that all carriers are captured in the contact barrier region. It is found that in this case the frequency dependence of the capacitance and of the barrier resistance is such that both these quantities are proportional to $\omega^{-2}$. This is the limiting case. Usually, only some carriers are captured in the electrodes and the degree of carrier capture is different under different conditions, and different frequency dependences are obtained. The exponents (indices) of the hyperbola may assume the values $^1/_2$, $^3/_2$, and can be different for the capacitance and for the resistance of the layer.

On the other hand, it has been reported that p = 0.5. This applies to the frequency dependence of the brightness, obtained by Diemer [44] for a single crystal of ZnS, and to measurements of the imaginary component of the permittivity carried out by Ince and Outley [79] on a large sample of polycrystalline zinc sulfide activated with copper.

These results are in very good agreement with the calculations of Shockley [35], who used the diffusion theory of rectification in an infinite single crystal to obtain a theoretical hyperbolic dependence of the resistance and capacitance of a p−n junction on the frequency. The exponent (index) of the hyperbola was found to be $^1/_2$.

b. Generality of the Proposed Circuit. From the preceding section, it is clear that a simple equivalent circuit which allows for the frequency dependence of the pure resistance is in good agreement with our experimental data. Therefore, it was of interest to determine the applicability of this circuit to the results of other workers. However, it was found that, with the exception of Zalm's paper [40], there were no published results of simultaneous measurements of the frequency dependences of the brightness and power. On double logarithmic scale, Zalm's dependences are represented satisfactorily by straight lines, and their slopes can be used to determine the exponent (index) of the hyperbola p. This exponent is found to be 0.78 from the measurements of tan δ and 0.75 from the measurements of the brightness. Such an agreement is very good. There is another interesting point about Zalm's results: when the frequency is increased from 20 cps to 10 kc, the influence of the electrodes and the shunting effect of the capacitance are completely absent, which is due to the low series resistance R of the electrodes. To achieve this, the capacitors should have the smallest possible area (here, 1 cm²).

Among the numerous experiments carried out by Zalm, the following is worth noting: He investigated the frequency dependence of the brightness of a capacitor in which the green-emitting electroluminescent phosphor ZnS: Cu, Al ($10^{-3}$ and $8 \cdot 10^{-4}$ g-atom/g-mole of Cu and Al, respectively) was suspended in tricresylphosphate. In such a sample, the ZnS grains are coupled electrically in the dc case because of the relatively low resistivity of tricresylphosphate. The frequency dependence of the brightness of this capacitor is shown in Fig. 13a. It differs considerably from the usual results for ZnS in an insulating medium because the brightness maximum lies at very low frequencies (less than 50 cps) so that the increase in the brightness, which in other capacitors extends from 20 cps to several kilocycles, is not observed at all. This experiment shows very clearly the effect of the medium in which an electroluminescent phosphor is suspended on the frequency dependence of the brightness. Zalm interpreted this experiment as a completely new frequency dependence of the brightness: the brightness was almost constant at low audio frequencies.

The experimental results given in Fig. 13a are replotted in Fig. 13b on a double logarithmic scale. We see that the curve is now of the same type as that for our capacitor No. 1 (cf. Fig. 12b) but it is strongly shifted in the direction of low frequencies. This was to be expected. Since the resistivity of the binder (tricresylphosphate) was low, the parameter $a$ of the capacitor, which was proportional to the resistance $R_0$ of the heterogeneous layer, should also be considerably smaller, but the decrease in $a$ shifts the brightness maximum in the direction of the low frequencies (cf. Fig. 11B).

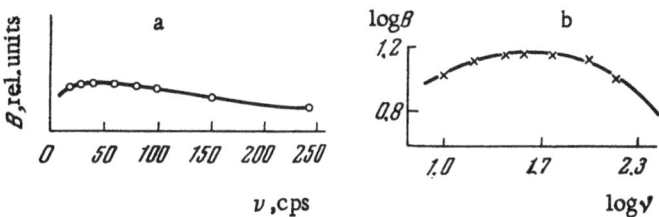

Fig. 13. Frequency dependence of the brightness of ZnS:Cu, Al suspended in tricresylphosphate: a) according to Zalm [40]; b) on a double logarithmic scale.

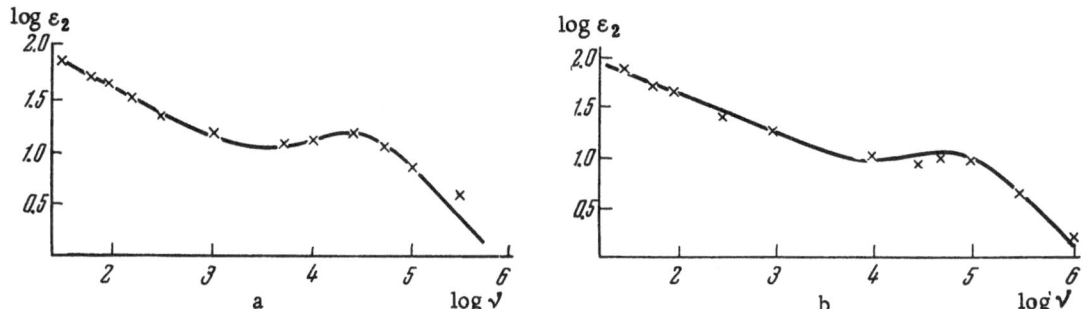

Fig. 14. Comparison of calculations: a) using circuit No. 5 (continuous curve) and circuit No. 13 (crosses) for p = 0.5; b) using circuit No. 5 for p = 0.6 and experimental data of Ince and Outley (crosses).

The frequency dependence of the brightness of fluorescent lamps made by the Sylvania Corporation has been investigated by Matossi [83]. At U = 380 V, $\nu > 7$ kc, the brightness is found to decrease. No explanation has been found for this effect. An approximate calculation on the basis of the equivalent circuit proposed in the present paper shows that, due to the influence of the resistance of the electrodes at 10 kc, the brightness of the lamp should be 10% less than the maximum brightness. This is in good agreement with Matossi's experiments [83].

Ince and Outley [79] measured the frequency dependence of the imaginary component of the complex permittivity $\varepsilon_2$ and calculated the active losses in a capacitor using circuit No. 13 (Fig. 8) on the assumption of a frequency dependence of two circuit elements: $C_0$ and $R_0$. Both these elements are assumed to be proportional to $1/\sqrt{\omega}$, in agreement with Shockley's theory [35]. We repeated the calculation, using the circuit employed by Ince and Outley [79] and the values of the parameters given in their paper, and compared the results with their calculations. The results agreed at all frequencies except at $3 \cdot 10^5$ cps. We then found that the resultant theoretical curve could be derived from a simpler circuit No. 5 (Fig. 8), which was used in the present investigation on the assumption that only the resistance $R_0$ depended on the frequency ($R_0 = R_0^* / \nu^P$, where p = 0.5). This theoretical curve is shown continuous in Fig. 14a; the crosses in the same figure represent the theoretical calculation of Ince and Outley based on circuit No. 13. Thus, the theoretical curve obtained by Ince and Outley [79] on the basis of a circuit with six parameters can also be derived from a circuit with three parameters.

It follows from Fig. 7 that the theoretical calculation of Ince and Outley does not satisfactorily represent their experimental results at $10^4 < \nu < 10^5$ cps. Figure 14b shows the theoretical curve (continuous line) calculated using our circuit. The values of $R_0^*$, R, and $C_0$ were taken from [79] and we assumed that p = 0.6 and not 0.5, as in [78]. This curve is in better agreement with the experiments of Ince and Outley, which are represented by crosses in the same figure.

Generally speaking, it is surprising that Ince and Outley did not find any extrema in their experimental or theoretical curves.

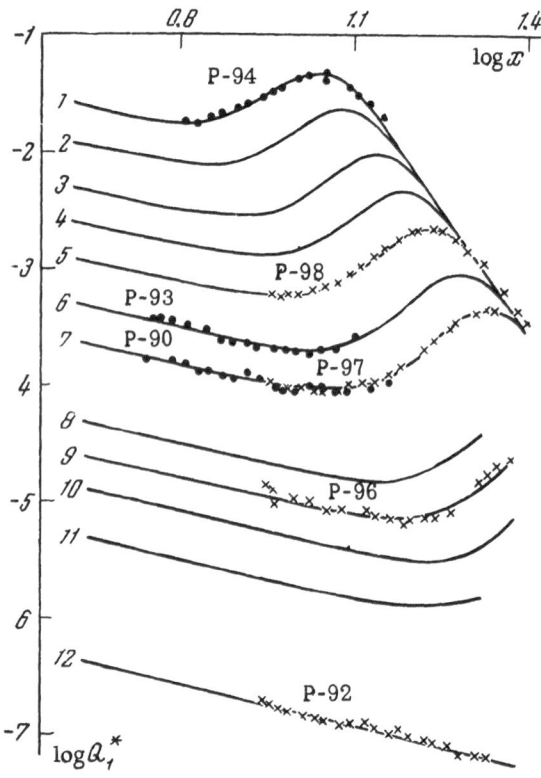

Fig. 15. Frequency dependence of the energy absorbed in one period. The continuous curves are the results of the calculation of $Q_1^*$; points and crosses are the experimental data.

c. Determination of the Parameters of a Capacitor. The experimentally determined frequency dependence of the energy Q absorbed in one period (Fig. 12a) can be used to determine all four parameters of the chosen equivalent circuit if the investigated range of frequencies includes a low-frequency linear region, a minimum, and a maximum. The parameter p, representing a heterogeneous layer, is found from the slope of the low-frequency linear region. If this value of p is used to calculate, by means of Eq. (III.14), the family of $Q_1^* = Q_1^*(x)$ curves which depend on the dimensionless parameter $a$ found from Eq. (III.17), we can find the value of $a$ corresponding to a given experimental curve by making the latter curve coincide with one of the theoretical curves. This fixes the curve relative to the dimensionless frequency axis and to the axis of dimensionless energy absorbed in one period. Then, Eqs. (III.19) and (III.17) can be used to determine the other parameters.

It should be stressed that the accuracy of the determination of the parameter $a$ is small: nevertheless, $R_0^*$, $C_0$, and R can be calculated with an accuracy up to 10% if p is found accurately. This can be done because the parameter p can be found simultaneously from the frequency dependence of Q and from the frequency dependence of the brightness plotted on a double logarithmic scale.

We shall now consider the possibility of determining p under unfavorable conditions. For example, in the range of frequencies investigated, a capacitor may have no low-frequency linear region in the dependence $Q(\nu)$ plotted using the coordinates $\log Q$, $\log \nu$ (cf. capacitors P-94, P-97, P-98 in Fig. 15). This is observed for capacitors with a high electrode resistance R. Sometimes the value of p of such capacitors cannot be de-determined either from the slope of the low-frequency region of the frequency dependence of the brightness plotted using the coordinates $\log B$, $\log \nu$. Then, p can be calculated indirectly, the characteristics having been determined of another capacitor with a low value of R but containing the same heterogeneous layer, since p is the characteristic of this layer itself. Figure 15 shows as a function of frequency the experimental results (crosses and points) for the energy absorbed in one period in two series of capacitors: first series (points) P-90, P-93, and P-94; second series (crosses) P-92, P-96, P-97, and P-98. All the capacitors in one series have the same dimensions, the same thickness, and the same composition of the heterogeneous layer (they were deposited at the same time) and differ only in the resistance of the transparent electrode, which increases in each series as the capacitor number goes up. In other words, all the capacitors in one series should have the same parameter p, which is governed only by the heterogeneous layer. The continuous lines in Fig. 15 represent the theoretically calculated $Q_1^*(x)$ curves for p = 0.85 (found experimentally for capacitors P-92 and P-90) for 12 values of the dimensionless parameter $a$: $10^7$, $2 \cdot 10^7$, $5 \cdot 10^7$, $10^8$, $2 \cdot 10^8$, $5 \cdot 10^8$, $10^9$, $5 \cdot 10^9$, $10^{10}$, $2 \cdot 10^{10}$, $5 \cdot 10^{10}$, $5 \cdot 10^{12}$ (curves 1-12, respectively, in Fig. 15).

It follows from Fig. 15 that the experimental results for other capacitors in the same range of frequencies fit well the theoretical curves corresponding to different but definite values of $a$, i.e., the remaining three parameters can be found for these capacitors.

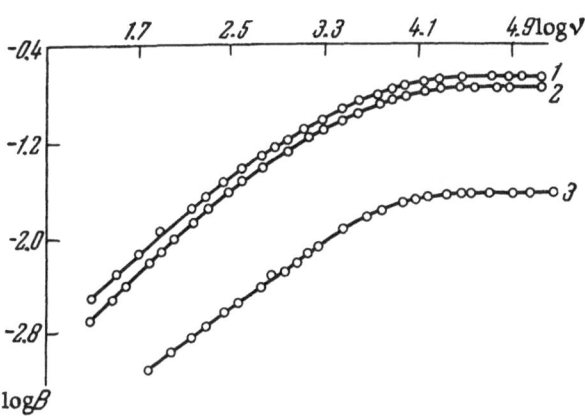

Fig. 16. Frequency dependence of the brightness B.

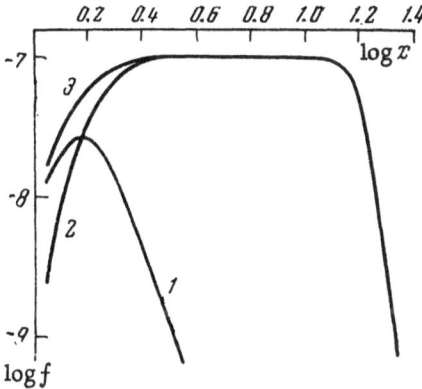

Fig. 17. Frequency dependence of the dimensionless energy absorbed only by barriers (curve 1), only by the interior (curve 2), and by barriers and the interior (curve 3) for dimensionless parameters $p = 0.85$, $a = 10^8$, $\varkappa_\infty = 10^7$.

## § 5. Luminescence of the Barriers and Luminescence of the Interior

In § 3 of the present chapter, we have considered the frequency dependence of the average brightness of capacitors, which shows that the brightness passes through a maximum. However, it is widely known that the brightness becomes saturated when the frequency is increased [39, 42, 83]. This is observed for capacitors with low electrode resistances R. Typical results obtained by us for two capacitors based on the green-emitting electroluminescent phosphor in polystyrene (curves 1 and 2) and on the blue-emitting phosphors in ÉP-096 resin (curve 3) are given in Fig. 16. To explain this frequency dependence, it is necessary to go back to the equivalent circuit [105].

So far, we have considered the equivalent circuit of an electroluminescent capacitor in which the heterogeneous layer resistance is found using Eq. (III.13). This formula is clearly approximate, since $R_0$ increases without limit as $\nu \to 0$, while it is known that, under dc conditions, capacitors have a low but finite conductivity. However, at very high frequencies ($\nu \to \infty$), $R_0$ decreases without limit, which is also untrue, since, even at the very highest frequencies, the losses in the heterogeneous layer itself are finite. Thus, the range of frequencies in which Eq. (III.13) is applicable has upper and lower limits. Therefore, it would be more correct to represent the resistance of the heterogeneous layer by the following formula:

$$R_0 = \frac{R_0^*}{(\alpha + \nu)^p} + R_\infty. \qquad \text{(III.26)}$$

The formula (III.26) differs from (III.13) by two new parameters $\alpha$ and $R_\infty$. However, the frequency bands in which these parameters are important are so far apart that it is, in practice, more convenient to consider separately the cases of high and very low frequencies. At high frequencies,

$$R_0 = \frac{R_0^*}{\nu^p} + R_\infty, \qquad \text{(III.27)}$$

i.e., only one new parameter $R_\infty$ is introduced and the physical meaning of this parameter can be easily established. It is known that high-resistivity transition layers are formed at a boundary between two components, and that the resistivity of these layers decreases as frequency is increased, while the volume resistivity of a crystal remains constant. Therefore, when the frequency is increased, the term $R_0^*/\nu^p$, which represents the barriers, may become comparable with or even smaller than the term $R_\infty$, which represents the interior; we may then assume that $R_0 = R_\infty = \text{const}$. In agreement with the definition of Q as the energy absorbed in one period, the dimensionless energy $Q_2^*$ for a heterogeneous layer of resistance $R_0$, defined by Eq. (III.27), is

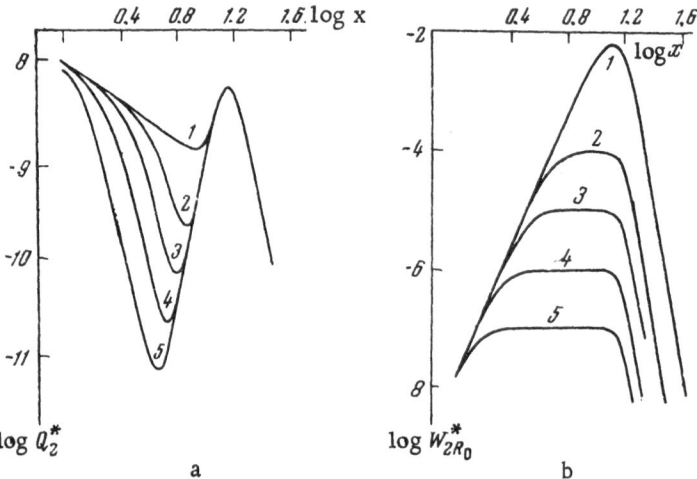

Fig. 18. Dimensionless frequency dependences, for five values of the parameter $\varkappa_\infty$, of the following quantities: a) dimensionless energy absorbed in one period, $Q_2^*$ in accordance with Eq. (III.28); b) dimensionless useful absorbed power $W_{2R_0}$, in accordance with Eq. (III.30). Values of $\varkappa_\infty$: 1) 99; 2) $10^4$; 3) $10^5$; 4) $10^6$; 5) $10^7$.

$$Q_2^* = \frac{1 + \varkappa_\infty + ax^{\frac{p}{p-1}} + \left(x + \frac{\varkappa_\infty}{a}x^{\frac{1}{1-p}}\right)^2}{\left[\left(1 + \varkappa_\infty + ax^{\frac{p}{p-1}}\right)^2 + \left(x + \frac{\varkappa_\infty}{a}x^{\frac{1}{1-p}}\right)^2\right]x^{\frac{1}{1-p}}}, \qquad (III.28)$$

where

$$\varkappa_\infty = \frac{R_\infty}{R}, \qquad (III.29)$$

and $Q_2^*$ is defined by Eq. (III.15). The subscript "2" is used because the value of $R_0$ is now different.

Since not all the energy absorbed by an electroluminescent capacitor can be transformed into light, we shall consider the useful absorbed power. We thus have three cases: 1) the energy is absorbed usefully only by the resistance $R_0^*/\nu P$, which represents the barriers; 2) the energy is absorbed usefully only by a constant resistance $R_\infty$, which represents the interior; 3) the energy is absorbed usefully by the total resistance $R_0$.

If the useful absorption of energy takes place in the barriers as well as in the interior (third case), then

$$W_{2R_0}^* = \frac{\varkappa_\infty + ax^{\frac{p}{p-1}}}{\left(1 + x_\infty + ax^{\frac{p}{p-1}}\right)^2 + \left(x + \frac{\varkappa_\infty}{a}x^{\frac{1}{1-p}}\right)^2}, \qquad (III.30)$$

where

$$W_{2R_0}^* = \frac{2W_{2R_0}R}{|U|^2} \qquad (III.31)$$

is the useful dimensionless power. For the first two cases, we obtain expressions which differ from (III.30) only in the numerator, which is $ax^{p/(p-1)}$ in the first case and $\varkappa_\infty$ in the second.

All these three cases are represented in Fig. 17, which shows that the useful absorption of energy by the barriers alone (curve 1) cannot explain the saturation of the brightness. The two other cases predict saturation but, nevertheless, the absorption of energy in the interior alone (curve 2) must be rejected because, in this case, the slope in the low-frequency region is twice as steep as that observed experimentally. The third interpretation — the useful absorption of the energy takes place in the barriers as well as in the interior — gives the correct frequency dependence at low frequencies and predicts saturation at high frequencies. Therefore, we shall consider it in some detail.

Figure 18 shows families of theoretical curves $Q_2^*$ and $W_{2R_0}^*$ for $p = 0.85$, $a = 10^8$. It follows from Fig. 18 that as $R_\infty/R$ increases, the maximum in the $W_{2R_0}^*$ broadens and the fall begins at lower frequencies. At very high values of $\varkappa_\infty$, the $W_{2R_0}^* = f(x)$ curve has a wide plateau (curves 4 and 5). The $Q_2^*$ curve at low values of $\varkappa_\infty$ and low frequencies has one decreasing linear region (curve 1), but when $\varkappa_\infty$ is increased, a kink appears (curves 2-5), which is followed by a more rapid decrease in $Q_2^*$ as the frequency is increased.

Thus, the frequency dependence of $W_{2R_0}^*$ clearly shows three main regions: 1) $R_0^*/\nu^P$ (increase in the brightness); 2) $R_\infty$ (constant brightness); 3) R (decrease in the brightness). If $R_\infty \lesssim R$, the second region is not very prominent. This should be observed for capacitors with large values of R. When $\varkappa_\infty = 0$, Eq. (III.30) transforms into (III.21).

Comparing the family of $W_{2R_0}^*$ curves with the family of $Q_2^*$ curves, we can draw the following conclusions: all three regions of the frequency dependence of $W_{2R_0}^*$ correspond to regions in the $Q_2^*$ curve with characteristic slopes, but the region at the lowest frequencies, where $R_0^*/\nu^P$ plays the dominant role, is sometimes hardly noticeable in the $Q_2^*$ curve, while it is still quite clear in the $W_{2R_0}^*$ curve (curve 4). Thus, all three regions of the frequency dependence of $Q_2^*$ and $W_{2R_0}^*$ are more clearly in evidence for higher values of $\varkappa_\infty$, provided the parameter $a$ is also large. At low values of $a$, calculations show that the boundaries between these regions become so indefinite that it is doubtful whether one can isolate the regions in which influence of $R_0^*/\nu^P$ or $R_\infty$ is dominant. Consequently, the $Q_2^*(x)$ and $W_{2R_0}^*(x)$ curves of some capacitors yield little information on the frequency dependence of the resistance, i.e., the determination of the parameters of the capacitor becomes very complicated. Thus, the frequency dependence of the useful absorbed power $W_{2R_0}^*$, which is being compared with the brightness of an electroluminescent capacitor on condition that $R_0$ is given by Eq.(III.27), has a rising region at low frequencies and a plateau at high frequencies. Consequently, this dependence can explain qualitatively the frequency dependence of the brightness of capacitors with low electrode resistances R (Fig. 16). We were unable to carry out a quantitative comparison because the frequency range (20 cps-200 kc) was insufficiently wide: the fall in brightness lay at higher frequencies and without this fall we could not determine $\varkappa_\infty$. We estimated that, for the results given in Fig. 16, $\varkappa_\infty > 10^2$.

We shall now consider very low frequencies, at which

$$R_0 = \frac{R_0^*}{(\alpha + \nu)^p} . \tag{III.32}$$

Under these conditions, the dependence of the useful power $W_{3R_0}$ (the subscript "3" is introduced because of a change in $R_0$) on the frequency $\nu$ is given by the formula

$$W_{3R_0} = \frac{|U|^2 R_0^* (\alpha + \nu)^p}{2R^2 \left\{ \left[ \frac{R_0^*}{R} + (\alpha + \nu)^p \right]^2 + (2\pi\nu C_0 R_0^*)^2 \right\}} , \tag{III.33}$$

from which it follows that when $\nu \to 0$, the useful power tends to a constant value

$$\frac{|U|^2 R_0^* \alpha^p}{2R^2 \left( \alpha^p + \frac{R_0^*}{R} \right)^2} .$$

However, if $\alpha \ll \nu$, this formula transforms into Eq. (III.21).

The lowest frequency at which our measurements were carried out was 20 cps, but there are published data for the frequency range $10^{-2}$ cps $\leq \nu \leq 10^2$ cps [87]. If we compare the results of these measurements with those expected on the basis of the proposed equivalent circuit, we find that down to $\nu = 1$ cps, Eq. (III.21) is still valid. According to this formula, the increase in the brightness as the frequency rises is slightly sublinear. At $\nu < 1$ cps, the brightness is reported to decrease more rapidly [87], which does not agree with Eq. (III.33). This disagreement can be explained as follows: At very low frequencies, the distribution of the fields between the components of the heterogeneous system changes considerably. In alternating currents, the fields are distributed inversely proportionally to the permittivities of the components [cf. Eq. (I.28)], which usually differ considerably, whereas, in a direct current, the distribution of the fields is inversely proportional to the conductivities [cf. Eq. (I.26)], which can differ by several orders of magnitude. The conductivity of ZnS is higher than the conductivity of the dielectric binder and, therefore, the fields in ZnS will decrease as $\nu \to 0$, i.e., the brightness will decrease more rapidly than at higher frequencies. Thus, at very low frequencies the formula (III.33) is inapplicable to heterogeneous systems because of a change in the nature of the field distribution between the components of the capacitor layer.

In experiments on single crystals there should be no such change in the fields, and, therefore, when the frequency is reduced the brightness should tend to a constant value, which is indeed observed experimentally [65]. Until now, no one had explained this.

This discussion of the frequency dependence of the brightness at low frequencies is also confirmed by the well-known experimental observation that the brightness of zinc sulfide electroluminescent phosphors is always lower for dc excitation than for ac excitation.

CHAPTER IV

# Dependence of the Characteristics of an Electroluminescent Capacitor on Voltage

In the preceding chapter, a relatively simple equivalent circuit was used to derive formulas for the total energy absorbed by an electroluminescent capacitor and for the energy which can be transformed into light. The agreement of these formulas with the experimental frequency dependences can be regarded as satisfactory. Therefore, it is of interest to consider the dependence of the parameters on the voltage U. Anticipating the results, we can mention here that these formulas do not agree with the experimentally determined dependences on U. This is because they are obtained on the assumption that the energy yield of a phosphor is constant. It is known that this yield depends strongly on the voltage (§ 1). An approximate allowance for this dependence makes it possible to explain the brightness characteristics (§ 2) and a more rigorous allowance gives a correct voltage dependence not only of the brightness but also of the optical yield of capacitors (§ 4).

## § 1. Dependence of the Brightness on the Voltage

The brightness of an electroluminescent capacitor increases very rapidly as the external exciting voltage is increased. This dependence can be described approximately by a power law [39, 83, 106-108]:

$$B \propto U^n, \tag{IV.1}$$

where B is the electroluminescence brightness, U is the external voltage, and n is a constant which lies in the range 2 < n < 8. However, this dependence can be described more accurately by an exponential law:

$$B \propto e^{-\frac{b}{\sqrt{U}}} \tag{IV.2}$$

or

$$B \propto e^{-\frac{U_0}{U}}, \qquad (\text{IV}.3)$$

where $U_0$ and b are constants.

The laws (IV.2) and (IV.3) can easily be related to the corresponding models of the process of excitation of an electroluminescent phosphor. The Destriau effect in insulated electroluminescent powders is associated, as shown in Chapter I, with the mechanisms of impact and tunnel excitation of the lattice (or activator). In both cases, the brightness is proportional to the ionization density, whose dependence on the field intensity is given by the formulas (I.11)-(I.13). However, depending on the nature of the barrier across which the field is concentrated, the relationship between the field intensity in the barrier and the potential difference is different for different mechanisms [cf. Eqs. (I.19), (I.23)]. If the excitation in a Mott—Schottky barrier and in a p—n junction is due to the tunnel or impact ionization, the brightness obeys (IV.2). In other cases of impact ionization, we should expect a voltage dependence of the brightness of the type given by Eq. (IV.3). When the field is concentrated in a Rose barrier [109], the impact ionization in weak fields and the tunnel ionization are represented by a dependence of the brightness on U in accordance with (IV.3). Impact ionization in a Rose barrier in strong fields should result in an even higher power of U, which is not confirmed experimentally.

The formulas (IV.2) and (IV.3) have a sign of proportionality because in a detailed calculation of the process of ionization and luminescence of ZnS different assumptions result in different actual formulas; experimenters describe their data using a number of exact mathematical formulas each of which includes an exponential dependence on U. We shall demonstrate this by several examples. Howard [110] assumes that the value of E is governed by the concentration of shallow donors, that the rate of liberation of carriers from deep donors is finite, and that the excitation of the activator is of the impact type; he further postulates that the donor levels all lie at the same depth. This gives the following formula for the brightness:

$$B = DU^{9/2}\left(1 - \frac{3\sqrt{U}}{b}\right)\exp\left(-\frac{b}{\sqrt{U}}\right), \qquad (\text{IV}.4)$$

which is in good agreement with Howard's experimental data and the results of Destriau and Ivey [84].

Taylor [111] has used the expression for the probability of electron acceleration, obtained by Seitz, and a model in which the field is concentrated at exhaustion barriers, to deduce a voltage dependence of the brightness in the form

$$B = B_0 U^n \exp\left(-\frac{b}{\sqrt{U}}\right), \qquad (\text{IV}.5)$$

where n is governed by the distribution of the levels.

The same formula with n = 0 has been used by Lehmann [112], Thornton [113], and Fok [58], whereas Schwertz et al. [114, 115] have found that the brightness of phosphors suspended in a dielectric or those chemically deposited fits Eq. (IV.5) very well for n = 1. Goldberg [88] found that the brightness changed in his experiments by only two orders of magnitude, and, therefore, that Eq. (IV.5) was equally valid with n = 0 and n = 1. The same could be deduced when the brightness varied by three orders of magnitude [111, 116]. This variety of the formulas employed is due to the fact that, usually, the ranges of the brightness and the exciting voltage are narrow, so that the weak dependence on U represented by the preexponential term is unimportant. Zalm, Diemer, and Klasens [59] varied the brightness by a factor of $10^8$ and even their experimental results were described satisfactorily by the following theoretical formulas:

$$B = DU^{1/2}\exp\left(-\frac{b}{\sqrt{U}}\right), \qquad (\text{IV}.6)$$

$$B = DU^{-5/4}\exp\left(-\frac{b}{\sqrt{U}}\right), \qquad (\text{IV}.7)$$

and by Eq. (IV.5) with n = 0. However, in addition to Eq. (IV.2), the formula

$$B = DU^n e^{-\frac{U_0}{U}} , \qquad\qquad\qquad (IV.8)$$

proposed by Destriau [93] is also used. The constant n can assume different integral values for different samples; these values range from zero to three [83, 85, 93, 117].

Thus, most investigators have used the formula (IV.2), although the weak dependence on the voltage represented by the pre-exponential term may differ from case to case.

Without considering in detail the frequency dependence of the pre-exponential term, we must draw attention to the coefficient b. This coefficient is different for capacitors prepared in different ways. For capacitors containing a fractionated electroluminescent phosphor, the value of b increases when the crystallite dimensions are reduced [112, 118, 119] and it is different for different activators in the same phosphor [59]. Prolonged operation of a capacitor increases the value b even when the excitation conditions are constant [113].

Several investigators have reported that the coefficient b depends on the frequency. For example, the dependence of the brightness on the voltage has been measured at two exciting field frequencies [41, 111, 112, 120]; b has been found to increase with the frequency. Lehmann [72] measured b at three fixed frequencies, the extreme frequencies differing by a factor of 100. He found that b changes markedly only at high frequencies, while at low frequencies it can be regarded as constant. Fuller information on the frequency dependence of b is not yet available, and there is as yet no interpretation of this phenomenon.

Since, in a given substance, electroluminescence may be due to quite different mechanisms, depending on the positions of local levels and on the structure of the local field, the problem of the validity of formulas (IV.2) and (IV.3) under particular experimental situations has not yet been solved. Nevertheless, attempts have been made to find the limits of validity of these laws. Lehmann [72], supported by experimental data, has suggested that the law of emission by single particles is that given by (IV.3), while (IV.2) should be considered as an integral radiation law of the whole system of particles of various dimensions. Mathematical calculation of the brightness of the electroluminescence of a system of particles (II.1), where each particle emits in accordance with the law (IV.3), shows that, in the case of moderate and weak voltages, the emission is in agreement with Eq. (IV.2), but in strong voltages it obeys (IV.3).

The transition from the dependence (IV.2) to (IV.3) has also been observed [121, 122] when the external voltage is increased, but these results have been interpreted differently. The field in a barrier is proportional to $\sqrt{U}$ as long as the excitation region is smaller than the dimensions of the crystal. However, when the external voltage is increased, the barrier widens and may eventually extend over the whole crystal; then the field in the barrier will vary proportionally to the applied voltage. The value of the external voltage at which this transition from one law to another is observed, and the dimensions of crystallites in an electroluminescent phosphor, have been used [122] to estimate the true field intensity in the barrier. It has been found that $E \approx 1.4 \cdot 10^5$ V/cm.

## § 2. Approximate Allowance for the Efficiency of an Electroluminescent Phosphor [123]

In the preceding section we considered several formulas which approximate quite satisfactorily the experimentally observed voltage dependence of the electroluminescence brightness. Each of these formulas has an exponential term which represents a very strong dependence of the brightness on the voltage. This reflects the proportionality of the number of radiative recombination events to the ionization density and represents an approximate allowance for the efficiency of an electroluminescent phosphor. We have introduced the term "efficiency," which will be used later as a characteristic of the energy yield of an electroluminescent phosphor, but these two concepts are not identical. The use of the term "efficiency" has been motivated by the following considerations: To find the energy yield of an electroluminescent phosphor, it is necessary to know the energy emitted and absorbed. The emitted energy can easily be measured, but it is practically impossible to determine the absorbed energy if the experiments are carried out on heterogeneous systems, because the dielectric binder takes part in the absorption of energy. This means that the experimentally determined value of the energy absorbed by a capacitor represents the sum of the absorption by the dielectric and the phosphor. Therefore, the ratio of the emitted to the absorbed energy does not give the energy yield of the phosphor but only a value which represents it in some measure and which we shall call the "efficiency."

Thus, in that range of voltages in which the absorption of energy is proportional to $U^2$, the efficiency of an electroluminescent phosphor is approximately proportional to the density of ionization.

In our experiments, the dependence of the brightness on the voltage was described by Eq. (IV.2) better than by Eq. (IV.3). Hence, it follows that the density of ionization in the process of excitation and the efficiency of a phosphor K, proportional to this density, have the form

$$K \propto \exp\left(-\frac{b_i}{E}\right),\tag{IV.9}$$

where $b_i$ and E represent an exhaustion barrier. If we consider an array of barriers connected in series, it is necessary to replace these quantities by their integrals, and, therefore, we find that

$$K \propto \exp\left(\frac{b}{\sqrt{U}}\right),\tag{IV.10}$$

where U is the external voltage. However, the latter expression is valid only in the case when the whole external voltage drop is concentrated in the barriers in the heterogeneous layer.

In §3 of the preceding chapter, it was shown that this is true only at low frequencies. At high frequencies, the absorption by the electrodes becomes important. Consequently, it is necessary to determine that fraction of the external voltage which represents the potential drop across the barriers. If the resistance of a heterogeneous layer is defined by Eq. (III.13), the whole voltage applied to the heterogeneous layer is concentrated in the barriers. The expression for the efficiency then assumes the form

$$K \propto \exp\left(-\frac{b^* \sqrt{|Z_{tot}|}}{\sqrt{U|Z_{hl}|}}\right),\tag{IV.11}$$

where $Z_{tot}$ is the total impedance of a capacitor and $Z_{hl}$ is the impedance of the heterogeneous layer.

Knowing the efficiency K and the energy absorbed by the heterogeneous layer $W^*_{1R_0}$, we can obtain an expression for the radiant flux F of an electroluminescent capacitor

$$F = K(U, x)W^*_{1R_0}.\tag{IV.12}$$

Since the brightness B is directly proportional to the radiant flux F, the validity of the expression for K (IV.11) can be established by considering the behavior of the brightness.

If the parameter b represents the phosphor, then, other conditions being equal, capacitors with different dielectric binders should have the same value of b. Figure 19 confirms this for capacitors in which the following four dielectrics of approximately the same thickness have been used: polystyrene with nitrocellulose (triangles), ÉP-096 resin (crosses), ML-92 resin (points), polyethylene terephthalate (circles). The frequency of the exciting field in this experiment was 500 cps. The straight lines in Fig. 19 are shifted along the ordinate axis in order to obtain coincidence.

Secondly, when the external field frequency is increased, the experimentally measured value of b = $b^* \sqrt{|Z_{tot}|/|Z_{hl}|}$ should increase. For a detailed verification of this conclusion, it is necessary to calculate

$$\sqrt{\frac{|Z_{tot}|}{|Z_{hl}|}} = \sqrt[4]{\frac{\left(1 + ax^{\frac{p}{p-1}}\right)^2 + x^2}{\left(ax^{\frac{p}{p-1}}\right)^2}}.\tag{IV.13}$$

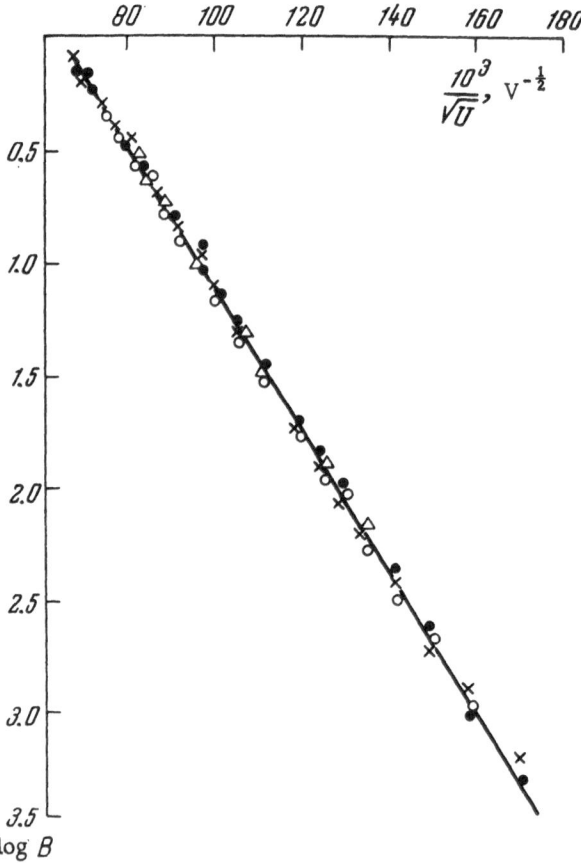

Fig. 19. Dependence of the brightness B on U for four types of dielectric binder (represented by different types of point).

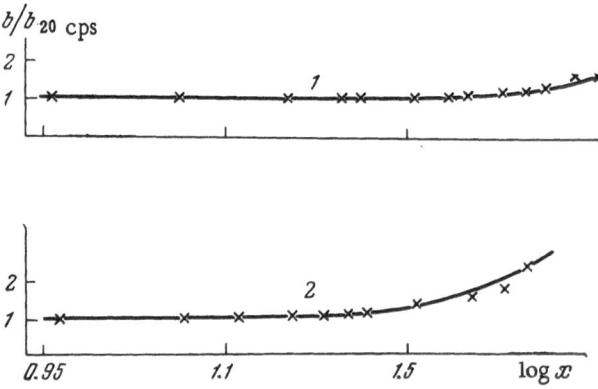

Fig. 20. Comparison of the experimental frequency dependence of the normalized index of the exponential function $b/b_{20\,cps}$ (crosses) with the dependence given by Eq. (IV.13) (continuous curves) for capacitors P-97 (curve 1) and P-98 (curve 2).

Here, $Z_{tot}$ and $Z_{hl}$ are expressed in terms of $a$ and $p$, which are the dimensionless parameters of the capacitor introduced in an earlier chapter; p is governed solely by the properties of the heterogeneous layer, while $a$ depends on all the components of the capacitor and is calculated using Eq. (III.17). Calculations by means of Eq. (IV.13) are easy for the capacitors investigated because the parameters p and $a$ for these capacitors can be found from independent energy measurements (cf. §4, Chapter III).

In Fig. 20, the continuous curves show the results of a calculation while the crosses represent the normalized experimental values of $b/(b)_{20\,cps}$ for capacitors P-97 and P-98 in which polystyrene is used as the binder. The parameters p and $a$ determine the shape of the curve, while the position of the curve along the frequency axis $\nu$ is determined by the parameter $\gamma$. Allowance for all these three parameters produces good agreement with experiment.

We finally find that the efficiency of an electroluminescent phosphor is given by

$$K \propto \exp\left(-\frac{b^*}{\sqrt{U}}\sqrt{\frac{|Z_{tot}|}{|Z_{hl}|}}\right).$$

Knowing the value of the useful absorbed power $W^*_{1R_0}$ and the efficiency of an electroluminescent phosphor, we find that the brightness $B_1$ is

$$B_1 \propto W^*_{1R_0}\exp\left(-\frac{b^*}{\sqrt{U}}\sqrt{\frac{|Z_{tot}|}{|Z_{hl}|}}\right). \qquad (IV.14)$$

The subscript "1" is introduced, as before, because $R_0$ is defined by Eq. (III.13). Here, the second factor represents the strong dependence of the efficiency on the voltage, which is characteristic of the phosphor itself. The frequency dependence of this factor is due to the fact that the phosphor represents only a part of a capacitor, and the voltage across the phosphor decreases when the frequency is increased because of the shunting effect of the capacitance $C_0$. The formula (IV.14) is valid if the phosphor itself does not exhibit a frequency dependence of the efficiency. In this case, the dependence of the dimensionless brightness

$$B^*_1 = A_1\frac{2W_{1R_0}R}{|U|^2 a} \qquad (IV.15)$$

on the dimensionless frequency x has the form

$$B^*_1 = \frac{A_1 x^{\frac{p}{p-1}}}{\left(1+ax^{\frac{p}{p-1}}\right)^2+x^2}\exp\left[-\frac{b^*\sqrt[4]{\left(1+ax^{\frac{p}{p-1}}\right)^2+x^2}}{\sqrt{Uax^{\frac{p}{p-1}}}}\right], \qquad (IV.16)$$

where $A_1$ and $b^*$ are constants.

The effect of the parameter $a$ on this dependence is illustrated in Fig. 21a, where the continuous curves show, for the sake of comparison, the results of a calculation of the useful absorbed power $W^*_{1R_0}$. The calculation has been carried out for p = 0.85 and three values of the parameter $a$ ($10^7$, $5 \cdot 10^7$, $2 \cdot 10^8$). The linear parts of both functions $f$ are made to coincide by shifting them along the ordinate. The introduction of an exponential factor does not change the form of the curve at low frequencies ($|Z_{tot}|/|Z_{hl}| = 1$), but results in a more rapid fall at high frequencies. Hence, we see the brightness agrees with $W^*_{1R_0}$ only at low frequencies. The parameter p affects the form of the $B^*_1 = B^*_1(x)$ curve in the same way as it affects $W^*_{1R_0}$ (cf. Fig. 11A). The third dimensionless parameter, $b^*/\sqrt{U}$, alters the form of the curve slightly, but it shifts the brightness maximum along the frequency axis (Fig. 21b). The values of $b^*/\sqrt{U}$ are given alongside the curves.

Comparison with experimental values shows that Eq. (IV.16) has a number of advantages compared with Eq. (III.21):

1. Equation (IV.16) gives a more accurate dependence of the brightness on the voltage. According to (III.21), the voltage dependence of the brightness should be quadratic, while the experimental dependence is

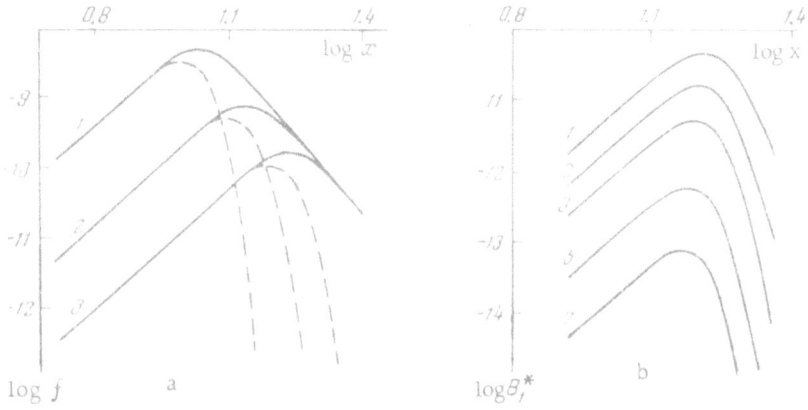

Fig. 21. Influence on the brightness $B_1^*$ of the parameters occurring in Eq. (IV.16). a) Dimensionless frequency dependence of $W_{1R_0}^*$ (continuous curves) in accordance with Eq. (III.21) and a similar dependence of $B_1^*$ (dashed curves) according to Eq. (IV.16). Values of a: 1) $10^7$; 2) $5 \cdot 10^7$; 3) $2 \cdot 10^8$. b) Influence of the dimensionless parameter $b^*/\sqrt{U}$ on $B_1^*$ in accordance with Eq. (IV.16).

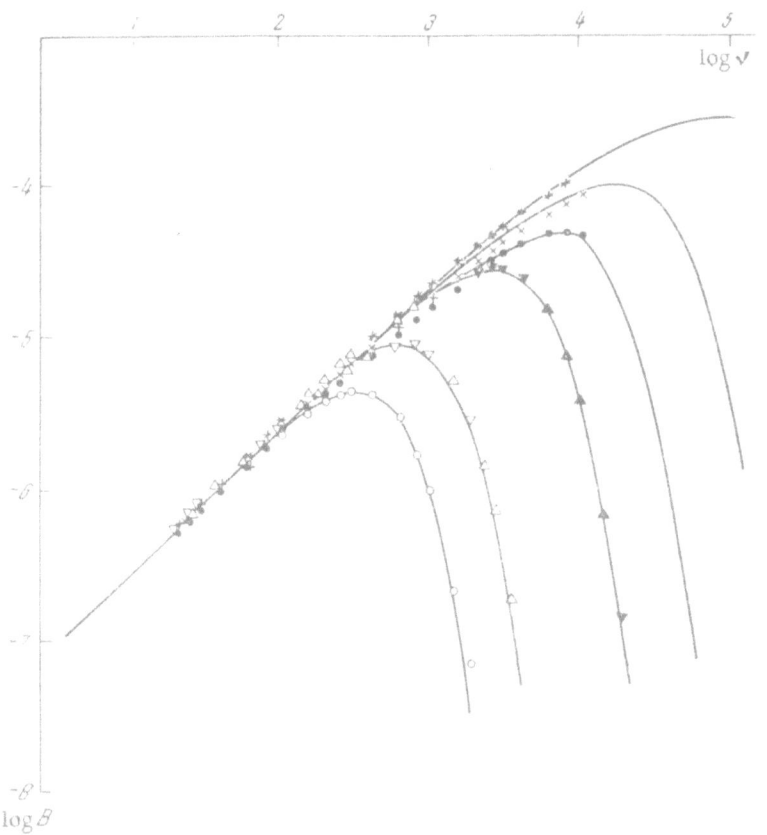

Fig. 22. Comparison of the experimental data (circles, triangles, black dots, crosses, stars) with calculations based on Eq. (IV.16) (continuous curves).

much stronger [83]. Usually, this dependence is described by Eq. (IV.2). According to Eq. (IV.16), the dependence of B on U is governed mainly by an exponential term of the same type as in Eq. (IV.2).

2. Equation (IV.16) explains the frequency dependence of the index of the exponential dependence of the brightness which is due to a redistribution of the external voltage between the components of an electroluminescent capacitor. This has been clearly demonstrated for capacitors P-97 and P-98 (Fig. 20).

3. Equation (IV.16) correctly reflects the frequency dependence of the brightness B. The agreement between experiments and Eq. (IV.16) with respect to this frequency dependence can be easily checked because the parameters p, $a$, and $\gamma$, which govern the form of the $B_1^* = B_1^*(x)$ curve and its position along the frequency axis, can be determined for each capacitor from independent energy measurements (cf. §4, Chapter III). The continuous curves in Fig. 22 represent the theoretical calculations. Stars, crosses, black dots, black triangles, open triangles, and circles represent the experimental results for capacitors with electrode resistances of 0.05, 0.54, 1.4, 4.9, 21, and 53 k$\Omega$, respectively. The agreement between the calculations and experimental results is good (cf. Fig. 12b).

4. Finally, Eq. (IV.16) shows that the frequency corresponding to the brightness maximum $x_m$ depends on the exciting field intensity.

An investigation of the function $B_1^*$ at an extremum gives a relationship between the frequency at which the brightness has its maximum value ($x_m$) and other parameters of the capacitor

$$\frac{b^*}{2\sqrt{\bar{U}}} \sqrt[4]{\frac{x_m^2 + \left(1 + ax_m^{\frac{p}{p-1}}\right)^2}{\left(ax_m^{\frac{p}{p-1}}\right)^2}} = \frac{pa^2 x_m^{\frac{2p}{p-1}} - p + (p-2)x_m^2}{x_m^2 + p + apx_m^{\frac{p}{p-1}}}. \tag{IV.17}$$

Hence, we see that the frequency of the brightness maximum is related to the value of the external voltage. The theoretical curves in Fig. 21b illustrate this dependence: when U is reduced, the maximum shifts toward lower frequencies. Such a shift has been observed experimentally for a very large number of electroluminescent capacitors. Figure 23a presents the results for the capacitor P-94 at U = 30, 50, 70, 100, 120, and 150 V (curves 1-6, respectively). It should be mentioned that not only the direction of the shift of the maximum, but also its value, is predicted correctly, while the function $W_{1R_0}^*$ does not predict any voltage dependence of the frequency of the brightness maximum. Figure 23b shows a similar dependence for the energy Q absorbed in one period, for the same voltages.

We have already mentioned in §1, Chapter III, that a shift of the brightness maximum along the frequency axis when the value of U is altered has already been observed [83] but has not yet been accounted for.

Thus, we may conclude that allowance for the efficiency of an electroluminescent phosphor in accordance with Eq. (IV.11) makes it possible to explain the frequency dependence of the brightness B of a capacitor which exhibits a maximum of B, and the dependence of the brightness of an electroluminescent capacitor on the external voltage across it.

All this applies to the case when the resistance of the heterogeneous layer $R_0$ is represented by Eq. (III.13). However, as shown in §5, Chapter III, the resistance $R_0$ is in general represented by Eq. (III.27). The question therefore arises: what is the form of the exponential term which governs the efficiency of an electroluminescent phosphor in the case represented by (III.27)? Since the behavior of the resistance $R_0$ is now complicated and its different components may play different roles in the process of electroluminescence, we cannot predict the role of the volume resistance $R_\infty$. Therefore, it is necessary to consider two expressions for the efficiency of an electroluminescent phosphor, which we shall denote by $K_2$ and $\mathcal{E}_2$:

1. The resistance $R_\infty$, as well as $R_0^*/v^p$, takes part in the electroluminescence process:

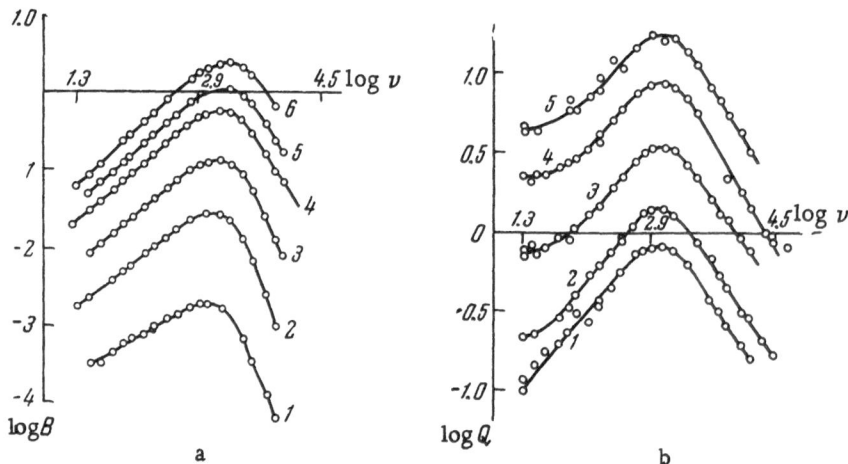

Fig. 23. Experimental frequency dependence of: a) the brightness B of capacitor P-94; b) the energy Q absorbed by capacitor P-98 in one period. The different curves represent different values of the external voltage U.

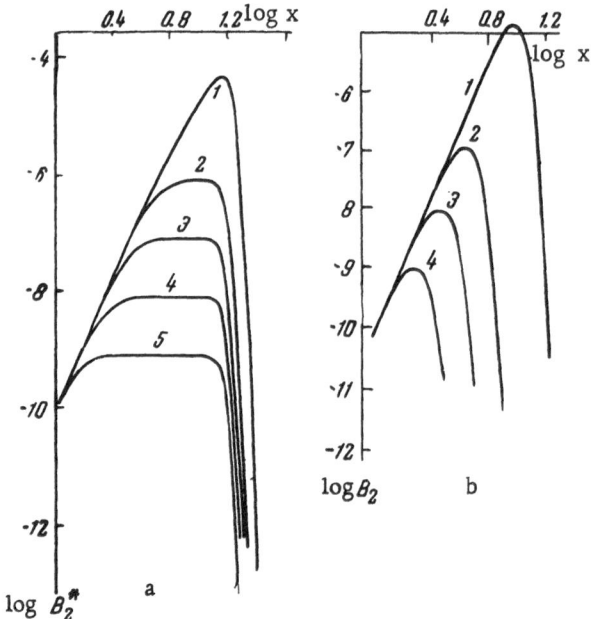

Fig. 24. Frequency dependence of the dimensionless brightness: a) $B_2^*$ calculated from Eq. (IV.18); b) $B_2^*$ in accordance with Eq. (IV.19).

$$K_2(U,\,x) = \exp\left[-\frac{b^* \sqrt[4]{\left(1 + \varkappa_\infty + ax^{\frac{p}{p-1}}\right)^2 + \left(x + \frac{\varkappa_\infty}{a}x^{\frac{1}{1-p}}\right)^2}}{\sqrt{U\left(\varkappa_\infty + ax^{\frac{p}{p-1}}\right)}}\right];$$ (IV.18)

2. The important components is only $R_0^*/\nu P$

$$\delta_2(U,\,x) = \exp\left[-\frac{b^* \sqrt[4]{\left(1 + \varkappa_\infty + ax^{\frac{p}{p-1}}\right)^2 + \left(x + \frac{\varkappa_\infty}{a}x^{\frac{1}{1-p}}\right)^2}}{\sqrt{U\,ax^{\frac{p}{p-1}}}}\right],$$ (IV.19)

where $b^*$ is a constant.

Because there are two possible expressions for the exponential term, the brightness $B_2^*$ will also be expressed in two different ways. We shall not give the appropriate formulas because they are very cumbersome and because they can be derived very easily using the formulas presented earlier and the expression for $W_{2R_0}^*$ given by Eq. (III.30). Figure 24a shows the frequency dependences of $B_2^*$. The calculation has been carried out for the following values of the parameters: $a = 10^8$, p = 0.85, $\varkappa_\infty = 99, 10^4, 10^5, 10^6, 10^7$ (curves 1-5, respectively).

The results with higher values of $\varkappa_\infty$ are very different. The plateau in the frequency dependence of $B_2^*$ is obtained only when the role of $R_\infty$ in the process of electroluminescence is as important as the role of the barrier resistance $R_0^*/\nu P$ because otherwise the $B_2^*(x)$ simply have a maximum. Thus, from the experimental data (Fig. 16) we may conclude that the efficiency of an electroluminescent phosphor is given by Eq. (IV.18) if the electrode resistance of a capacitor is small and if it is necessary to allow for the resistance of the interior of a crystal, $R_\infty$. This formula simplifies [cf. the second factor in Eq. (IV.16)] for $\varkappa_\infty = 0$, i.e., when the resistance $R_\infty$ can be neglected.

In summarizing our discussion of capacitors with low electrode resistances, we may conclude that the interior of a crystal plays an important part in the light emission properties, and that, under certain conditions, the interior may emit more strongly than the boundary regions.

Thus, an investigation of the frequency dependence of the electroluminescence, allowing for the efficiency of the phosphor, yields information on the behavior of the barriers and the interior of a crystal, and on their respective roles in the luminescence and ionization processes.

## §3. Dependence of the Absorbed Energy on the Voltage

The problem of the dependence of the energy absorbed by an electroluminescent capacitor on the voltage has been investigated much less than the voltage dependence of the brightness. Nevertheless, the dependence of the energy on the voltage is important not only from the practical point of view but also in relation to the processes taking place in an excited phosphor. The scarcity of investigations of this dependence is due to the difficulty of measuring the value of the actively absorbed energy, as mentioned in Chapter II. Therefore, in the present section we shall consider in detail the experimental results on the absorption, measured by methods which allow for the nonsinusoidal nature of the waveform of the current in electroluminescence [78, 101].

We must first recall that the active power absorbed by an electroluminescent capacitor has been calculated in § 3, Chapter III. The result obtained in that section has been considered from the point of view of the frequency dependence, and Eq. (III.14) has been found to be in good agreement with the experimental results. In investigations of the frequency dependences, the external voltage is assumed to be a parameter which, according to Eqs. (III.14) and (III.15), only affects the value of the absorbed energy and not its dependence on the frequency. Consequently, the extremal points of $Q(\nu)$ should lie at the same frequencies for different values

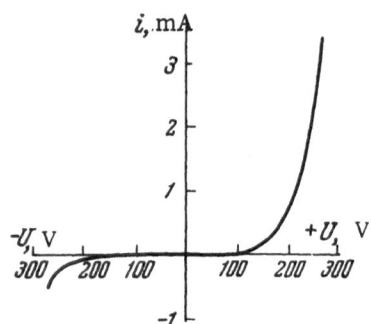

Fig. 25. Current-voltage characteristic of an electroluminescent capacitor. The sign of the voltage refers to the voltage applied to the lower electrode.

of U. Figure 23b shows the dependence Q($\nu$) for sample P-98 for U = 50, 70, 100, 150, and 200 V (curves 1-5, respectively). We can see that the maximum of Q is not affected by U, in contrast to the behavior of the brightness B (Fig. 23a). According to Eq. (III.15), the actively absorbed power W should depend quadratically on the voltage, whereas, in fact, this dependence is stronger. Usually, $W \propto U^n$, where n > 2, although at low voltages this dependence may be regarded as quadratic. Thus, we have an obvious discrepancy between the calculations and the experimental results. The reason for this discrepancy is as follows: The expression for the actively absorbed power has been obtained using Eq. (III.5). However, formula (III.5) is applicable only to systems which obey Ohm's law, and, as already mentioned [82], an electroluminescent capacitor behaves as a nonlinear element. This nonlinearity appears clearly in the processes of electrical energy absorption. A demonstration of the nonlinear relationship between the current and the voltage across an electroluminescent capacitor is given in Fig. 25. Here, U is the constant voltage applied to a sample and i is the constant current flowing through the sample. The current-voltage characteristic clearly shows the nonlinear relationship between the current and the voltage.

In our calculations of the absorbed energy, the nonlinearity of the system with respect to the voltage has not yet been allowed for and, therefore, the experimental results and theory disagree.

When U is increased, the nonlinearity of the system gets stronger and the deviations from Eq. (III.15) become greater. Thus, to describe the dependence of the absorbed energy on the voltage, it is necessary to allow for the nonlinearity of the system.

The first attempt to allow for this nonlinearity was made by Lehmann [97], who calculated the absorbed power from the formula

$$W = \omega U^2 C \cos\varphi, \tag{IV.20}$$

where C is defined by Lehmann as the complex capacitance, and $\varphi$ is the phase shift between the current through the sample and the applied voltage. The value of $C \cos\varphi$ increases when the external voltage is increased. From his experiments, Lehmann found an expression for $C \cos\varphi$ in terms of the electroluminescence brightness B and constants $k_0$ and k:

$$C \cos\varphi = k_0 + k\,(B/\omega)^{1/2}. \tag{IV.21}$$

Since the brightness B is a nonlinear function of the voltage U, in accordance with (IV.2), the second term in Eq. (IV.21) represents the absorption of energy by a nonlinear resistance. Lehmann did not give any physical interpretation of this result.

The absorption of the energy in a capacitor has been considered in [55] on the basis of the following model: Free electrons are generated mainly in a barrier, where the field has its maximum value, and they are all subjected to practically the same potential difference U. The total energy absorbed by all these electrons is $W_{tot} \approx eN_e U$, where $N_e$ is the number of free electrons. From the temperature dependence of the brightness waves, it has been concluded that the liberation of electrons from capture levels at room temperature is governed by multiphonon tunnel transitions whose probability is assumed to be proportional to $N_e$. When these assumptions are made, the power W is found to be

$$W \propto U^{3/2} \exp(b_1^0 U), \tag{IV.22}$$

where $b_1^0$ is a constant.

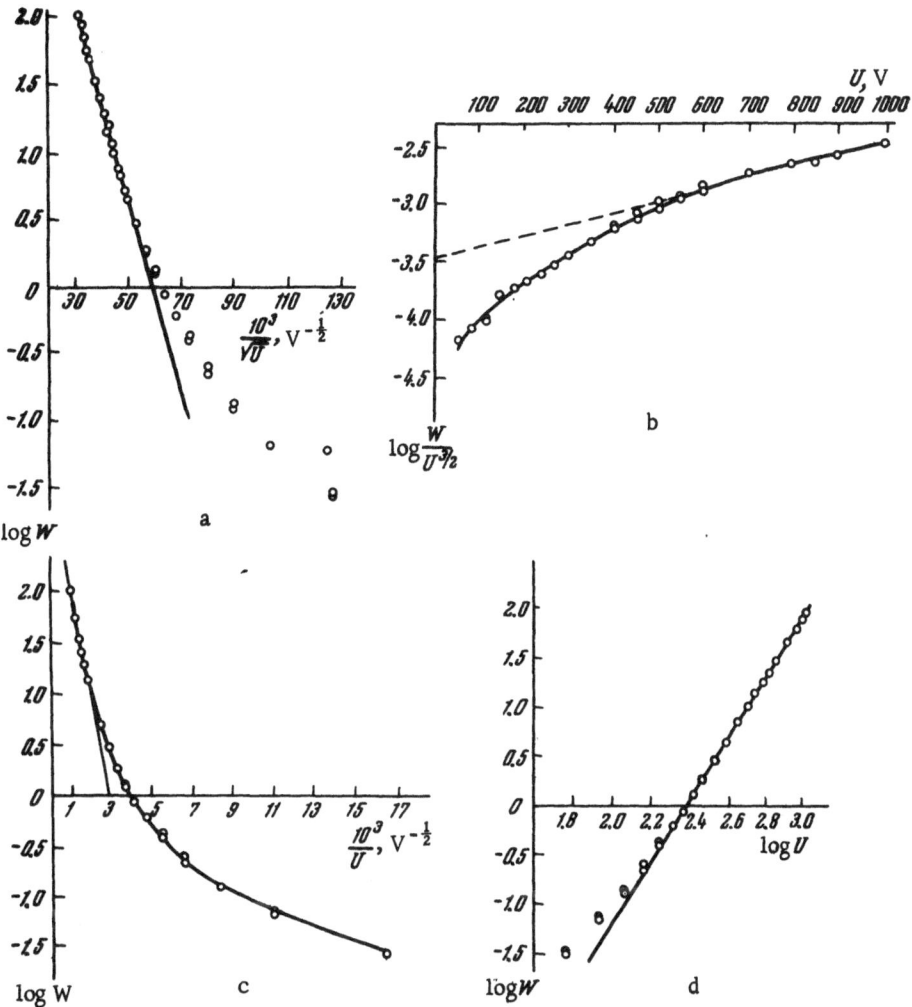

Fig. 26. Dependence of the power W on the voltage, in various coordinates.

Georgobiani, L'vova, and Fok [78] have proposed the following formula for the power:

$$W \propto U \left( e^{-\frac{b}{\sqrt{U}}} + A e^{b_1^0 U} \right), \qquad (IV.23)$$

and these authors suggest that the first term in parentheses can be neglected, so that the difference between (IV.22) and (IV.23) reduces to different pre-exponential terms. The different dependences of the pre-exponential terms on U are due to the different accuracies of the theories of the process, and, generally speaking, the exponent of U in the pre-exponential term may differ from $^3/_2$ or 1. Assuming that the absorbed energy is represented by Eq. (IV.22) or Eq. (IV.23), Georgobiani, L'vova, and Fok ignored the absorption of the energy by linear elements. Formula (IV.22) was found to well represent the experimental results of Georgobiani et al. for external voltages ranging from 100 to 400 V. However, measurements over a wider range of voltages (50–1000 V) showed that there were considerable deviations from Eq. (IV.22).

The dependence W = W(U) is shown in Fig. 26 using the following coordinates:†

$$\text{a) } \log W, \ \frac{10^3}{\sqrt{U}}; \quad \text{b) } \log \frac{W}{U^{3/2}}, \ U;$$

———————
† The experimental results presented in Figs. 26, 30, and 32a have been kindly supplied by E. Yu. L'vova.

$$c) \quad \log W, \ \frac{10^3}{U} ; \quad d) \quad \log W, \ \log U.$$

The straight lines in Fig. 26 represent the corresponding laws. Hence, we see that the dependence W = W(U) cannot be represented in these coordinates by a straight line over the full range of voltages investigated.

The absorption of the energy at low and high values of U is due to different processes and is therefore described by different formulas. At low values of U, we observe a quadratic dependence on the applied voltage, whereas at high values of U, the dependence W = W(U) can be described by several formulas capable of reflecting the physics of the process — for example, by Eq. (IV.22), as well as by

$$W \propto e^{-\frac{const}{U}}, \tag{IV.24}$$

or

$$W \propto e^{-\frac{const}{\sqrt{U}}}. \tag{IV.25}$$

It is therefore not surprising that the absorption by a nonlinear element is described by Eq. (IV.22) in [55] and by (IV.25) in [97].

To describe the process completely, it is necessary to include the absorption in linear and nonlinear elements.

Lehmann [97] has allowed for the absorption by linear and nonlinear elements using Eq. (IV.25). He has not given a physical meaning to the term $\exp(-const/\sqrt{U})$, but the nature of this voltage dependence suggests that it represents the absorption by tunnel-liberated electrons in one-phonon or no-phonon processes. It follows from the experimental results of Lehmann that the absorption of power cannot be represented satisfactorily by two terms.

If we allow additionally for the absorption of energy by electrons liberated by the tunnel effect from traps [the term $\exp(b_1^0 U)$ in Eq. (IV.26)], we then obtain

$$W \propto U^2 \left( 1 + d_2 e^{-\frac{const}{\sqrt{U}}} + d_1 e^{b_1{}^0 U} \right). \tag{IV.26}$$

This formula includes two parameters $d_1$ and $d_2$ which cannot be determined from the measurements of the brightness B = B(U) or from the measurements of W = W(U). These two free parameters make it possible to describe the experimental results, but the arbitrary nature of the selection of $d_1$ and $d_2$ is a disadvantage of this formula. Therefore, it would be interesting to carry out independent experiments to determine $d_1$ and $d_2$, but it has not yet been done.

The sum of two exponential functions in Eq. (IV.26) can be approximated by a single power function. It follows from Fig. 26 that this power function represents the experimental results over a wider range of voltages than do the formulas (IV.22), (IV.24), and (IV.25).

Bearing in mind what has been said before, we find that the absorbed power W is

$$W \propto U^2 (1 + dU^n), \tag{IV.27}$$

where n is a positive fraction and d is some constant. Formula (IV.27) describes the experimental results well over a wide range of values of U. The main disadvantage of this formula is that the power-function approximation masks the physical meaning of the nonlinear behavior of a capacitor:

$$d_2 e^{-\frac{const}{\sqrt{U}}} + d_1 e^{b_1{}^0 U} = dU^n. \tag{IV.28}$$

## § 4.  Energy Yield of an Electroluminescent Capacitor

The practical value of investigations of the energy yield of electroluminescence [124] is self-evident because one of the principal applications of electroluminescent capacitors is as light sources. The maximum light yield, obtained under laboratory conditions for zinc sulfide phosphors emitting green luminescence, does not exceed 18 lm/W in alternating fields [112] and 1-2 lm/W in dc fields [125]. Mass-produced capacitors usually have a yield of 2-5 lm/W [126].

The energy-yield measurements are equally important from the theoretical point of view in connection with the processes taking place in a phosphor. The value of 18 lm/W is higher than the limiting values of $\eta_{max}$ obtained theoretically on the basis of various models [127-129]. This shows that these models are inadequate. There is as yet no generally accepted view about the process which governs the dependence of the energy yield on the frequency and voltage, and the dominant process has not even been determined yet.

The energy yield of electroluminescence is the ratio of the energy emitted by an electroluminescent phosphor to the energy absorbed by this phosphor. The energy yield of a phosphor is, as a rule, higher than the energy yield of an electroluminescent capacitor because of the losses in the dielectric binder and in the electrodes.

On the basis of the calculations given in the preceding chapter and in § 2 of the present chapter, the frequency dependence of the energy yield of an electroluminescent capacitor $\eta$ is found to be given by

$$\eta = \frac{\varkappa_\infty + ax^{\frac{p}{p-1}}}{1 + \varkappa_\infty + ax^{\frac{p}{p-1}} + \left(x + \frac{\varkappa_\infty}{a}x^{\frac{1}{1-p}}\right)^2} \exp\left[-\frac{b^*\sqrt[4]{\left(1 + \varkappa_\infty + ax^{\frac{p}{p-1}}\right)^2 + \left(x + \frac{\varkappa_\infty}{a}x^{\frac{1}{1-p}}\right)^2}}{\sqrt{U\left(\varkappa_\infty + ax^{\frac{p}{p-1}}\right)}}\right]. \quad \text{(IV.29)}$$

Formula (IV.29) is best considered by looking at Fig. 27, which shows this dependence for several values of all three parameters: $\varkappa_\infty$, $a$, and p. Figure 27a shows the curves for the following values of $\varkappa_\infty$: 99, $10^4$, $10^5$, $10^6$, $10^7$ (curves 1-5, respectively) for $a = 10^8$, $b^*/\sqrt{U} = 4.75$, and p = 0.85. The curves in Fig. 27a consist of three distinct regions. The yield is constant in that range of frequencies in which only the resistance of the heterogeneous layer is important. The yield begins to decrease when the shunting effect of the capacitance $C_0$ becomes important and the pre-exponential term begins to decrease. The very rapid fall at the highest frequencies is due to the exponential factor and is related to a redistribution of the voltage between the electrodes and the heterogeneous layer.

Variation of the parameter $a$ does not change the form of the curves but simply shifts them along the frequency axis. The larger the value of $a$, the higher are the frequencies at which the energy yield begins to fall. This is illustrated in Fig. 27b, which is plotted for three values of $a$: $10^4$, $10^6$, $10^8$ (curves 1-3, respectively) for p = 0.85 and $\varkappa_\infty$ = 99.

Figure 27c shows curves plotted for three values of p: 0.2, 0.5, and 0.75 (curves 1-3, respectively). It can be seen from this figure that an increase in the parameter p extends the region in which $\eta$ is constant. The variation of p affects the rate of fall of the energy yield at high frequencies, i.e., it alters the shape of the curve.

Formula (IV.29) is valid for heterogeneous electroluminescent capacitors at audio and ultrasonic frequencies. This follows from the good agreement between the experimental data and the formulas deduced for the absorbed and emitted energy, and from Fig. 28, where the continuous curves represent the calculated values and the various symbols represent the experimental values of $\eta$ for a number of capacitors with different electrode resistances. The continuous curves in Fig. 28 have been calculated using the parameters found by the method described in § 4 of Chapter III.

The dependence of $\eta$ on the external voltage is, according to Eq. (IV.29), the same as the dependence for B = B(U), and is mainly governed by the same exponential factor. The results of measurements of the brightness and light yield, presented in Fig. 29, indicate that this conclusion is valid for weak fields. As the external

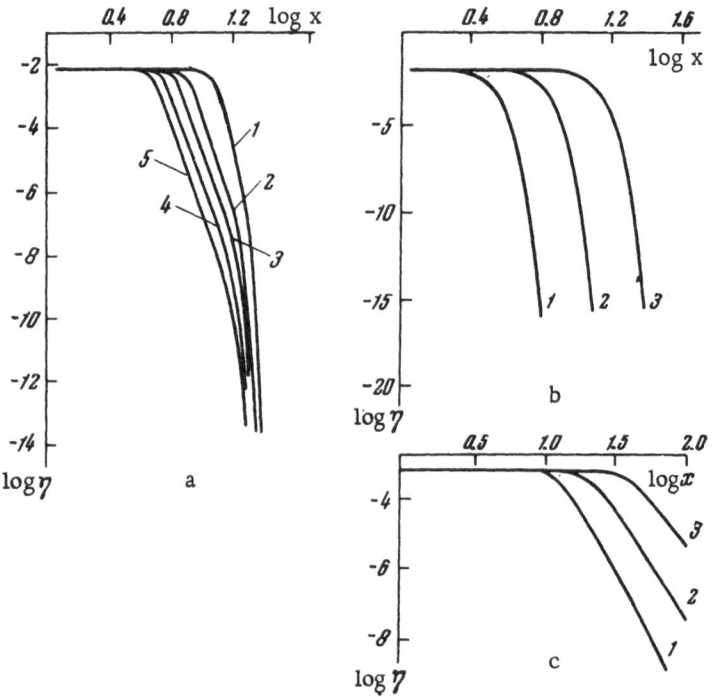

Fig. 27. Influence of the parameters $\varkappa_\infty$, $a$, and $p$ (represented by figures a, b, and c) on the frequency dependence of $\eta$ for an electroluminescent capacitor (calculated value).

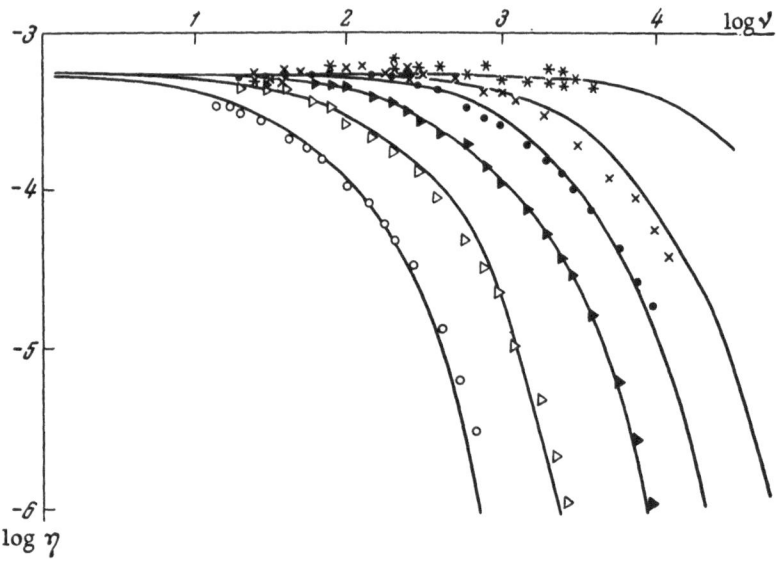

Fig. 28. Frequency dependence of the energy yield of various capacitors. The continuous curves represent the calculations based on Eq. (IV.29); the various types of point represent the experimental data.

voltage is increased, the deviations from (IV.29) become increasingly larger: the brightness rises and the light yield passes through a maximum. A similar dependence of the energy yield on the voltage has been reported elsewhere [56, 78]. Thus, in strong fields, i.e., when the luminescence brightness is high, Eq. (IV.29) does not correctly predict the dependence of $\eta$ on U. The reasons for this are obvious. Equation (IV.29) has been derived on the assumption that an electroluminescent capacitor obeys Ohm's law. However, as shown in § 3 of the present chapter, this is not true.

In order to obtain a formula for $\eta$ valid over a wide range of voltages, it is necessary to allow for the non-linearity of an electroluminescent capacitor. To do this, it is best to consider the process of electroluminescence at a fixed frequency of external exciting field.

The dependences of the absorbed and emitted energies on the voltage across a capacitor are known from experiments. Knowing these dependences, we can easily find the voltage dependence of the energy yield (or light output) of an electroluminescent capacitor. However, we shall show later that it is much more difficult to find a mathematical expression for this dependence which would have a clear physical meaning.

First, we shall consider the formulas for $\eta = \eta(U)$ which have already been proposed.

At low voltages, $\eta(U)$ is described satisfactorily by Lehmann's empirical formula [97]:

$$\eta = \frac{B/\omega}{aU^2\left[k_0 + k\left(\frac{B}{\omega}\right)^{1/2}\right]}.$$ (IV.30)

In Fig. 29, taken from Lehmann's paper [97], the dashed curve is the theoretical dependence calculated from Eq. (IV.30), while the points represent the experimental values of the brightness and light yield $\eta$.

Georgobiani [55] has suggested the following expression for the energy yield $\eta$:

$$\eta = \frac{a}{U^{3/2}\exp\left(b_1^0 U + \frac{b}{\sqrt{U}}\right)},$$ (IV.31)

where $b_1^0$ and b are constants which can be determined experimentally. Comparison of the calculated values with the experimental results (crosses) is shown in Fig. 30. We can see that the experimental and theoretical dependences are of the same nature but that there is no quantitative agreement at low values of U.

Thus, neither Eq. (IV.30) nor (IV.31) gives quantitative agreement with experiment when the parameters which occur in these formulas are determined independently. Therefore, it seemed interesting to determine the behavior of these formulas when the parameters which occur in them are varied widely in order to determine the influence of these parameters on the form of the curve and on the position of an extremum. An analysis of Lehmann's formula over a wide range of the parameters that occur in it has shown that it is not possible to simultaneously satisfy two conditions:

1) agreement between the shapes of the experimental and theoretical curves;

2) agreement between the value of $U_{max}$ at which the maximum of energy yield is observed. However, it is possible to select values of the parameters which satisfy one of these requirements. The same can be said of Eq. (IV.31).

Since the approximation of the voltage dependence of the brightness by the $\exp(-b/\sqrt{U})$ formula can be regarded as satisfactory, it is evident that the discrepancy between the experimental data and the theoretical results for $\eta(U)$ should be attributed to an insufficiently accurate theoretical allowance for the dependence W = W(U).

In § 3 of the present chapter, it has been shown that this dependence is best represented by Eq. (IV.27).

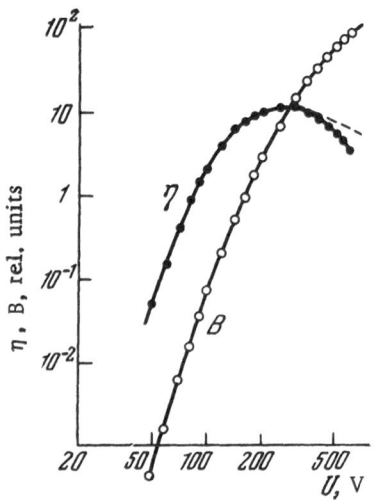

Fig. 29. Dependence of the optical yield $\eta$ and of the brightness B of an electroluminescent capacitor on the external voltage U (according to Lehmann).

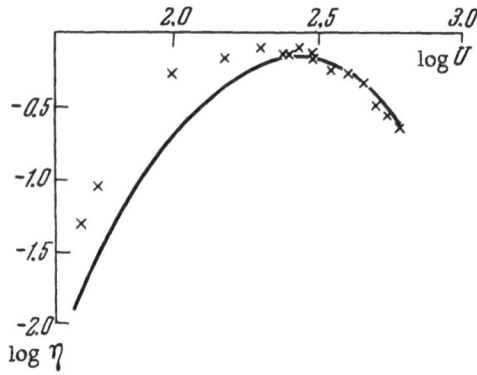

Fig. 30. Comparison of the experimental results for $\eta = \eta(U)$ with the calculations based on Eq. (IV.31). Crosses represent the experimental data.

Then, we find that the energy yield is

$$\eta = \frac{\text{const } e^{-\frac{b}{\sqrt{U}}}}{U^2 (1 + d U^n)} . \qquad (IV.32)$$

This equation has three parameters: b, n, and d.

The parameter b can be found from independent measurements of the brightness. Its influence on the form of the curve can be seen from Fig. 31a, where curves 1-3 have been plotted for b = 30, 61, and 92, respectively; $d = 10^{-3}$, n = 1.5. An increase in b shifts the maximum in the direction of higher voltages.

The parameter n has practically no effect on the shape of the curves, but it shifts the maximum in the direction of lower voltages when it is increased. Curves 1-4 in Fig. 31b are plotted for n = 1.3, 1.5, 1.6, and 1.7, respectively. The values of the parameters b = 92 and d = $10^{-3}$ are common for all curves in Fig. 31b. The parameter n can be determined from the measurements of the dependence of W on U.

The parameter d cannot be determined by independent measurements. Nevertheless, it is interesting to find its influence on the shape and position of the curves (Fig. 31c). The values b = 92 and n = 1.5 have been used to plot curves 1-5 in Fig. 31c, where d has the following values: $10^{-1}, 10^{-2}, 10^{-3}, 10^{-4}$, and $10^{-5}$. Higher values of d do not affect the shape or position of the curve with respect to the abscissa, i.e., in this case, the unity in Eq. (IV.32) can be ignored, and the variation of d simply shifts the curve along the ordinate. At very low values of d, the same effect is also observed; d does not affect the shape or position of the curve.

From Eq. (IV.32) we can find the condition for an extremum of $\eta$ which includes all three parameters d, b, and n:

$$b + db U_{max}^n - 4 \sqrt{U_{max}}$$
$$- 2 d (n + 2) U_{max}^{n+0.5} = 0. \qquad (IV.33)$$

Since $U_{max}$, b, and n are known from experiments, the value of d can be calculated from Eq. (IV.33).

If this is done and Eq. (IV.32) is used to find the dependence $\eta = \eta(U)$ and to compare it with the experimental results, it is found that good agreement between the calculated and experimental results is obtained over the whole range of values of U. Figure 32a (crosses) and Fig. 32b (circles) show the results obtained for two samples; the continuous curves represent the calculations using Eq. (IV.32). The curves can be made to coincide simply by shifting them along the log $\eta$ axis.

Figure 33 presents the experimental results obtained by Lehmann [112] for samples containing a fractionated phosphor. The granulometric position affected not only the value of $\eta$ max but also the nature of the curve. The coarser the grain, the flatter is the $\eta = \eta(U)$ curve at low values of U. These experimental results can be explained by any of the proposed formulas for $\eta$: (IV.31), (IV.30), and (IV.32). Lehmann also established [112] that when the grain size of the phosphor was increased, the parameter b, representing the dependence of the

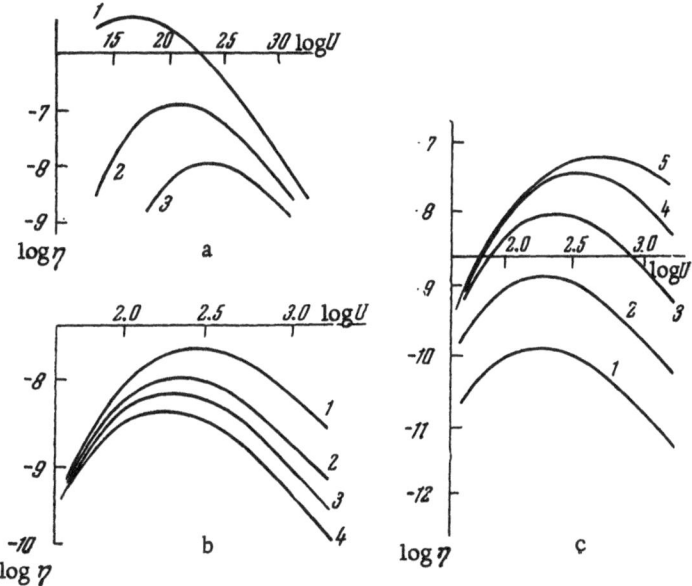

Fig. 31. Dependence $\eta = \eta(U)$ according to Eq. (IV.32): a) for various values of the parameter b; b) for various values of n; c) for various values of d.

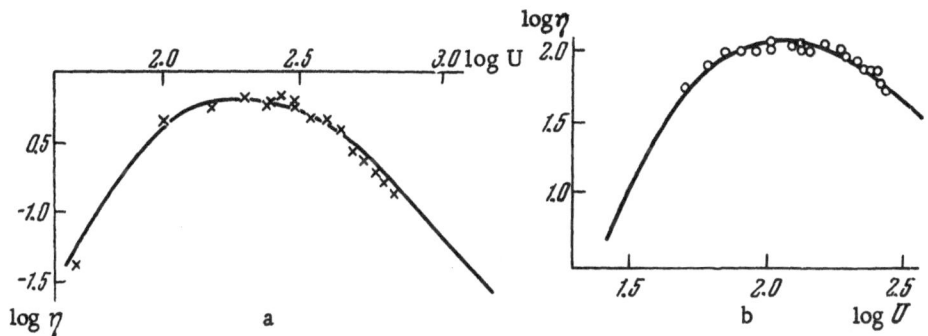

Fig. 32. Comparison of the experimental results for two samples with the calculations based on Eq. (IV.32).

brightness on voltage, decreased. The reduction in b should result in a decrease in the curvature of $\eta = \eta(U)$ to the left of the maximum, which was indeed observed experimentally.

Thus, summarizing this discussion, we may conclude that Eq. (IV.29) describes the frequency dependence of the energy yield of electroluminescent capacitors and Eq. (IV.32) describes the voltage dependence of this yield. A formula valid over a wide range of frequencies for the energy yield and the voltages will thus have the form

$$\eta = \frac{\text{const}}{U^2(1+dU^n)} \frac{\varkappa_\infty + ax^{\frac{p}{p-1}}}{1+\varkappa_\infty + ax^{\frac{p}{p-1}}+\left(x+\frac{\varkappa_\infty}{a}x^{\frac{1}{1-p}}\right)^2} \times$$

$$\times \exp\left[-\frac{b^* \sqrt[4]{\left(1+\varkappa_\infty + ax^{\frac{p}{p-1}}\right)^2+\left(x+\frac{\varkappa_\infty}{a}x^{\frac{1}{1-p}}\right)^2}}{\sqrt{U\left(\varkappa_\infty + ax^{\frac{p}{p-1}}\right)}}\right].$$

(IV.34)

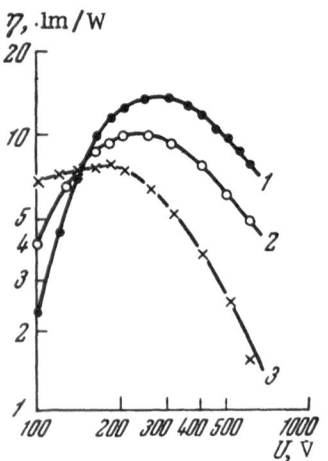

Fig. 33. Dependence of the optical yield $\eta$ on the external voltage U for electroluminescent capacitors with different ZnS grain dimensions: 6, 10, and 20 $\mu$ (curves 1-3, respectively).

Using Eq. (IV.34), we can describe all the available experimental results on the energy yield of electroluminescent capacitors: the maximum of $\eta$ at moderate voltages, the constancy of $\eta$ in a certain range of audio frequencies, the unavoidable fall of $\eta$ at high frequencies, and the fall of $\eta$ at low frequencies, which is observed only occasionally. We shall consider the last effect in detail. This phenomenon was first observed by Zalm [82]. It was sometimes observed in our samples, but only when the samples exhibited a strongly nonsinusoidal current. Since the nonsinusoidal nature of the current became more pronounced at low frequencies, this resulted in a very weak dependence of the parameters d and n on $\nu$, which led to a certain fall in $\eta$ at a fixed value of the external exciting voltage.

§ 5.  Distribution of the Brightness Across the Surface of a Capacitor

We have demonstrated that the brightness of luminescence depends very much on the voltage across the heterogeneous layer and that this fraction is smaller for points further from the contacts. If an electroluminescent capacitor is in the form of a rectangular slab with contacts along the shorter side and with one electrode of considerable resistance, its properties are best described by the equivalent circuit shown in Fig. 34. At any moment, the value of the potential at one electrode is constant, but across the other electrode there is a potential gradient and, consequently, a gradient of the brightness along the longer side of the slab. To find the potential, it is necessary to solve the problem of the propagation of an electromagnetic wave in a lossy medium. The solution of Maxwell's equations for such a system has the form

$$V\,(x,\,t) = e^{i\omega t}(Ae^{-\gamma x} + Be^{\gamma x}). \tag{IV.35}$$

The expression in parentheses is the sum of the incident and reflected waves. The parameter $\gamma$ is a complex quantity ($\gamma = a + ib$) and is expressed in terms of the parameters of the system as follows:

$$a = \left[\left(\frac{R}{R_0}\right)^2 + (\omega C_0 R)^2\right]^{1/4} \cos\left[\frac{\tan^{-1}(\omega C_0 R_0)}{2}\right], \tag{IV.36}$$

$$b = \left[\left(\frac{R}{R_0}\right)^2 + (\omega C_0 R)^2\right]^{1/4} \sin\left[\frac{\tan^{-1}(\omega C_0 R_0)}{2}\right]. \tag{IV.37}$$

If we only consider the quasi-stationary case, then A and B can be found from the boundary conditions

$$A + B = U \text{ for } x = 0 \tag{IV.38}$$

and

$$\frac{\gamma}{R}(-Ae^{-\gamma x} + Be^{\gamma x}) = 0 \text{ for } x = L. \tag{IV.39}$$

Hence, we can find the values of A and B. Substituting these values into Eq. (IV.35), we find

$$V_x = \frac{U\cosh[\gamma(L-x)]}{\cosh[\gamma L]}. \tag{IV.40}$$

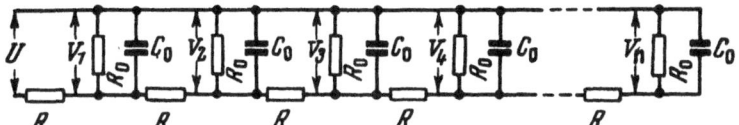

Fig. 34. Equivalent circuit of an electroluminescent panel.

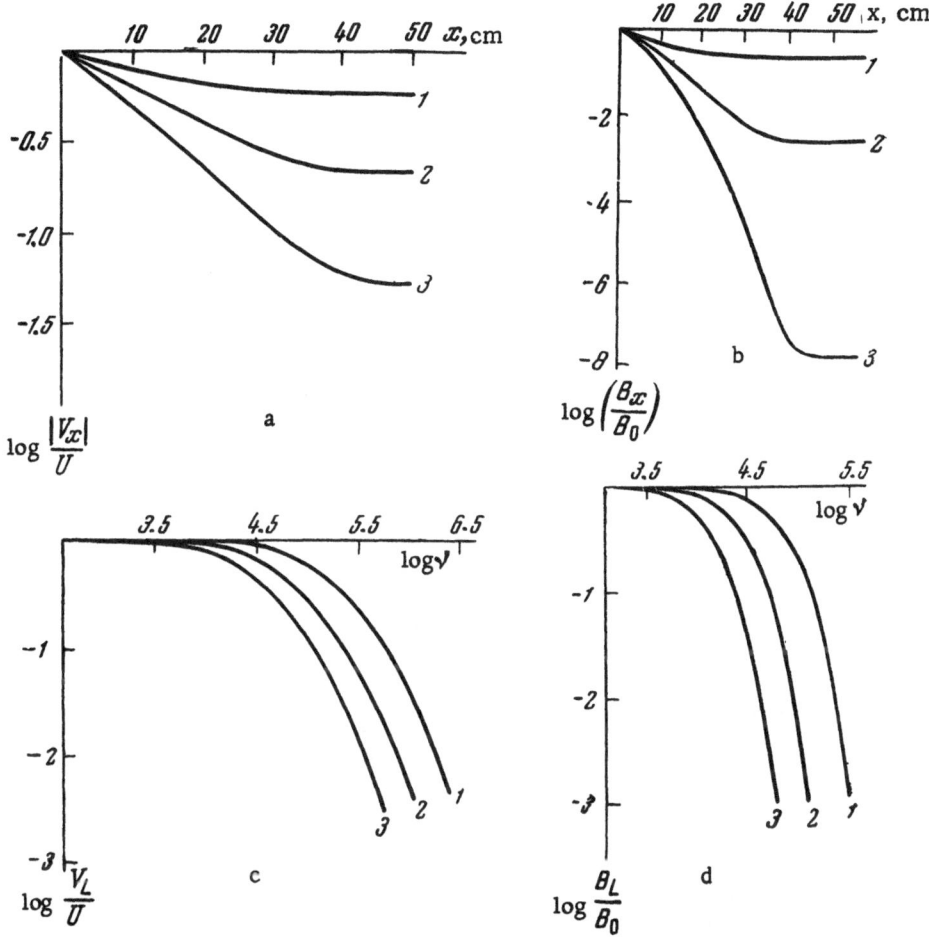

Fig. 35. Theoretical calculations of: a) the distribution of the potential difference across the panel; b) the distribution of the luminescence across the panel; c) the frequency dependence of the potential difference at the far end of the panel; d) the frequency dependence of the brightness of the far end of the panel.

The modulus of the function of the complex variable is, in this case,

$$|V_x| = U \sqrt{\frac{\cosh[2a\,(L-x)] + \cos\,[2b\,(L-x)]}{\cosh\,[2aL] + \cos\,[2bL]}} \,. \tag{IV.41}$$

Here, L is the length of the sample and x is the distance from a contact. The potential drop at the end of the sample at a distance L from a contact will be represented using a quantity $V_L$:

$$V_L = \frac{\sqrt{2}\,U}{[\cosh(2aL) + \cos\,(2bL)]^{1/2}} \,. \tag{IV.42}$$

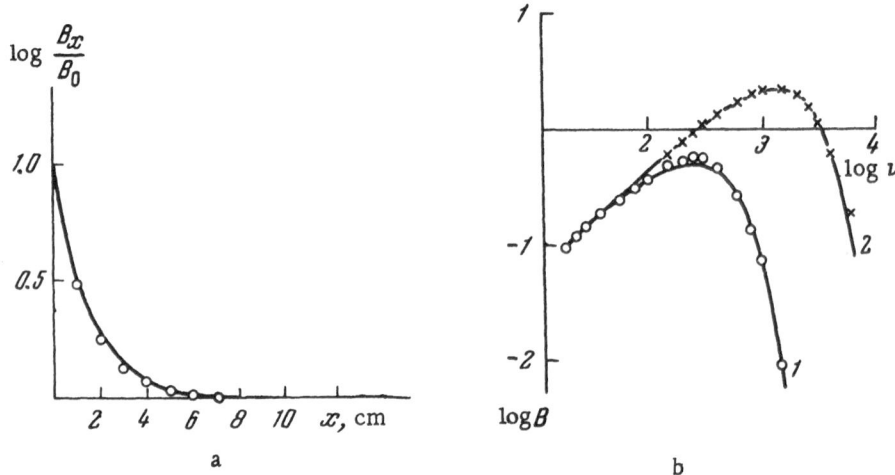

Fig. 36. Comparison of the experimental data (points) with the calculated results
(continuous curves) for samples (a) S-295 and (b) K-10.

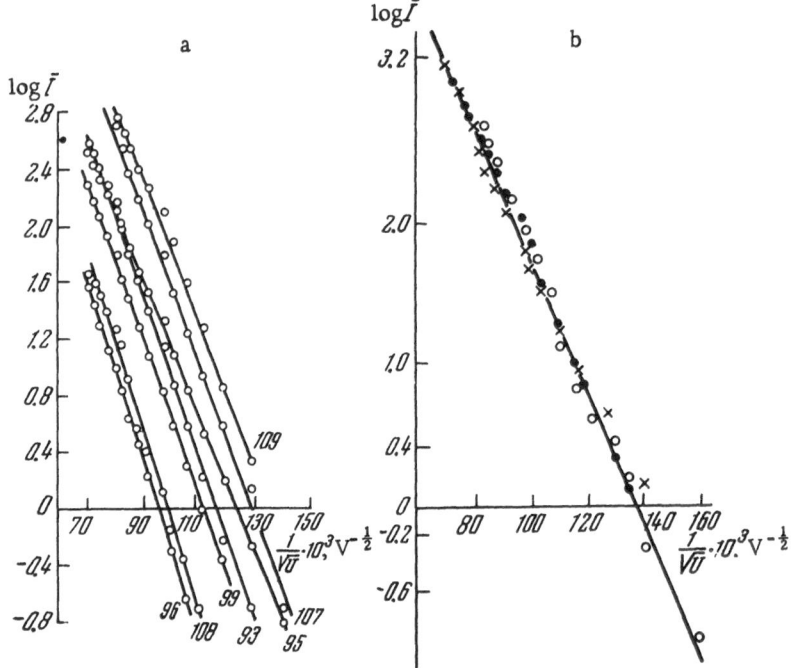

Fig. 37. Dependence of the rectified current $\bar{I}$ on the voltage: a) for capacitors
with resin ML-92; b) for capacitors with three different dielectric binders.

Knowing the distribution of the potential across the capacitor surface, we can use Eq. (IV.12) to calculate the brightness at any point. Figures 35a and 35b show the results of a theoretical calculation of the dependence of $|V_x|/U$ on the distance from a contact x and the corresponding distribution of the normalized brightness potential $B_x/B_0$, where $B_x$ is the brightness at a distance x from a contact and $B_0$ is the brightness at the contact. These results are given for three values of the electrode resistance (100, 300, and 800 $\Omega$ per square — curves 1, 2, and 3, respectively, in Figs. 35a and b). The first of these values is typical of capacitors on glassy substrates; the other two values are typical of capacitors on steel and Plexiglas (Perspex) substrates. The calculations were carried out for the following values of the parameters: $\nu$ = 20 kc, p = 0.95, $R_0^*$ = 4 $\cdot 10^9$ $\Omega$, $C_0$ = 0.5 $\cdot 10^{-9}$ F (the values of $R_0^*$ and $C_0$ are given per centimeter of length), U = 120 V, $b^*$ = 60 $V^{\frac{1}{2}}$. It is evident from Fig. 35b that a panel 0.5 m long subjected to a 20-kc field will emit light only at the contact if R = 300 and 800 $\Omega$. If the electrode resistance is R = 100 $\Omega$ per square, the luminescence from the half of the panel which is further from the contact will be five times weaker than that from the nearer half.

Panels of this type cannot be used at high frequencies. In order to determine the upper limit of the frequency at which luminescence is uniform over the whole surface, we must look at Figs. 35c and 35d. Calculations of $V_L/U$ and $B_L/B_0$ ($B_L$ is the brightness at the far end of the sample) were carried out using the previous values of the parameters for a sample 0.5 m long. The resistance of the electrodes was assumed to be 20, 50, and 100 $\Omega$ per square for curves 1, 2, and 3, respectively. It follows from Fig. 35d that the frequency limit for the uniformity of the brightness is not higher than 10 kc. It is easy to check this calculation using samples with high values of the electrode resistance R since, in this case, the influence of the electrodes becomes significant at relatively low frequencies and short lengths. Figure 36a shows the results of a comparison of the experimental data obtained for a sample S-295 (a phosphor EL-510$^m$ mixed with a resin ÉP-96 and subjected to a signal of U = 160 V and $\nu$ = 15 kc; R = 2 k$\Omega$ per square), with calculations carried out specially for this sample using parameters determined from the energy measurements by the method described earlier. The calculations are represented by a continuous curve and the experimental values are represented by points; the experimental brightness has been normalized to the brightness of the contact strip. The agreement between the calculated and experimental values is satisfactory. Figure 6b shows the frequency dependence of the brightness of the contact end of a sample K-10 (crosses) and of the far end (circles) for U = 150 V and an electrode resistance R = 14 k$\Omega$ per square for a panel 10 cm long. The continuous curves 1 and 2 represent the calculated values. This problem was first dealt with by Ivey [130], who solved it in the linear approximation. His results were therefore only in rough agreement with the experimental findings.

CHAPTER V

# Rectification of the Current by an Electroluminescent Capacitor

The investigation of the dependence of the average brightness of electroluminescence on the external exciting field is one of the most widely used methods to study the electroluminescence mechanism, but it is not the only method. "Brightness waves," energy absorption processes, and the frequency dependence of the absorbed and emitted energy all yield some information on the electroluminescence mechanism.

In the present chapter, we shall describe how the phenomenon of current rectification can be used to investigate the excitation process [131]. If an electroluminescent capacitor is connected in the circuit shown in Fig. 5, the magnetoelectric instrument will record a current in the circuit. Since the capacitor is excited with an alternating current, the presence of a dc component indicates that the capacitor has rectifying properties. Therefore, the electroluminescent capacitor is represented as a detector in the circuit of Fig. 5. We shall consider the dependence of the rectified current on the applied voltage and its frequency without considering at first the nature of rectification.

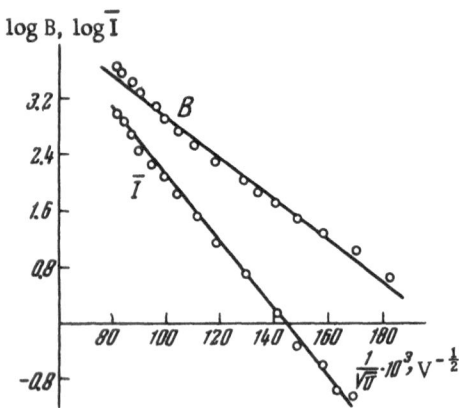

Fig. 38. Voltage dependences of the brightness B and of the rectified current $\bar{I}$ for the same capacitor.

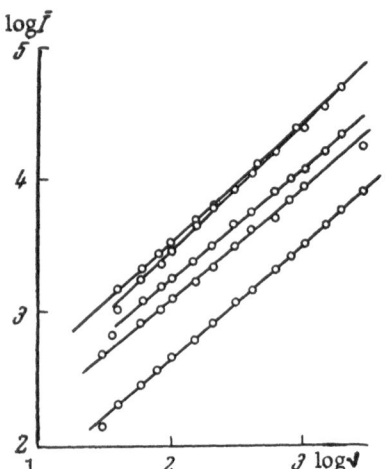

Fig. 39. Frequency dependence of the rectified current for several sample (U = 150 V).

## §1. Dependence on the Voltage

Dependence of the rectified current on the voltage applied to a capacitor is shown in Fig. 37a for capacitors of various thicknesses bound with the ML-92 resin. The numbers alongside the curves are the capacitor numbers. The ordinate represents the logarithm of the rectified current and the abscissa $10^3/\sqrt{U}$. It follows from Fig. 37a that the experimental results are described satisfactorily by the formula

$$\bar{I} = Ae^{-\frac{\beta}{\sqrt{U}}}. \qquad (V.1)$$

Figure 37b shows, in relative units, the dependence of the rectified current $\bar{I}$ on the applied voltage U for capacitors made from three different dielectrics (the points represent the results for ML-92, the crosses those for ÉP-096, and the circles those for polystyrene with nitrocellulose). It follows from Fig. 37b that B is the same for all the dielectrics.

The voltage dependences of the electroluminescence brightness B and of the rectified current $\bar{I}$ are described by the same law, but, for all the investigated samples, the index of the exponential function in the expression for $\bar{I}$ is larger than the corresponding quantity in the expression for B.

It follows from Fig. 38 that the straight line representing the rectified current $\bar{I}$ has a steeper slope. The identical natures of the voltage dependences of $\bar{I}$ and B can be explained by the fact that the current-voltage and brightness characteristics of an electroluminescent capacitor are governed primarily by the same process.

The nature of this process can be determined by considering the dependence of $\bar{I}$ on the external voltage, because nonequilibrium carriers, which contribute to the current, also affect the value of the brightness.

If minority carriers are generated by transitions through a barrier, then the dependence of the current on the voltage is given by the formula (I.16), which disagrees with the law $\bar{I} = \bar{I}(U)$ found in our experiments. The experimental results agree with Eqs. (I.11) and (I.12) provided the relationship between the field in a barrier and the potential difference across it is given by Eq. (I.19).

Thus, the observed voltage dependence can only be explained by tunnel and impact ionization in weak fields. Consideration of the actual fields in ZnS has shown [58] that the criterion of the weakness of the field is not satisfied over a wide range of voltages used in the measurements of electroluminescence. It therefore follows that the tunnel ionization is the only possible process.

## §2. Frequency Dependence of the Rectified Current

Figure 39 shows, on a double logarithmic scale, the frequency dependence of the rectified current for several polystyrene capacitors subjected to an exciting voltage of 150 V. This type of dependence is typical of capacitors with low electrode resistances. If the electrode resistance is high, the rectified current has a maximum followed by a rapid decrease.

To account for the frequency dependence of the rectified current, we used the equivalent circuit for an electroluminescent capacitor employed earlier [131, 132]. It consists of a capacitance $C_0$ and a resistance $R_0$ connected in parallel to represent a phosphor layer in a dielectric, with the electrode resistance R connected in series with these two elements. Let the resistance $R_0$ depend on the frequency in accordance with the law (III.13):

$$R_0 = \frac{R_0^*}{\nu^p},$$

where $R_0^*$ and p are constants as defined in §2 of Chapter III. In this circuit, the rectification is attributed to the resistance $R_0$. If an alternating voltage U is applied to such a capacitor, the alternating voltage $U_0$ across the rectifying element $R_0$ is given by

$$U_0 = U \frac{|Z_{hl}|}{|Z_{tot}|}. \tag{V.2}$$

The detection law gives the relationship between the constant emf $\delta_1$ and the alternating voltage $U_0$:

$$\delta_1 = f\left(U \frac{|Z_{hl}|}{|Z_{tot}|}\right). \tag{V.3}$$

Then, the constant current $\bar{I}$ is governed by the values of the resistances $R_0$ and R:

$$\bar{I}_1 = \frac{f\left(U \frac{|Z_{hl}|}{|Z_{tot.}|}\right)}{R_0 + R}. \tag{V.4}$$

At low frequencies, we can neglect the influence of the electrodes and of the capacitance $C_0$.

At such frequencies, the value of the rectified current $\bar{I}_1$ is governed by the constant emf, which is given by an exponential term of the type represented by Eq. (V.1), and by the resistance of the barriers, given by Eq. (III.13):

$$\bar{I}_1 = \text{const} \frac{\exp\left(-\frac{\beta^*}{\sqrt{\bar{U}}}\right)}{R_0^*} \nu^p. \tag{V.5}$$

In this range of frequencies, the tangent of the angle of slope of the straight line $\bar{I} = \bar{I}(U)$, plotted using the coordinates $\log \bar{I}$ and $10^3/\sqrt{U}$, will be denoted by $\beta^*$. At such frequencies, $\beta = \beta^*$. The relationship between $\beta$ and $\beta^*$ over a wide range of frequencies is given by Eq. (V.11). It follows from Eq. (V.5) that the tangent of the angle of slope of the straight line $\bar{I}_1 = \bar{I}_1(\nu)$, plotted on a double logarithmic scale, is equal to p. This has been confirmed by experiments carried out at high voltages. However, if we measure the frequency characteristics for various values of the external voltage, we find discrepancies in the behavior of $\bar{I}$, W, and B. Measurements of W and B indicate that the parameter p is practically independent of U and usually close to unity, but always p < 1 (cf. Fig. 23a). For many samples, the variation in p is found to lie within the limits of the experimental error. Measurements of $\bar{I}$ show that the tangent of the angle of slope of $\bar{I} = \bar{I}(\nu)$, plotted using the coordinates $\log \bar{I}$, $\log \nu$, exhibits a strong dependence on the voltage. We shall denote this tangent by $p_{\bar{I}}$. This is illustrated in Fig. 40, where a double logarithmic scale is used to plot the $\bar{I}(\nu)$ curves for a range of values of the external voltage applied to a sample P-4: 25, 30, 40, 50, 60, 80, 100, 160 V. It follows from Fig. 40 that the strong frequency dependence of the rectified current at high values of U gradually disappears as U is reduced. The measurements reported in Fig. 40 were carried out on 16 samples and they all gave the same results.

On the other hand, experiments show that the parameter $\beta$ depends on the frequency. Figure 41 shows the dependence $\bar{I} = \bar{I}(U)$ for two fixed frequencies. The line 1 refers to a frequency of 50 cps and the line 2 to one of 5 kc. The second line is shifted along the ordinate axis. It follows from Fig. 41 that $\beta$ increases with the frequency [cf. Eq. (V.1)].

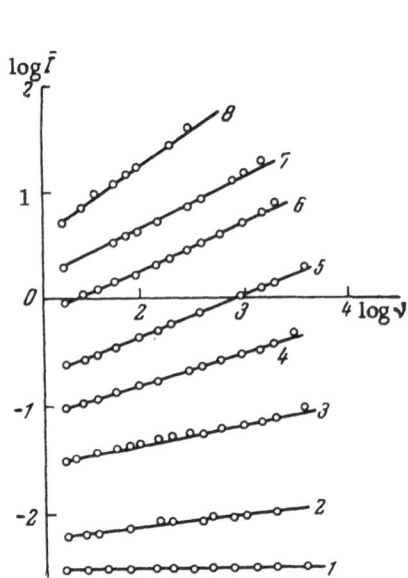

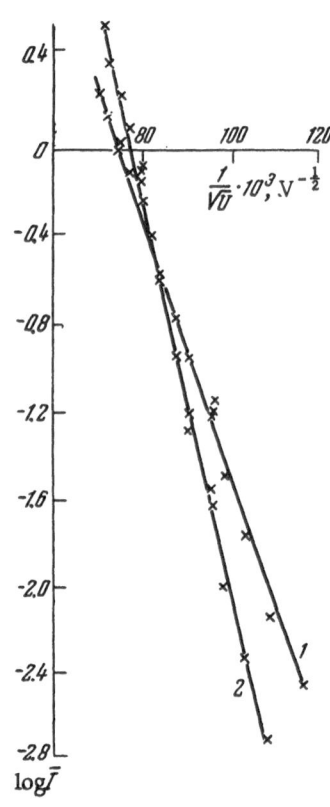

Fig. 40. Frequency dependence of the rectified current $\bar{I}$, plotted on a double logarithmic scale for various values of the external voltage.

Fig. 41. Dependence $\bar{I} = \bar{I}(U)$ for two frequencies of the exciting field.

At low frequencies ($|Z_{hl}| / |Z_{tot}| = 1$ and $\beta = \beta^{*}$), the frequency dependence of this parameter is of the type shown in Fig. 42. The crosses show the values of $\beta^{*}$ for an electroluminescent panel P-4. These crosses are joined by a straight line, which is represented by

$$\beta^{*} = \beta_0 + \xi \log v, \tag{V.6}$$

where $\beta_0$ is a constant,† and $\xi$ is the tangent of the angle of slope of the straight line.

If the expression for $\beta^{*}$ of Eq. (V.6) is substituted into Eq. (V.5) and the logarithms are taken of both sides, we obtain

$$\log \bar{I} = -\frac{\beta_0 + \xi \log v}{\sqrt{\bar{U}}} 0.434 - \log R_0^{*} + p \log v + \log C. \tag{V.7}$$

Grouping the terms depending on the frequency, we find

$$\log \bar{I} = -0.434 \frac{\beta_0}{\sqrt{\bar{U}}} - \log R_0^{*} + \left( p - 0.434 \frac{\xi}{\sqrt{\bar{U}}} \right) \log v + \log C. \tag{V.8}$$

---

† The value of $\beta_0$ can be easily calculated from Eq. (V.6). The experimental data presented in Fig. 42 give $\beta_0 = 51 \pm 2\,V^{\frac{1}{2}}$ for sample P-4. It is interesting to note that $\beta_0$ agrees well with the value of b, which is $53 \pm 2\,V^{\frac{1}{2}}$ for sample P-4 and is constant in the range of frequencies defined by the condition that the impedance of the barriers must be the dominant term in the total impedance of an electroluminescent capacitor.

The above equation means that if the frequency dependence of the rectified current is measured at different values of the external voltage, we should observe a dependence on U of the tangent of the angle of slope of the straight line plotted using the coordinates $\log \bar{I}$, $\log \nu$. In other words, $p_{\bar{I}}$ is related to the voltage U by

$$p_{\bar{I}} = p - \xi \frac{0.434}{\sqrt{\bar{U}}}. \qquad (V.9)$$

The value of $\xi$ varies considerably from sample to sample. The crosses in Fig. 43 indicate the values of $p_{\bar{I}}$ plotted as a function of $1/\sqrt{U}$ for sample P-4. The heterogeneous layer in this sample consisted of an electroluminescent phosphor ÉL-510 and a mixture of polystyrene and nitrocellulose, which was used as a dielectric binder. The dashed line in Fig. 43 represents a theoretical dependence plotted on the assumption that p = 0.95 (which was found from independent data on the frequency dependence of the brightness) and $\xi$ = 11 (found from the frequency dependence of $\beta^*$ shown in Fig. 42). It is evident from Fig. 43 that the agreement between the calculated and experimental data can be regarded as satisfactory over a wide range of voltages.† However, it must be stressed that the expression (V.9) is not valid over the whole range of voltages, since $p_{\bar{I}}$ does not become negative at any value of the external voltage. If we use $U_m$ to denote the maximum voltage at which $p_{\bar{I}} = 0$, the value of $p_{\bar{I}}$ is also equal to 0 for U < $U_m$.

The values of $U_m$ are different for different samples, but they are usually close to 20-50 V. When $U_m$ is applied, a sample luminesces very weakly. The crosses in Fig. 43 show the values of $p_{\bar{I}}$ obtained for sample No. 109, in which the heterogeneous layer consisted of ÉL-510 and a resin binder ML-92 (mixture of melamine-formaldehyde and glyptal resins). The theoretical line (shown continuous) is plotted in the same way as that for sample P-4.

It follows from Fig. 43 that at high voltages the values of $p_{\bar{I}}$ lie above the straight line representing Eq. (V.9). The value of $p_{\bar{I}}$ increases rapidly as U rises and becomes equal to p, which is found from the frequency dependences of the absorbed and emitted energies. A further increase in U causes electrical breakdown of the samples. In the prebreakdown field range, the rectified current becomes less stable and this makes it difficult to carry out the measurements.

Thus, we have found that the frequency dependence of the rectified current is due to the frequency dependences of two elements: the resistance of the heterogeneous layer and the constant emf across the barrier.

The frequency dependence of the resistance $R_0$, given by Eq. (III.13), has been derived from independent measurements of the absorbed and emitted energies, as discussed in the preceding chapters. Therefore, the experimental data just discussed provide additional independent confirmation of the proposed equivalent circuit. The physical meaning of the frequency dependence of the resistance $R_0$ has been considered before; the constant emf has been discussed in § 4 of the present chapter.

The formula (V.5) and the formulas that follow it are valid only in that range of frequencies in which the impedance of the barriers is the dominant component of the total impedance of an electroluminescent capacitor. However, Eq. (V.4) easily yields and expression for $\bar{I}_1$ which is valid over a wide range of frequencies:

$$\bar{I}_1 = \frac{A}{R_0^* \nu^{-p} + R} \exp\left(-\frac{\beta^*}{\sqrt{U \frac{|Z_{hl}|}{|Z_{tot}|}}}\right). \qquad (V.10)$$

Here, $\beta^*$ is given by Eq. (V.6) and

$$\beta^* \sqrt{\frac{|Z_{tot}|}{|Z_{hl}|}} = \beta. \qquad (V.11)$$

---

† It should be remembered that the calculations yield not only the slope, but also the absolute positions of the theoretical lines.

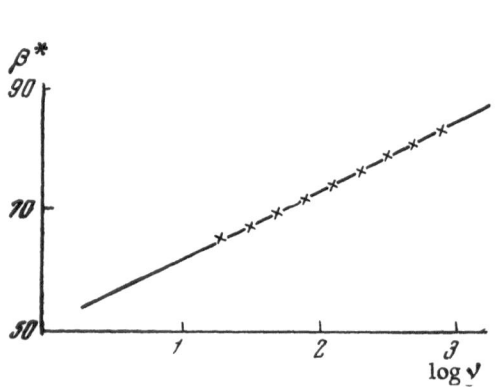

Fig. 42. Frequency dependence of $\beta^*(V^{\frac{1}{2}})$ for sample P-4 (crosses represent experimental data).

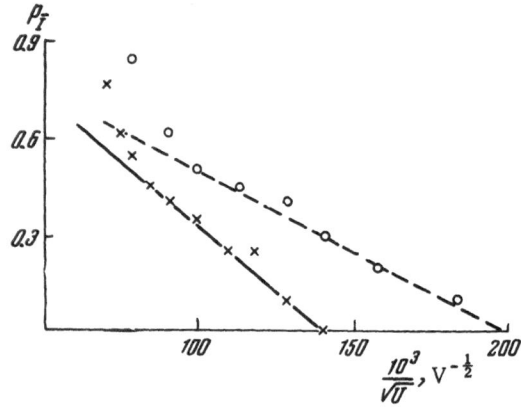

Fig. 43. Dependence of $p_{\bar{I}}$ on the voltage for samples P-4 (circles) and No. 109 (crosses). The continuous and dashed curves represent the calculated values.

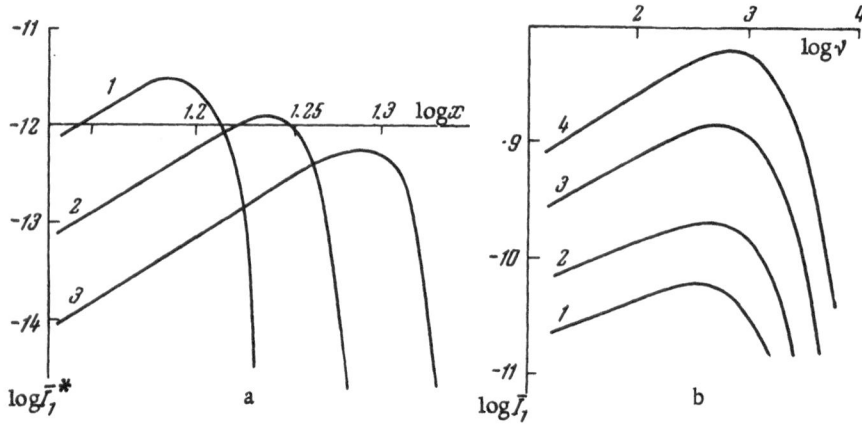

Fig. 44. Frequency dependence of the rectified current for various values: a) of the parameter $a$; b) of the parameter U.

The expression (V.10) has, in terms of dimensionless quantities, the following form:

$$\bar{I}_1^* = \frac{1}{1 + ax^{\frac{p}{p-1}}}\exp\left[-\frac{\beta^*}{\sqrt{U}}\sqrt[4]{\frac{\left(1 + ax^{\frac{p}{p-1}}\right)^2 + x^2}{\left(ax^{\frac{p}{p-1}}\right)^2}}\right]. \tag{V.12}$$

In the above equation, $\overrightarrow{\mathbf{I}_1^*}$ represents the dimensionless value of the rectified current, which is proportional to $\bar{I}_1$.

The family of $\mathbf{I}_1^* = \bar{I}_1^*(x)$ curves depends on five parameters: p, $a$, $\beta_0$, $\xi$, and U. For most of the investigated samples, the value of p is slightly less than unity. The influence of the parameter $a$ is demonstrated in Fig. 44a. The curves are plotted for the following values of the parameters: p = 0.95, $\varkappa = 10^5$, $\beta_0 = 50$, and $\xi = 10$, for $a = 3\cdot10^{24}$, $3\cdot10^{25}$, and $3\cdot10^{26}$ (curves 1-3, respectively).

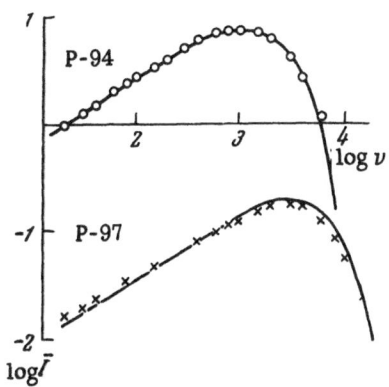

Fig. 45. Comparison of the experimental data (circles and crosses) with the calculated values of $\bar{I}$. The continuous curves represent the calculated values.

The values of the parameters $\beta_0$ and $\xi$ are different for different capacitors. The value of $\beta_0$ for our capacitors was 40-50 $V^{\frac{1}{2}}$ and the value of $\xi$ was usually about 10, but could be higher.

The change in the frequency dependence $\overline{I_1^\bullet} = \overline{I_1^\bullet}(U)$ due to a change in the external voltage U may be very considerable (Fig. 44b). Curves 1-4 in Fig. 44b are calculated for different values of the external voltage across a capacitor (U = 50, 65, 100, and 150 V). It follows from Fig. 44b that a change in U considerably alters the slope of the low-frequency region.

The general nature of the frequency dependence of the rectified current can be explained as follows. At low frequencies, there is a linear region whose slope is equal to $p\bar{I}$ in accordance with Eqs. (V.8) and (V.9). In this range of frequencies, $|Z_{hl}| / |Z_{tot}| = 1$. When the frequency is increased, $|Z_{hl}| / |Z_{tot}|$ decreases, i.e., the index of the exponential function [cf. Eq. (V.11)] increases and, therefore, the current begins to decrease. The increase in the index of the exponential function is due to a redistribution of the voltage between the components of an electroluminescent capacitor.

The formula (V.12) has been checked over a wide range of frequencies, including those at which the rectified current reaches its maximum value. For this purpose, the frequency dependence of the absorbed power has been measured for a number of capacitors and the proposed equivalent circuit has been used to calculate the parameters of these capacitors. Using such parameters, theoretical curves have been plotted for the frequency dependence of the rectified current. Figure 45 demonstrates the degree of agreement between the experimental results (circles and crosses) and the theoretical (continuous) curves for the investigated capacitors. The agreement with experiment is good at low frequencies and somewhat poorer at high frequencies, but the shapes of the curves are correct.

The experimental results presented in Fig. 45 and the corresponding theoretical curves have been plotted assuming that the resistance of the heterogeneous layer is given by Eq. (III.13). However, as shown in § 5 of Chapter III, this formula is approximate and is applicable only if the electrode resistance is relatively high. If the electrode resistance is low, we should use the expression (III.27) for $R_0$ at audio and ultrasonic frequencies. To obtain an expression for the rectified current under these conditions, it is necessary to determine which

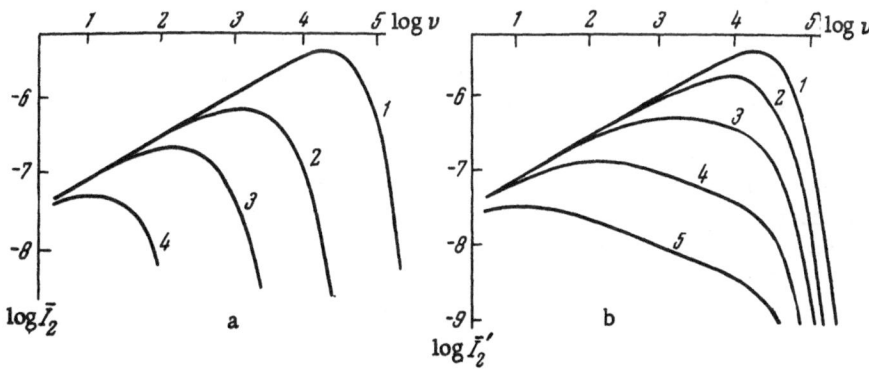

Fig. 46. Frequency dependence of the rectified current for two possible definitions of the constant emf: a) according to Eq. (IV.19); b) according to Eq. (IV.18).

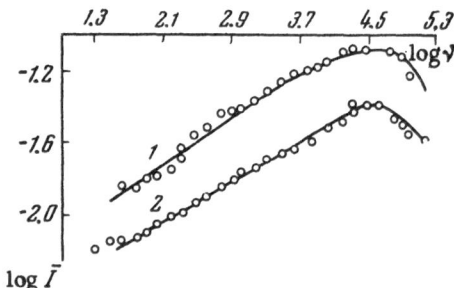

Fig. 47. Frequency dependence of $\bar{I}$ for capacitors with a low electrode resistance R (two capacitors).

element produces rectification. Since now $R_0$ consists of two resistances (barrier and interior), we cannot a priori predict the role of the interior in the rectification process. Therefore, it is necessary to consider two expressions for the constant emf.

If the rectification takes place only at the barriers, the emf is given by Eq. (IV.19). If the role of the volume in the rectification is comparable with that of the barriers, the emf is given by Eq. (IV.18). In this case, the frequency-independent $b^*$ should be replaced in Eqs. (IV.18) and (IV.19) with $\beta^*$, whose dependence on the frequency $\nu$ is given by Eq. (V.6). Both cases are presented in Fig. 46 for p = 0.95, $\gamma = 12.1$, $a = 2.34 \cdot 10^{26}$, $\varkappa = 7.41 \cdot 10^5$, and values of $\varkappa_\infty$ equal to 0, $10^2$, $10^3$, $10^4$, $10^5$ (curves 1-5, respectively). Figure 46a represents the emf found using Eq. (IV.19) and Fig. 46b represents the emf found using Eq. (IV.18). The frequency dependence of $\bar{I}$ is quite different in these two figures. If the constant emf appears only at the barriers, the frequency dependence $\bar{I}$ has a well-defined maximum. In this case, the rise of the current is followed, after a maximum, by a rapid fall. However, if the interior and the barriers play comparable roles in the rectification, then the frequency dependence of the rectified current has a very flat maximum which sometimes resembles a plateau (curve 3 in Fig. 46b). In any case, when $\varkappa_\infty > 10^2$, the rise of the current $\bar{I}$ is replaced by a very slow fall which becomes rapid only at very high frequencies. Therefore, it seemed interesting to determine the frequency dependence of the rectified current over a wide range of frequencies using capacitors which exhibited saturation of the brightness at high frequencies. The results of the measurements for two capacitors are given in Fig. 47 (cf. Fig. 16). The rise of the current is found to be followed by a rapid fall. This result is fairly reliable although the scatter of the experimental points is fairly large. When the frequency is increased, the rectified current $\bar{I}$ varies in the same way as for capacitors which have a large value of the resistance R; consequently, the constant emf of an electroluminescent capacitor is given by Eq. (IV.19), and the frequency dependence of the rectified current for capacitors with low electrode resistances has the following form:

$$\bar{I}_2^* = \frac{1}{1 + \varkappa_\infty + ax^{\frac{p}{p-1}}} \exp\left[ -\frac{\beta^*}{\sqrt{\bar{U}}} \sqrt[4]{\frac{\left(1 + \varkappa_\infty + ax^{\frac{p}{p-1}}\right)^2 + \left(x + \frac{\varkappa_\infty}{a} x^{\frac{1}{1-p}}\right)^2}{\left(ax^{\frac{p}{p-1}}\right)^2}} \right], \quad (V.13)$$

where $\bar{I}_2^*$ is the dimensionless value of the rectified current. The subscript "2" is used to distinguish the cases when $R_0$ is defined by Eq. (III.27) and by Eq. (III.13).

Thus, the results of investigations of the frequency dependence of the rectified current permit us to draw two important conclusions: 1) the constant emf appears only at the barriers; 2) the process of volume redistribution of the charge plays an important role at the barriers.

If the rectification is due to the tunnel effect, which has no inertia at audio frequencies, then the frequency dependence of the rectified current $\bar{I}$ indicates that another process takes place at the barriers and is responsible for the frequency dependence of the current. This process is probably associated with the volume redistribution of the charge, which is different at different frequencies. It must be stressed that the same process (volume redistribution of the charge) governs the frequency dependences of three quite different effects: rectification of the current, absorption of the energy, and emission of the energy by an electroluminescent capacitor.

## § 3.  Nature of the Rectification

First, we must consider which element of an electroluminescent capacitor is responsible for the rectification. Our capacitors had different electrodes: one electrode was made of aluminum, the other of $SnO_2$ film. Therefore, the question arises: is the rectification effect associated with the electrode material? To answer this question, we prepared capacitors with identical electrodes: both aluminum or both conducting glass; we found that the rectifying properties of the capacitors were still exhibited in full. Therefore, we could conclude that the rectification of the current was mainly due to components other than the electrode material.

From the results given in Fig. 37b, we may conclude that the dielectric is also an inactive element as far as the rectification is concerned, but more accurate information was obtained as follows: Capacitors were prepared from the binder alone without a phosphor. These capacitors did not rectify the current. To carry out a more reliable comparison, we prepared pairs of capacitors (one containing a mixture of ZnS and a dielectric binder and the other containing only the dielectric binder), which, in the absence of electroluminescence, should have very similar electrical properties. When the samples were excited even with fairly high dc voltages, there was no electroluminescence, but weak alternating fields were sufficient to produce excitation. The selected pairs of capacitors had very similar values of the loss-angle tangent and of the capacitance when they were subjected to an alternating voltage of 10 V of 5-kc frequency; a constant current i through the samples, under constant external voltage of 100 V, was also similar. The results of these measurements are given in Table 1.

When these capacitor pairs were subjected to an alternating voltage sufficiently strong to generate electroluminescence, the capacitors containing ZnS produced a rectified current, but no rectification was observed in the capacitors free of ZnS. The minimum current measured by our apparatus was $10^{-10}$ A; therefore, we could assume that, if the capacitors consisting of pure dielectric produced a rectified current, this current was $10^4$ times smaller for the polystyrene capacitors and $10^2$ times smaller for the ML-92 resin capacitors than for the corresponding capacitors with ZnS. Thus, we could conclude that the rectification effect was associated with the presence of zinc sulfide.

The question then arose: how important was the presence of electroluminescent ZnS for the rectification effect? We prepared capacitors in which electroluminescent zinc sulfide was replaced with a photoluminescent phosphor of the FKP-03k grade, which had the same luminescence centers as the electroluminescent phosphor ÉL-510. The rectification effect was absent in capacitors with FKP-03k not only in darkness but also during photoexcitation. The photoexcitation was applied in order to equalize the conductivities of the crystallites in FKP-03k and ÉL-510, because the value of the rectified current could have depended strongly on the conductivity. The FKP-03k phosphor differed from ÉL-510 by the presence of a conducting phase enriched with copper, which was associated directly with the electroluminescence. The presence of a conducting phase at the ZnS boundaries suggested that the rectification could be due to tunnel transitions of electrons from the conducting phase through thin layers of the dielectric. Therefore, using the same method as before, we prepared samples in which ZnS was replaced with aluminum dust. These samples did not rectify the current.

Thus, the whole series of these experiments (see also § 2 of the present chapter) indicated that the rectification was due to effects taking place at conducting phase—ZnS boundaries. We shall consider this point in more detail.

The electroluminescence of zinc sulfide copper-activated phosphors [40] is observed only when copper is introduced in concentrations not less than $5 \cdot 10^{-3}\%$. The amount of copper in excess of the equilibrium value is precipitated in the form of a second phase. This phase may be in the form of compounds of the $Cu_xS$ type, where $1 \leq x \leq 2$. Compounds of this type [133] are usually binary systems ($Cu_2S$ + CuS) or $Cu_2S$ with an excess of sulfur. Most investigators [40, 75, 134] are of the opinion that the second phase on the surface of ZnS is $Cu_2S$, since CuS is less stable and decomposes into sulfur and $Cu_2S$ when heated. Generally speaking, an excess of sulfur impedes the decomposition of copper sulfide.

The asymmetry of the heterogeneous layer in an electroluminescent capacitor may be established during its preparation and may possibly be associated with the deposition of ZnS particles on a substrate because of the

TABLE 1

| Sample | $i$, A | $\tan \delta$ | $C$, pF | $\bar{I}$, A |
|---|---|---|---|---|
| Capacitor with polystyrene and phosphor ÉL-510 | $60 \cdot 10^{-8}$ | $7 \cdot 10^{-2}$ | $25 \cdot 10^{3}$ | $3 \cdot 10^{-6}$ |
| Capacitor with polystyrene | $6 \cdot 10^{-8}$ | $2 \cdot 10^{-2}$ | $42 \cdot 10^{3}$ | — |
| Capacitor with ML-92 resin and phosphor ÉL-510 | $5 \cdot 10^{-8}$ | $5 \cdot 10^{-2}$ | $25 \cdot 10^{3}$ | $1 \cdot 10^{-8}$ |
| Capacitor with resin ML-92 | $2.5 \cdot 10^{-8}$ | $10^{-2}$ | $8 \cdot 10^{3}$ | — |

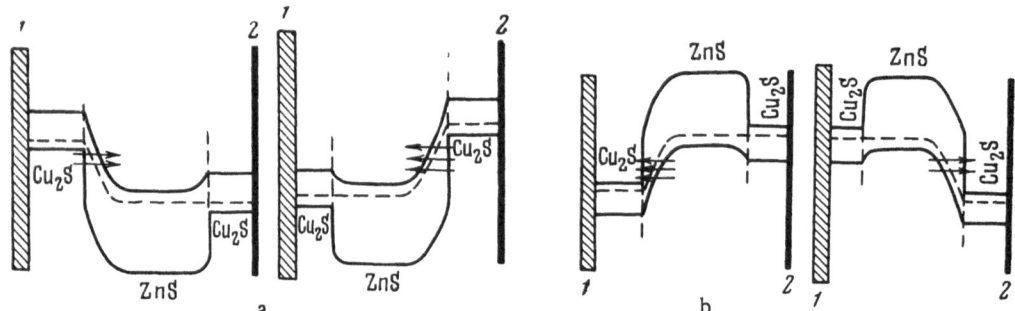

Fig. 48. Tunnel currents during successive half-periods of the exciting field for two types of conduction in ZnS: a) n-type ZnS; b) p-type ZnS.

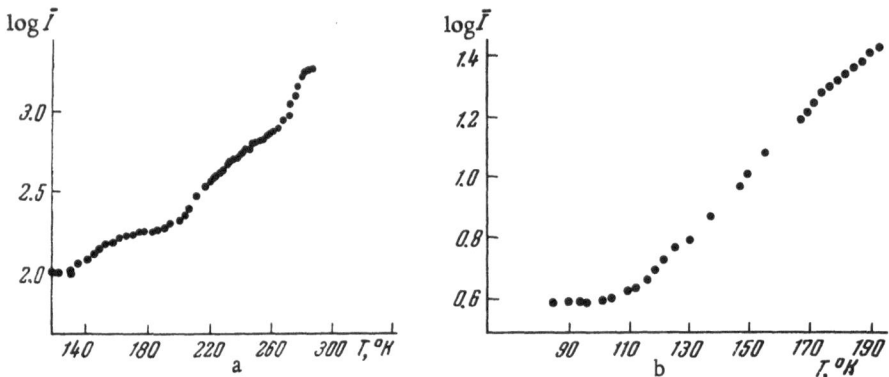

Fig. 49. Temperature dependence of the current $\bar{I}$ for two capacitors.

high specific gravity of these particles. This is supported by the fact that the direction of the constant emf is the same for different substrates: glass, steel, aluminum, i.e, for very different electrodes (in the first case, $SnO_2$ is the lower electrode, whereas, in the second and third cases, this compound is the upper electrode). In all these cases, the direction of the constant emf is such that the electrode which is lower during the preparation becomes negative.

Zinc sulfide particles are of irregular shape. During the preparation of a capacitor (spraying by means of an atomizer, precipitation on a substrate) they may acquire an ordered distribution. For example, sharp corners may be oriented in the direction of one of the electrodes. The field will then be concentrated at such sharp points (cf. § 2, Chapter I) and some barriers will be subjected to stronger fields, since the concentration of the field at the barriers will be added to the concentration of the field at the sharp points. If these points are oriented to some extent, the tunnel current in one half-period of the alternating exciting field will not be equal to the tunnel current in the second half-period. The tunnel currents during consecutive half-periods of the field are shown in Fig. 48 for two types of conduction of ZnS (Fig. 48a represents the n-type conduction case, Fig. 48b the p-type case). The lower electrode is shown shaded and is indicated by the number 1; 2 is used to denote the upper electrode at which a positive emf appears.

Figure 48 is much simplified not only because it completely excludes the influence of the dielectric, but also because of the continuous joining of the bands. The thermal width of the forbidden band of $Cu_2S$ is 0.6 eV [133] and that of ZnS is $\Delta = 3.7$ eV. Thus, the nature of the joining of the bands (Fig. 1) may differ considerably from case to case, depending on $\varphi_1 - \varphi_2$ and the nature of the conduction in ZnS. However, even such a simplified scheme allows us to draw the following conclusion: If the lower electrode is charged negatively, the sharp points must be oriented in the direction of the upper electrode in the case of n-type ZnS, while in the case of p-type ZnS, the field should be concentrated most strongly at the lower electrodes. There are no reliable data on the type of conduction in ZnS electroluminescent phosphor powders. Usually, such powders are regarded as n-type semiconductors [39]. In such a case, the polarity of the constant emf would indicate that the field is concentrated most strongly at the upper electrode. The same result is obtained from the dc current-voltage characteristic (Fig. 25).

## § 4. Temperature Dependence of the Rectified Current

In § 1 of the present chapter, we considered the voltage dependence of the rectified current and found that this dependence could be accounted for satisfactorily by the tunnel effect. The temperature dependence of the rectified current was investigated in order to determine the nature of the electron−phonon interaction during the process of tunnel leakage [135].

The experimentally measured complex temperature dependence of the rectified current $\bar{I}$ is shown in Fig. 49 for capacitors in which the heterogeneous layer consisted of an electroluminescent phosphor ZnS : Cu and a mixture of polystyrene with nitrocellulose. The investigated samples were heterogeneous systems and, therefore, the rectified current $\bar{I}$ could be considered in relation to the tunnel effect mechanism only in that range of temperatures where $\tan \delta$ and $\varepsilon$ of the binder were constant. We measured the capacitance and the loss-angle tangent of capacitors filled only with polystyrene and nitrocellulose. These measurements showed that $\varepsilon$ and $\tan \delta$ of the binder in our electroluminescent capacitors were constant in the range $77 < T < 170°K$. In this range of temperatures, the dependence of $\bar{I}$ on T indicates a one-phonon nature of the tunnel transitions. As the temperature rises, $\bar{I}$ increases exponentially beyond a region in which it is practically constant.

In the case of direct tunnel transitions, the temperature dependence of $\bar{I}$ should be, according to Eq. (I.1), a weakly rising function over a wide range of temperatures. This type of temperature dependence is obtained for the emission from electrodes (I.10), for the ionization of local levels (I.9), and for the Zener effect with scattering by charged impurities (I.8). Consequently, these mechanisms must be rejected. There is no point in comparing our results with those expected for multiphonon tunnel transitions (I.6) and for the impact ionization (I.13), since they do not yield the experimentally observed voltage dependence. Moreover, in the impact ionization case, the current should decrease as the temperature rises. Thus, we are left with only one mechanism: one-phonon tunnel transitions (I.2), which give a temperature dependence such as that shown in Fig. 49.

However, in a system as complex as an electroluminescent capacitor, we cannot say a priori which electron transitions make the main contribution to the rectified current, although from the general agreement between the results in Fig. 49 and Eq. (I.2) we may conclude that they are band−band transitions. However, several types of transition are possible: 1) from the valence to the conduction band of ZnS; 2) from the valence band of $Cu_xS$ to the conduction band of ZnS; 3) from the electrode metal to the conduction band of ZnS.

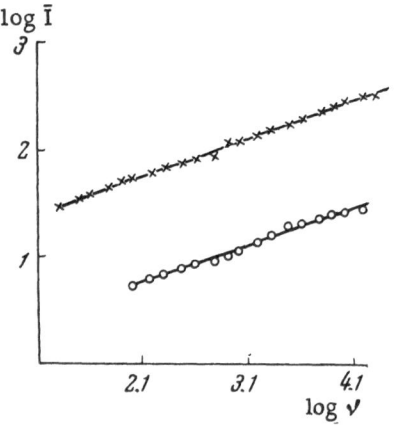

Fig. 50. Frequency dependence of $\bar{I}$ at T = 290°K (crosses) and 120°K (circles).

This last possibility, which represents electron transitions from a metal electrode to the conduction band of ZnS, must be rejected on the basis of the experiments described in § 3 in the present chapter, which were carried out on capacitors with identical electrodes.

We shall now try to find which of the possible transitions does in fact take place. In this connection, we must draw attention to the following observation. At $T < T_0$, where $T_0$ is defined by Eq. (I.4), we observe a constant number of tunnel-generated pairs $n_0$,

$$n_0 \propto \exp\left[-\frac{4\sqrt{2m^*}}{3e\hbar E}(\Delta + \hbar\omega)^{3/2}\right]. \qquad (V.14)$$

The rectified current $\bar{I}$, which we shall regard as proportional to the number of tunnel-generated pairs, is also constant at low temperatures, and, therefore,

$$\bar{I} \propto \exp\left(-\frac{\beta^*}{\sqrt{U}}\right). \qquad (V.15)$$

The latter formula is integral, since the index of the exponential function includes the voltage applied to the whole capacitor. It has been shown in Chapter III that, over a fairly wide range of audio frequencies, the whole voltage applied to the heterogeneous layer is concentrated at the barriers. This also applies to measurements over a fairly wide range of temperatures, as shown in Fig. 50. Hence, if we know the number of barriers $\mathfrak{M}$,

$$U = \mathfrak{M}V, \qquad (V.16)$$

where V is the voltage across one barrier. The rectified current through one barrier is proportional to

$$\exp\left\{-\frac{\beta_{\mathfrak{M}}^*}{\sqrt{V}}\right\} = \exp\left\{-\frac{\beta_{\mathfrak{M}}^*\sqrt{\mathfrak{M}}}{\sqrt{U}}\right\} = \exp\left\{-\frac{\beta^*}{\sqrt{U}}\right\}. \qquad (V.17)$$

Here, $\beta_{\mathfrak{M}}^*$ represents one barrier and all $\mathfrak{M}$ barriers are regarded as identical. The above equation is correct because barriers are connected in series. Since the field intensity across a barrier is proportional to the square root of the applied voltage (cf. Eqs. (I.19) and (I.4)], the low-temperature region where the current is constant can be represented by

$$\frac{\beta^*}{\sqrt{U}} = \frac{(\Delta + \hbar\omega)^{3/2}}{3kT_0\sqrt{\Delta}}. \qquad (V.18)$$

Hence, the height of the barrier through which a tunnel transition takes place is found to be

$$\Delta = \frac{3kT_0\beta^*}{\sqrt{U}} - \frac{3}{2}\hbar\omega. \qquad (V.19)$$

The quantities U, $\beta^*$, $T_0$, and $\hbar\omega$ which occur in Eq. (V.19) are known from experiments (we shall deal with $\hbar\omega$ later) and, therefore, we can find the value of $\Delta$. According to our results, $\Delta = 0.35 \pm 0.1$ eV. Judging by the height of this barrier to be overcome by electrons, we can reject the following possible transitions:

1) from the valence band of ZnS to the conduction band of ZnS, since, in this case, the height of the barrier should be 3.7 eV;

2) from the valence band of p-type $Cu_2S$ to the conduction band of n-type ZnS;

3) from the valence band of n-type $Cu_2S$ to the conduction band of p-type ZnS.

In the latter two cases, the barrier height should have been not less than 0.67 eV, which represents the thermal width of the forbidden band of $Cu_2S$.

It is interesting to note the following: while the type of conduction of $Cu_2S$ is unimportant in the formation of a blocking layer, i.e., it does not affect the possibility of concentrating the electric field, the height of the barrier to be overcome by electrons does depend on the type of conduction of $Cu_2S$. The barrier between ZnS and $Cu_2S$ (Fig. 1) for different types of conduction is certainly greater than the value obtained by us, while the barrier at the boundary between the same substances having the same type of conduction (Fig. 1) may be equal to the value obtained here.

It has been shown in [136] that the second phase on the surface of ZnS exhibits metallic conduction, i.e., if this phase represents $Cu_2S$, it must be so heavily doped that it has the properties of a strongly degenerate semiconductor. Since no special doping with other elements has been carried out, this may mean that there is a strong local departure from the stoichiometric composition.

Thus, if there is a region where $\bar{I}$ is constant, then, on the basis of Eq. (V.19), we can estimate the energy height of the barrier. On the other hand, we can use Eq. (I.4) to estimate the value of the field in the barrier from the temperature at which this region ends. This estimate is interesting because the fields cannot be measured directly in such complex heterogeneous samples as our electroluminescent capacitors. Assuming that $\Delta = 0.35$ eV, we find that $E \approx 10^6$ V/cm. If, other conditions being equal, we assume that $\Delta = 3.7$ eV, then $E \approx 3 \cdot 10^6$ V/cm. In estimating the fields, we have assumed that $m^* = m$. In fact, this is not correct [137]. Usually, $m^* < m$. Therefore, the true fields are evidently weaker than the calculated fields. However, even if the effective mass is 4 times smaller than m, the values of the field still amount to half the calculated values.

Estimates of the fields in the exhaustion barriers have been obtained before. In a paper cited earlier [102], it is concluded that the electroluminescence begins in fields of $5 \cdot 10^4$ V/cm, while in fields of $1.4 \cdot 10^5$ V/cm, the exhaustion barrier spreads over the whole volume of a ZnS crystallite. However, experiments on the quenching of electroluminescence by a magnetic field [47, 48] (cf. § 2, Chapter I) indicate that local fields should be stronger than $2 \cdot 10^5$ V/cm, and this agrees well with our estimates.

We shall now consider the nature of the frequency dependence of the field. The field is concentrated in a barrier and the concentration process is characterized by a relaxation time. At low frequencies, the process of field concentration is more complete than at high frequencies. Thus, for the same external voltage across a sample, we may expect the fields in the barrier layers to be weaker at higher frequencies. The values of the power exponent should be larger at high frequencies, which is indeed observed experimentally (Fig. 42); the experimental results presented in Fig. 42 were obtained at room temperature. This relationship between the field and frequency should also apply at lower temperatures. Our previous estimates indicate a field of $E = 10^6$ V/cm; this applies to an external field frequency $\nu = 50$ cps. At higher frequencies, we could not determine $T_0$, since the region where $\bar{I}$ was constant was below liquid nitrogen temperature. This way, a reduction in E due to an increase in the frequency caused a proportional fall in $T_0$. For the majority of the investigated samples, $T_0 \approx 120°K$ and the region where the current was constant lay at lower temperatures. Thus, a drop in E by 20-30% reduced $T_0$ so much that we could not observe the region where the current was constant. Consequently, we were unable to check the frequency dependence of the field by independent temperature measurements. However, a frequency dependence of the type predicted can be regarded as firmly established from the temperature measurements, which were carried out on a large number of samples.

Thus, the field E which governs the rectified current depends on the frequency (Fig. 42). In the preceding sections, we have established a close relationship between the rectification of the current and the excitation of the electroluminescence. However, b, which is the power exponent in the expression for the brightness (Fig. 20), is independent of the frequency at low frequencies where, nevertheless, there is a dependence of $\beta$ on $\nu$. Hence, it follows that the field which causes electroluminescence is independent of the frequency. This discrepancy is possibly due to the fact that, for a current to pass right through a sample, a potential drop of 0.4 eV is required, while about 3 eV is needed for the ionization of luminescence centers.

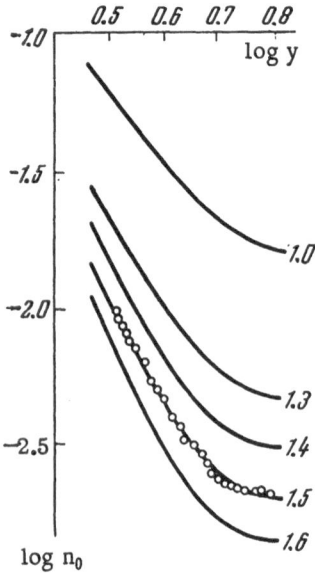

Fig. 51. Temperature dependence of the rectified current. The circles represent experimental values. The curves are calculated using Eq. (V.26).

Knowing the value of the field in a barrier, we can estimate other characteristics of the barriers formed in the cathode regions of ZnS crystallites. If we assume that the density of the positive space charge is constant in the region where this charge exists, then, on the basis of Eq. (I.20), we can obtain the following expression for the barrier width:

$$L = \left( \frac{\varepsilon V}{2\pi e N} \right)^{1/2} \qquad (\text{I.20'})$$

and the maximum field intensity in a barrier is

$$E = \frac{2V}{L}. \qquad (\text{I.20})$$

Here, $\varepsilon$ is the permittivity, N is the number of ionized donors in 1 $cm^3$, and V is the voltage across the barrier. Then, the concentration of ionized donors N is given by:

$$N = \frac{\varepsilon E^2}{8\pi e V}. \qquad (\text{V.21})$$

The value of E was found earlier in the present paper; according to [138], $\varepsilon = 7$ and the value of V can be estimated from Eq. (V.16) knowing $\mathfrak{M}$. The value of $\mathfrak{M}$ can be estimated from the thickness of the layer and the granulometric composition of the electroluminescent phosphor. In our case, $\mathfrak{M} = 4\text{-}8$. Using these numerical values, we find from Eq. (V.21) that $N \approx 10^{17}$ $cm^{-3}$. Allowance for a possible error in the determination of E shows that N can be 4 times smaller than the calculated value. The value for the concentration of ionized donors which is thus obtained ($10^{17}$ $cm^{-3}$) is quite reasonable because, although the introduction of copper in amounts of $10^{-4}\text{-}10^{-5}$ g-atom/g-mole (which is the solubility limit for Cu in ZnS) corresponds to a donor level concentration of $10^{18}\text{-}10^{17}$ $cm^{-3}$, some of these donors may not be ionized.

The width of an exhaustion barrier L may be estimated using Eq. (I.20'), making the assumptions referred to earlier. This width is found to be $2 \cdot 10^{-5}$ cm. Allowance for the effective mass may increase this value by a factor of 2.

All these estimates refer to the temperature range where the current is constant: $\bar{I} = \bar{I}(T) = \text{const}$. At temperatures of about 100°K, the rectified current begins to increase exponentially. In this range of temperatures, we can use Eq. (I.2) to estimate the energy of the phonons taking part in the tunnel transitions. If the field intensity in a barrier is not high, so that we can neglect the second term in the square brackets of Eq. (I.2), we can determine $\hbar\omega$ from the slope of the experimental curve representing the dependence $\bar{I} = \bar{I}(T)$, plotted using the coordinates $\log \bar{I}$, $1/T$. However, in the case of our capacitors the field E is usually such that we need to include both terms.

If we use this method to find $\hbar\omega$, we obtain a value of the phonon energy which is too low, and, moreover a value which depends on the field intensity at which the experiment is carried out. Therefore, we shall determine the phonon energy as follows:

We shall rewrite Eq. (I.2) using $N_k$ from Eq. (I.3) and E from Eq. (I.4). Then

$$n_0 = F \left[ \frac{1}{\exp\left( \frac{\hbar\omega}{kT} \right) - 1} \left( 1 + e^{-\frac{\hbar\omega}{kT_0}} \right) + e^{-\frac{\hbar\omega}{kT_0}} \right], \qquad (\text{V.22})$$

where F can be regarded as independent of temperature. We shall introduce dimensionless parameters G, y, and $y_0$ by means of the formulas

$$G = \frac{\hbar\omega}{0.043} \,, \tag{V.23}$$

and

$$y = \frac{0.043}{kT} \,, \tag{V.24}$$

$$y_0 = \frac{0.043}{kT_0} \cdot \tag{V.25}$$

Substituting these parameters into Eq. (V.22), we obtain

$$n_0 = F \left[ \frac{1}{e^{Gy} - 1} (1 + e^{-Gy_0}) + e^{-Gy_0} \right]. \tag{V.26}$$

The choice of the factor 0.043 is arbitrary. We have used this value because this is the energy of the longitudinal optical phonons reported in [139] for G close to unity. For a fixed value of $y_0$, Eq. (V.26) gives a one-parameter family of curves. This family is shown in Fig. 51 for several values of G (continuous lines). The numbers alongside the curves are the values of G. The experimental curve in Fig. 49b agrees best with the theoretical curve for G = 1.5, after shifting it along the ordinate. Hence, the energy of the phonon taking part in the tunnel transitions is 0.065 ± 0.004 eV.

From [139, 140] it is known that the energy of the longitudinal optical phonons in ZnS is $4.3 \cdot 10^{-2}$ eV, while that of the transverse optical phonons is $3.4 \cdot 10^{-2}$ eV. Thus, if ZnS phonons take part in the transition, we must assume that a transition requires the participation of two transverse optical phonons [141]. However, they may be $Cu_xS$ phonons, whose energies are not known. The following point must be stressed: A tunnel transition does not take place inside ZnS alone but from one substance to another. Since the dependence of the crystal momentum on the energy is not known exactly for these substances, we cannot predict a priori what phonons will be necessary to compensate for the difference between the crystal momenta resulting from this transition. All this applies to temperatures below 170°K, since at T > 170°K the permittivity ε of the dielectric binder (polystyrene + nitrocellulose) increases, and at T > 250°K, tan δ begins to rise rapidly. Thus, summarizing our discussion, we may conclude that at 77 < T < 170°K the behavior of the rectified current can be accounted for by a constant emf which results from a phonon-assisted tunnel transition for E(T) = const.

It is very difficult to interpret the rise of the rectified current in the temperature range 180 < T < 250°K. On the basis of the most elementary considerations, we may assign it to a change in the permittivity of the binder at these temperatures: the rise of ε of the dielectric binder results in an increase of the field in ZnS, because the fields in the components of a heterogeneous system are inversely proportional to the permittivities of the components. $\bar{I}$ is an exponential function of the field and, therefore, we should observe an exponential rise of the rectified current. However, experiments on various dielectrics (with different temperature dependences of ε) have shown that it is not possible to account for the rise of $\bar{I}$ simply by a change in ε of the binder. It is very likely that the behavior of $\bar{I}$ at these temperatures is due to phonon-assisted tunnel transitions under the action of a uniformly increasing field in an exhaustion barrier.

## Conclusions

We have investigated the dependence of the energy absorbed and emitted by an electroluminescent capacitor and the dependence of the rectified current on the voltage and frequency of the external exciting field. To explain the results obtained, we have proposed an equivalent circuit for the electroluminescent capacitor in the form of the capacitance and resistance of the heterogeneous layer connected in parallel and an electrode resistance in series with the other two elements. The capacitance of the heterogeneous layer and the electrode resistance are assumed to be independent of the frequency. The resistance of the heterogeneous layer itself has two components. One is frequency dependent and is interpreted as the resistance of the barriers; the other ($R_\infty$) is independent of frequency and is attributed to the volume resistance. Good quantitative agreement is obtained between the frequency dependence of the energy absorbed by such a capacitor and calculations based on this equivalent circuit. Measurements of the frequency dependence of the actively absorbed energy make it possible to determine the parameters of the equivalent circuit, which can be used in the calculation of other characteristics of the electroluminescent capacitor.

This equivalent circuit not only explains the frequency dependence of the actively absorbed energy, but it also accounts for the frequency dependence of the electroluminescence brightness and of the current rectified by the capacitor.

It has been established experimentally that, at very low frequencies, the electroluminescence brightness is constant and that it begins to increase rapidly when the frequency is increased, reaching a brightness maximum or saturation at a frequency of several kilocycles or several tens of kilocycles.

The proposed circuit predicts this frequency dependence for the useful absorbed power $W_{R_0}$, which is related to the electroluminescence brightness if the energy yield of the electroluminescent phosphor is constant. The circuit makes it possible to clearly interpret the various parts of the frequency dependence. At very low frequencies, the resistance of the capacitor can be assumed to be constant, which corresponds to a region of constant brightness. At low audio frequencies, the main contribution to the brightness is made by barriers whose resistance increases when the frequency is increased; this is accompanied by an increase in the brightness. Saturation of the brightness is observed only when the volume resistance of an electroluminescent phosphor is higher than the electrode resistance. It is usually observed at high audio and ultrasonic frequencies, at which the luminescence of the interior is the dominant component of the total electroluminescence. For capacitors with high-resistance electrodes, the quantity $R_\infty$ in the equivalent circuit, which represents the interior, can be neglected, and we then have a brightness maximum instead of saturation. The fall in the brightness at high frequencies results from a redistribution of the external voltage between the elements of a capacitor. The impedance of the heterogeneous layer filling a capacitor decreases when the frequency is increased not only because of the shunting effect of the capacitance, but also because of a drop in the value of the barrier resistance. Consequently, when the frequency is increased, a growing fraction of the external voltage becomes concentrated at the electrodes, and the useful voltage across the heterogeneous phosphor decreases and the brightness falls off.

Since the measurements of the brightness were carried out independently of the measurements of the active power, they could be used to check the correctness of the proposed equivalent circuit. Having determined the parameters of the circuit from the measurements of the active power, we could calculate the frequency dependence of $W_{R_0}$. Comparison of the experimental and calculated values show that the nature of the dependences was the same. Another independent confirmation of the correctness of the equivalent circuit was provided by the growth of the rectified current when the frequency was increased. We shall consider this later in more detail.

Thus, the equivalent circuit and the assumption that the system is linear with respect to the voltage makes it possible to explain the frequency dependence of the absorbed energy and correctly predicts the nature of the frequency dependence of the emitted energy. However, a real electroluminescent capacitor is basically a nonlinear system, as indicated by its current-voltage characteristic. This affects the voltage dependences of the absorbed and emitted power. The dependence of the absorbed power on the voltage is stronger than $U^2$; the dependence of the brightness on the voltage is exponential. The nonlinearity of these characteristics of the electroluminescent capacitor is due to the phosphor itself. The processes which take place in an electroluminescent phosphor crystal during excitation and de-excitation are complex and many.

However, in spite of the complexity of the electroluminescence, we must point out one process which plays a decisive role; this is the ionization in a barrier, the probability of which process is exponential.

Therefore, even in the simplest formula proposed for the efficiency of an electroluminescent phosphor, we find a term which is proportional to the ionization density. We have shown that the inclusion of just this term makes it possible to account for the dependence of the electroluminescence brightness on the voltage. It is necessary to allow for this term when the frequency dependence of the brightness is considered at high frequencies.

At low frequencies, when the whole voltage is applied to the heterogeneous layer filling a capacitor, the exponential term is constant and does not affect the form of the curve. When the frequency is increased, the voltage is redistributed and the fraction of the voltage across the heterogeneous layer decreases. This alters the value of the exponential term: it decreases when the frequency is increased, resulting in a more rapid decrease in the brightness. When the exponential term is taken into account, the frequency dependences of the brightness found by calculation and measured experimentally agree very well. Such a comparison has been carried out for frequencies of 20 cps $\leq \nu \leq$ 200 kc.

It has been shown that an increase in the exciting field intensity results in an increase in the importance of other phenomena which take place in electroluminescent zinc sulfide: the absorption of the electric field energy by electrons which tunnel through a barrier or which are liberated by the field from deep traps, etc. Allowance for such processes makes it possible to explain the experimentally observed dependence of the absorbed power on the voltage. The expression for the efficiency of an electroluminescent phosphor becomes very complex [cf. Eq. (IV.34)]. Formula (IV.34) is in good agreement with the experimental results.

A new method is proposed for investigating the processes which take place in an electroluminescent capacitor with a heterogeneous layer, based on the investigation of the rectified current characteristics. An investigation of the rectifying properties of a series of specially prepared capacitors has shown that this is due to a definite orientation of the electroluminescent phosphor grains. The dependence of the rectified current on the external voltage indicates a relationship between the mechanism of the establishment of a constant emf and the tunnel effect in strong fields. Investigations of the temperature dependences, carried out in order to determine the nature of the electron−phonon interaction, suggest that the tunnel transitions of carriers at $77 < T < 170°K$ are assisted by the lattice phonons.

The experimental data, together with the theory of one-phonon tunnel transitions, show that the height of the energy barrier through which the tunnel effect takes place is $\Delta = 0.35 \pm 0.1$ eV and the electric field intensity in the barrier, which governs the rectification of the current, is $E = 10^6$ V/cm. Two transverse optical phonons, probably from ZnS ($2\hbar\omega = 0.065$ eV), take part in the process. Assuming that the field is concentrated in Mott−Schottky exhaustion barriers, it is found that the width of such barriers is $2 \cdot 10^{-5}$ cm and the concentration of ionized donors which establishes such barriers is about $10^{17}$ cm$^{-3}$.

The frequency dependence of the rectified current in the range from 20 cps to 200 kc is governed by two factors: the frequency dependence of the constant emf and the frequency dependence of the barrier resistance. The frequency dependence of the emf, which appears only at the barriers, is related to the processes of field concentration and voltage redistribution between the elements of an electroluminescent capacitor. At high frequencies, the voltage across the barriers decreases and the number of tunneling carriers decreases; this governs the value of the constant emf. At low frequencies, the constant current increases when the frequency is increased, as expected on the basis of the proposed equivalent circuit, in which it is assumed that the barrier resistance depends on the frequency.

The frequency dependence of the rectified current calculated on the basis of this equivalent circuit for the capacitor parameters found from the energy measurements is in good agreement with the experimental results over a wide range of frequencies.

The author takes this opportunity to thank her scientific director, M. V. Fok, for his constant interest, help in the investigation, valuable comments, and advice.

## Literature Cited

1.   O. V. Losev, Telegr. i telef. bez provodov, 18:61 (1923).
2.   G. Destriau, J. Chem. Phys., 33:620 (1936).
3.   J. R. Haynes and H. B. Briggs, Phys. Rev., 86: 647 (1952).
4.   R. Newman, W. C. Dash, R. N. Hall, and W. E. Burch, Phys. Rev., 98: 1536 (1955).
5.   R. Newman, Phys. Rev., 100:700 (1955).
6.   I. K. Vereshchagin and V. S. Teslyuk, Izv. vuzov, seriya fiz., No. 6, 114 (1958).
7.   K. W. Böer and U. Kümmel, Ann. Physik, 14 :341 (1954).
8.   G. A. Wolf, I. Adams, and J. W. Mellichamp, Phys. Rev., 114:1262 (1959).
9.   A. W. Smith, Am. J. Phys., 27:591 (1959).
10.  A. Fischer, Z. Physik, 149:107 (1957).
11.  R. Braunstein, Phys. Rev., 99:1892 (1955).
12.  S. Larach and R. E. Shrader, Phys. Rev., 102:582 (1956).
13.  G. A. Wolff, R. A. Hebert, and J. D. Broder, Phys. Rev., 110:1144 (1955).
14.  D. H. Smith, Elec. Eng., 33(397):164 (1961).
15.  H. K. Henisch, Brit. J. Appl. Phys., 12:660 (1961).
16.  W. W. Piper and F. E. Williams, Solid State Physics, 6:95 (1958).
17.  W. W. Piper and F. E. Williams, Phys. Rev., 98:1809 (1955).
18.  H. K. Henisch, Electroluminescence. [Russian translation, Izd. Mir (1964)].
19.  A. G. Chynoweth, Progr. Semiconductors, 4:95 (1960).
20.  C. M. Zener, Proc. Roy. Soc. (London), A145:523 (1934).
21.  W. Franz, Ann. Phys., 11:17 (1952).
22.  K. B. McAfee, E. J. Ryder, W. Shockley, and M. Sparks, Phys. Rev., 83:650 (1951).
23.  L. V. Keldysh, Zhur. éksp. i teor. fiz., 34:962 (1958).
24.  R. H. Bube, Photoconductivity of Solids, 1960. [Russian translation, IL (1962)].
25.  P. S. Serebrennikov, Radiotekhnika i elektronika, 3(3):536 (1962).
26.  W. Franz and L. Tewordt, Halbleiterprobleme, 3:1 (1956).
27.  R. H. Fowler and L. Nordheim, Proc. Roy. Soc. (London), A119:173 (1928).
28.  W. Franz, Handbuch der Physik, Vol. 7, Berlin (1956).
29.  A. F. Ioffe, Physics of Crystals. Gosizdat, Moscow-Leningrad (1929).
30.  F. Seitz, Phys. Rev., 76:1376 (1949).
31.  B. I. Davydov and I. M. Shmushkevich, Zhur. éksp. i teor. fiz., 10:1043 (1940).
32.  V. A. Chuenkov, Fiz. tverd. tela, No. 2, 200 (1959).
33.  L. V. Keldysh, Zhur. éksp. i teor. fiz., 37:713 (1953).
34.  W. C. Dunlap, Jr., An Introduction to Semiconductors, 1957. [Russian translation, IL (1959)].
35.  W. Shockley, Electrons and Holes in Semiconductors, 1950. [Russian translation, IL (1959)].
36.  J. R. Haynes and W. C. Westphal, Phys. Rev., 101:1676 (1956).
37.  D. Rücker, Z. Angew Phys., 10:254 (1958).
38.  C. Z. van Doorn and D. De Nobel, Physica, 22: 338 (1956).
39.  W. W. Piper and F. E. Williams, Brit. J. Appl. Phys., Suppl. 4:39 (1955).
40.  P. Zalm, Philips Res. Rep., 11:353 (1956).
41.  A. Kremheller, J. Electrochem. Soc., 107:8 (1960).
42.  K. H. Butler and J. P. Waymouth, Brit. J. Appl. Phys., Suppl., 4:33 (1955).
43.  E. E. Loebner and H. Freund, Phys. Rev., 98:1545 (1955).
44.  G. Diemer, Philips Res. Rep., 10:194 (1955).
45.  V. E. Oranovskii and B. A. Khmelinin, Optika i spektroskopiya, 7, Brit. 542 (1959).
46.  D. R. Frankl, Phys. Rev., 111:1540 (1958).
47.  A. Levialdi and E. Guercigh, Compt. Rend., 257:852 (1963).
48.  A. N. Ince, Proc. Phys. Soc. (London), B67 :870 (1954).
49.  A. F. Ioffe, Doklady Akad. Nauk SSSR, 27 :547 (1940).

50. A. F. Ioffe, J. Phys. USSR, 10: 49 (1946).
51. W. Schottky, Z. Physik, 118: 539 (1942).
52. A. F. Ioffe, Physics of Semiconductors. Izd. Akad. Nauk SSSR, Moscow-Leningrad (1957).
53. A. Rose, Helv. Phys. Acta, 29: 199 (1956).
54. M. V. Fok, Uspekhi fiz. nauk, 72: 467 (1960).
55. A. N. Georgobiani, Trudy Fiz. Inst. Akad. Nauk, 23 : 3 (1963).
56. K. Maeda, J. Phys. Soc. Japan, 13: 1352 (1958).
57. B. M. Vul, Fiz. tverd. tela, 2: 2961 (1960).
58. A. N. Georgobiani and M. V. Fok, Optika i spektroskopiya, 10 : 188 (1961).
59. P. Zalm, G. Diemer, and K. A. Klasens, Philips Res. Rep., 10: 205 (1955).
60. G. I. Skanavi, Physics of Dielectrics (Weak Fields). GITTL, Moscow-Leningrad (1949).
61. G. G. Harman, Phys. Rev., 111: 27 (1958).
62. S. Roberts, J. Opt. Soc. Am., 42: 850 (1952).
63. A. G. Fischer, J. Electrochem. Soc., 109: 1043 (1962).
64. A. G. Fischer, J. Electrochem. Soc., 110: 733 (1963).
65. V. E. Oranovskii, E. I. Panasyuk, and B. T. Fedyushin, Inzh.-fiz. zh., 2: 40 (1959).
66. O. N. Kazankin, F. M. Pekerman, and L. N. Petoshina, Izv. Akad. Nauk SSSR, seriya fiz., 21: 721 (1957).
67. I. N. Orlov, Izv. Akad. Nauk SSSR, seriya fiz., 21: 731 (1957).
68. A. A. Cherepnev, Optika i spektroskopiya, 2: 770 (1957).
69. P. I. Gruder and D. A. Cusano, J. Opt. Soc. Am., 45: 493 (1955).
70. C. Feldman and M. O'Hara, J. Opt. Soc. Am., 47: 300 (1957).
71. N. A. Vlasenko and Yu. A. Popkov, Optika i spektroskopiya, 8: 81 (1960).
72. W. Lehmann, J. Electrochem. Soc., 107: 20 (1960).
73. G. S. Kozina and L. P. Poskacheeva, Optika i spektroskopiya, 8: 214 (1960).
74. O. N. Kazankin, F. M. Pekerman, and L. N. Petoshina, Optika i spektroskopiya, 7: 776 (1959).
75. V. N. Favorin and G. S. Kozina, Optika i spektroskopiya, 10: 91 (1961).
76. A. M. Bonch-Bruevich and O. S. Marenkov, Optika i spektroskopiya, 8: 855 (1960).
77. A. Luyckx, I. Vandewauwer, and S. Ries, Ann. Soc. Sci. Bruxelles, 72: 58 (1958).
78. A. N. Georgobiani, E. Yu. L'vova, and M. V. Fok, Optika i spektroskopiya, 13: 564 (1962).
79. A. N. Ince and C. W. Outley, Phil. Mag., 46: 1081 (1955).
80. W. Lehmann, J. Electrochem. Soc., 103: 24 (1956).
81. I. M. Vishenchuk, E. P. Sogolovskii, and B. I. Izvetskii, Cathode-Ray Oscillograph and Its Applications in Measurement Techniques. Fizmatgiz (1959).
82. P. Zalm, G. Diemer, and B. A. Klasens, Philips Res. Rep., 9: 81 (1954).
83. F. Matossi, J. Electrochem. Soc., 103: 34 (1956).
84. G. Destriau and H. F. Ivey, Proc. IRE, 43(12): 1911 (1955).
85. H. Gobrecht, D. Hahn, and H. E. Gumlich, Z. Physik, 136: 612 (1954).
86. G. R. Hoffman and D. H. Smith, J. Electron. and Control, 9(3): 161 (1960).
87. K.-S. K. Rebane and E. K. Tal'viste, Trudy IFA Akad. Nauk ESSR, No. 21, 257 (1962).
88. P. Goldberg, J. Electrochem. Soc., 106: 948 (1959).
89. W. Lehmann, Phys. Rev., 101: 489 (1956).
90. G. Destriau, Brit. J. Appl. Phys., Suppl., No. 4, 49 (1955).
91. D. Curie, J. Phys. Radium, 14: 672 (1953).
92. J. F. Weymouth, C. W. Jerome, and W. G. Gungle, Sylvania Technologist, 5: 54 (1952).
93. G. Destriau, Phil. Mag., 38: 700 (1947).
94. G. Destriau and I. Saddy, J. Phys. Radium, 6: 12 (1945).
95. A. Lempicki, D. R. Frankl, and V. A. Brophy, Phys. Rev., 104: 1238 (1957).
96. M. V. Fok, Optika i spektroskopiya, 11: 98 (1961).
97. W. Lehmann, Illum. Eng., 51: 684 (1956).
98. C. W. Jerome and W. G. Gungle, J. Electrochem. Soc., 100: 34 (1953).
99. M. V. Fok, Theory of Electroluminescent Image Converters. Izd. Sovetskoe radio (1961).

100.   A. N. Gubanov, Theory of the Rectifying Action in Semiconductors. GITTL (1956).
101.   Yu. P. Chukova, Optika i spektroskopiya, 1:339 (1963).
102.   J. R. MacDonald, Phys. Rev., 92:4 (1953).
103.   R. J. Friauf, J. Chem. Phys., 22:1329 (1954).
104.   A. B. Lidiard, Ionic Conduction in Crystals. [Russian translation, IL (1962)].
105.   Yu. P. Chukova, Optika i spektroskopiya, 18:1035 (1965).
106.   V. N. Favorin and L. P. Poskacheeva, Optika i spektroskopiya, 7:706 (1959).
107.   D. W. G. Ballentyne, J. Phys. Radium, 17:758 (1956).
108.   S. Nudelmann and F. Matossi, Phys. Rev., 98:238 (1955).
109.   G. F. Alfrey and K. N. R. Taylor, Helv. Phys. Acta, 30:206 (1957).
110.   B. T. Howard, Phys. Rev., 98:1544 (1955).
111.   J. B. Taylor, Brit. J. Appl. Phys., Suppl., 4:44 (1955).
112.   W. Lehmann, J. Electrochem. Soc., 105: 585 (1958).
113.   W. A. Thornton, J. Electrochem. Soc., 107: 895 (1960).
114.   F. A. Schwertz, J. J. Mazenko, and E. R. Michalick, Phys. Rev., 98:1133 (1955).
115.   F. A. Schwertz and R. E. Freund, Phys. Rev.,98:1134 (1955).
116.   G. F. Alfrey, Brit. J. Appl. Phys., Suppl., No. 4, 45 (1955).
117.   W. Lehmann, J. Electrochem. Soc., 103: 667 (1956).
118.   W. A. Thornton, J. Appl. Phys., 28: 313 (1957).
119.   P. Goldberg, J. Phys. Rev., 106: 34 (1959).
120.   P. Zalm, J. Phys. Radium, 17: 777 (1956).
121.   W. A. Thornton, Phys. Rev., 102: 38 (1956).
122.   V. S. Trofimov, Optika i spektroskopiya, 4:113 (1958).
123.   Yu. P. Chukova and V. D. Ligasova, Optika i spektroskopiya, 18: 846 (1965).
124.   E. Nad', Proc. Seventh. Conf. on Luminescence (Crystal Phosphors), Tartu (1959).
125.   H. F. Ivey, Electroluminescence and Related Effects. New York-London (1963).
126.   R. H. Weber, Illum. Eng., 58: 592 (1963).
127.   P. Zalm, Philips. Res. Rep., 11: 417 (1956).
128.   G. Neumark, Phys. Rev., 116:1425 (1959).
129.   F. F. Morehead, J. Electrochem. Soc., 107: 281 (1960).
130.   H. F. Ivey, IRE Trans. Electron. Devices, 335 (1959).
131.   Yu. P. Chukova and M. V. Fok, Zhur. tekhn. fiz., 35: 762 (1965).
132.   M. V. Fok and Yu. P. Chukova, Zhur. tekhn. fiz., 35: 2065 (1965).
133.   L. Eisenmann, Ann. Physik, 10:129 (1952).
134.   O. N. Kazankin, Author's Abstract of Dissertation, Leningrad (1964).
135.   M. V. Fok and Yu. P. Chukova, Zhur. tekhn. fiz., 35:1139 (1965).
136.   F.I. Vergunas and K. Sh. Enikeeva, Izv. Akad. Nauk SSSR, seriya fiz., 27:475 (1962).
137.   F. A. Kröger, Physica, 22:637 (1956).
138.   D. Curie, Luminescence in Crystals, 1960. [Russian translation, IL (1961)].
139.   J. J. Lambe, C. C. Klick, and D. L. Dexter, Phys. Rev., 103:1715 (1956).
140.   F. A. Kröger and H. I. G. Mayer, Physica, 20:1149 (1954).
141.   S. S. Mitra and R. Marschal, Proc. Seventh Internat. Conf. Semiconductor Phys. (Paris, 1964) p. 1085.